SECOND EDITION

LASER
SAFETY
Tools and Training

OPTICAL SCIENCE AND ENGINEERING

Founding Editor
Brian J. Thompson
University of Rochester
Rochester, New York

RECENTLY PUBLISHED

*Please visit our website **www.crcpress.com** for a full list of titles*

SECOND EDITION

LASER
SAFETY
Tools and Training

Edited by KEN BARAT

CRC Press
Taylor & Francis Group
Boca Raton London New York

CRC Press is an imprint of the
Taylor & Francis Group, an **informa** business

CRC Press
Taylor & Francis Group
6000 Broken Sound Parkway NW, Suite 300
Boca Raton, FL 33487-2742

First issued in paperback 2017

© 2014 by Taylor & Francis Group, LLC
CRC Press is an imprint of Taylor & Francis Group, an Informa business

No claim to original U.S. Government works

Version Date: 20130923

ISBN 13: 978-1-4665-8137-1 (hbk)
ISBN 13: 978-1-138-07200-8 (pbk)

Library of Congress Cataloging-in-Publication Data

Laser safety : tools and training / editor, Ken Barat. -- Second edition.
    pages cm -- (Optical science and engineering ; 141)
    Includes bibliographical references and index.
    ISBN 978-1-4665-8137-1 (hardback)
    1. Lasers--Safety measures. I. Barat, Ken.

TA1677.L3698 2013
621.36'60289--dc23   2013036752

Visit the Taylor & Francis Web site at
http://www.taylorandfrancis.com

and the CRC Press Web site at
http://www.crcpress.com

Dedication

This second edition of *Laser Safety: Tools and Training* is as much a surprise to others as myself.

It is true, family support makes the difference in writing a book, so thanks for being understanding, Pat (my wife), Emily and Leah (my daughters). I would also like to thank my mom and dad, Lil and Max Barat, who instilled in me many of the traits that have aided my success. I would be dishonest if I did not say it is the willingness of friends and colleagues to contribute chapters that really pushed this book forward. So to all the contributors, this book is your fault, and I really thank you for making this text possible, and Taylor & Francis—CRC Press for taking a chance with me again.

Every good laser safety officer adds to his or her education and understanding through contact with laser users and laser systems. Some professional associations have a greater impact than others. I am taking this opportunity to acknowledge the contribution to my laser knowledge by some exceptional laser users: Wim Leemans, Robert Schoenlein, Musa Ahmed, David Sliney, Oliver Gessner, Gabor Somorjai, Steve Fournier, Robert Kaindl, Jim Rockwell, Csaba Toth, Keith Kanz, Heinz Frei, Marcus Hertlien, and Joel Ager.

Contents

Contents

Introduction

This second edition of *Laser Safety: Tools and Training* is as much a surprise to others as myself. I hope the new chapters and updates are a welcome addition to your application and knowledge base.

The use of laser technology is on a Moore's law track, increasing and expanding into areas unthought of a few years ago. So how do we make the user and those around them laser safe? As one reads through the text, one will see that it is really a matter of the hazards evaluation method and the desire to think laser safety. This means to think beyond laser-protective eyewear. One can say if eyewear is required, then the laser setup is not as safe as possible.

A number of chapters and sections deal with components of laser work and laser safety from the optical table, power meter, training adults, optics, and so forth. The purpose of including these items in this text is that they are generally not discussed in the contents of laser safety or only in stand-alone text if at all. Take the power meter for example, a common item in laser labs and owned by many laser safety officers (LSO). But here we present information on how to choose and the best way to use them, information I hope any LSO or laser user will find useful. The overall goal of this text is to provide the fundamental means or road map for laser users to see how simple laser safety can be with some thought and the desire to have the safest system possible.

I hope readers will find this text useful, entertaining, and informative.

Ken Barat
Laser Safety Solutions

Editor

Ken Barat is the former laser safety officer (LSO) for the Lawrence Berkeley National Laboratory and National Ignition Facility Directorate at Lawrence Livermore National Laboratory. Presently he is continuing his passion for laser safety through private consulting, Laser Safety Solutions (lasersafetysolutions@gmail.com).

Ken is a fellow of the Laser Institute of America and senior member of IEEE. He has received Jim Rockwell awards (2009 and 2005) for Laser Safety Leadership and Education, and the Tim Renner User Service Award (2002) from the Advance Light Source as well as recognition awards from the Department of Energy and LSO Working Group. As former group leader of the nonionizing group of the Arizona Regulatory Agency, he successfully led the effort for statewide laser regulations. He is also one of the founders of the Bay Area Laser Safety Officer Society, the nation's oldest LSO networking group. He was among the first LSOs certified by the Board of Laser Safety.

Ken organized and directed the first eight LSO workshops, the premier LSO training workshops in the United States. He has given presentations on laser safety at national and international meetings and acted as a consultant and trainer for a wide variety of companies. He is a member of the United Kingdom's Laser Safety Forum and helped launch and instructed at both the Hong Kong University of Science and Technology first laser safety seminars and the Hong Kong Medical Society LSO training seminar. He is an active member on several American National Standards Institute committees.

Contributors

A number of chapters have been written by individuals, in my opinion and according to my knowledge, who are the best in their areas. I cannot thank them enough for their hard work and willingness to contribute to this book. So here is a little about all these special people.

Stephen Benson

Stephen Benson began his career in free electron lasers (FELs) on the Stanford SCA FEL. He commissioned the Mark III user facility at Stanford University and then at Duke University. He was the first to demonstrate third harmonic lasing in an FEL and assisted in efforts to achieve the first lasing at the second and fifth harmonics. Since 1982, he has been working on developing high-power FELs. The Jefferson Lab IR FEL presently has the FEL power record of 14.3 kW at 1.6 μm. He won the FEL prize in 2000 for his part in developing the first kilowatt average power FEL. He is an American Physical Society fellow. He is the laser safety supervisor for the FEL facility and developed the laser safety specifications for its laser personnel safety system.

Jerald A. Britten

Jerald A. Britten earned a bachelor of science from Michigan State University and master of science and PhD from University of Colorado, all in chemical engineering. Presently, he is an associate program leader for Advanced Optical Components in the National Ignition Facility (NIF) and Photon Sciences Program at Lawrence Livermore National Laboratory (LLNL). Since 1993, he has been developing the technology for manufacturing laser damage-resistant, large-aperture diffraction gratings for high-energy laser systems.

James Fisher

James Fisher is currently the vice president of Newport's Optical Components and Vibration Control Business with headquarters in Irvine, California. Before his current role, he led various technical, sales, and marketing groups within Newport and launched a number of new product platforms and patented products. He earned a bachelor's degree in engineering from the University of Illinois–Urbana and his MBA from the University of Southern California.

James Foye

James Foye is a training course developer for the NIF at the LLNL. He is a strong supporter of safety and has years of experience in training development.

Larry Green

With almost three decades of instrumentation experience and almost 10 years in the laser industry, Larry Green brings a unique perspective to measuring spatial beam profiles. He earned a bachelor's degree in engineering sciences from the State University of New York at Stony Brook, specializing in materials science, and a master of business from the Polytechnic Institute of New York. His chapter goes to the central issue of laser processing (Chapter 17). If you need a beam of a certain shape for your process, you should be measuring it quantitatively. Walter Deming said it all, "If you cannot measure it, you cannot control it."

John Hansknecht

John Hansknecht is the vice president of Laser Safety Systems, a company that specializes in Class IV engineered laser safety controls. He has been working with high-power laser systems since 1995 and has more than 25 years of experience in the Department of Defense nuclear engineering and the Department of Energy research laboratory environments. He is an engineer, entrepreneur, and inventor, currently holding one U.S. patent with a second patent pending. He was interested in writing a chapter for this book (Chapter 15) because he has seen so many people struggle with the engineered safety requirements for Class IV laser areas as outlined in American National Standards Institute (ANSI) Z136.1.

Patty Hunt

Patty Hunt works for the Thomas Jefferson Site Office in Newport News, Virginia, as the environmental programs manager. She earned a bachelor of science in environmental engineering from Old Dominion University and a master's degree in occupational health from Temple University. She is registered as a professional engineer in Virginia and is a certified industrial hygienist. Her latest interests are in environmental management of nanomaterials.

Damon Kopala

As a project engineer at Edmund Optics for 3 years, Damon Kopala is a valuable asset to anyone with questions on optical mounts. He has written articles in this area.

Jack Lund

Jack Lund is enjoying retirement following 50 years actively engaged in laser-related research and development since graduating from Western Illinois University in 1961. He was one of the original members of the Joint Army Laser Safety Team established in 1968 at the Frankford Arsenal in Philadelphia, Pennsylvania, to study laser bioeffects. His focus for the past 40 years has been on the effect of laser radiation on ocular tissue and the visual system.

Bill Molander

Bill Molander is an integral member of the National Ignition Team, as well as a member of its Laser Safety Working Group. He has influenced many as an instructor at Los Positas College.

Burt Mooney

As a graduate of San Jose City College Laser Program, Burt Mooney was an applications/sales engineer for 17 years at Molectron, 8 years as a power and energy meter specialist, and now sales development manager for Gentec-EO USA. His chapter on power and energy meters (Chapter 19) delves into the specifics on how one goes about selecting a sensor head for one's laser and/or lasers, and then matching that up to a display or PC data collection option. The chapter covers some of the history, looks at the present, and peeks at what the future holds.

Matt Vaughn

Matt Vaughn is a training developer at the LLNL. He is too modest to say much about himself.

Diana Warren

Diana Warren is the product manager responsible for New Focus tunable diode lasers and amplifiers, electro-optical modulators, and high-speed photoreceivers. New Focus, a Newport Corporation Brand, is based in Santa Clara, California. She has a background in laser spectroscopy and earned her PhD in chemistry from the University of Southern California.

Bill Wells

Bill Wells works for the Lawrence Berkeley National Laboratory as the safety compliance program manager. He earned both a bachelor and a master of science in biology from the University of Louisiana, Monroe, and a PhD in environmental health from the University of Cincinnati, Kettering Laboratory. He is a certified industrial hygienist and certified safety professional. His current professional interests include mishap investigation and environment, safety, and health system assurance.

David C. Woodruff

David C. Woodruff is an engineering physicist designing laser shutters for 25 years. He is the president of NM Laser Products Inc. The continuing evolution and sophistication of the laser shutter motivated him to write Chapter 20 and provide industry awareness of the technologies involved.

1 Laser Safety
Where Are We?

Ken Barat

CONTENTS

INTRODUCTION

No laser user wants to receive a laser eye injury or skin burn (well some physicians think of a minor laser finger burn as a rite of passage). This statement is like saying none of us want a speeding ticket, yet we all keep on exceeding the speed limit. So how is our laser safety *gene* doing? If we are honest with each other, the answer is we could be doing better, but that takes effort and not all laser users see the need for such effort.

A SAFETY CULTURE

A survey in 2005 stated that 70% of U.S. drivers wear their seat belts. While this statement seems far removed from laser safety, it really is not. For the survey results demonstrate a cultural shift that took years to take root. Now people expect drivers and front seat passengers to be wearing a seat belt. Law enforcement has started giving tickets rather than warnings for those who are not wearing their seat belts. Child safety seats are similar in usage today. One day bus riders will buckle up.

Getting back to laser safety, what the survey above represents is a cultural change. One that laser safety is in the midst of, group responsibility over individual safety. Another analogy is smoking; the concern over secondhand smoke exceeds the argument that it is my lungs and if I want to smoke I will. We will not even go into health-care costs related to smoking. For years one could enter a laser lab and if they were smoking, everyone would protest due to the concern of particulates on optics, with no concern to the individual's safety. Yet one could enter the same laser use area and not put eyewear on and few would say a word of warning or protest. The environment is one in which each person makes their own risk decision. If eyewear is required in a laser setting, all in the room or in hazard zone need to be wearing eyewear, and it is the responsibility of all in the area to make sure anyone entering puts on eyewear, not to leave it to a matter of personnel choice.

ANSI Z136.8 SAFE USE OF LASERS IN RESEARCH, DEVELOPMENT, AND TESTING

The publication of the Z136.8 standard is a positive step in a laser safety culture. Why? It is the first laser American National Standards Institute (ANSI) standard that addresses many of the real-world situations found in the research setting, regardless if it is academic or commercial product development setting. It needs to be adopted by any institution that is engaged in research.

LASER INCIDENT

What follows is the official report on a laser incident that occurred at a U.S. government research lab. At the end, you will find my interpretation of the incident, why it occurred, and how it could have been avoided. This case study demonstrates a number of important points that will be built on in later chapters of this book. Even a novice to lasers and laser safety, on review of this incident, will see areas for improvement and, more importantly, where a lack of safety awareness or concern contributed to the incident (Figure 1.1).

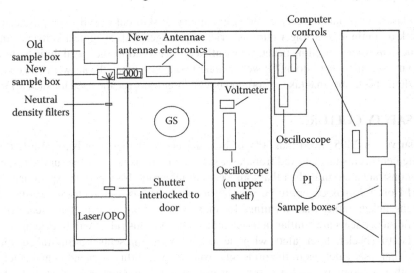

FIGURE 1.1   Laser lab setup.

ACCIDENT DESCRIPTION FROM INITIAL ACCIDENT REPORT (SOURCE: DEPARTMENT OF ENERGY, 2005)

At approximately 12:00 PM, Wednesday, January 19, 2005, a researcher sustained a *laser* injury to his right eye while working in Solar Energy Research Facility (SERF) Lab E-218.

The light involved in this incident is generated by a neodymium:yttrium aluminum garnet (YAG) laser–driven optical parametric oscillator (OPO). The light exiting the equipment housing has a wavelength of 750 nm and a power of 200 mW/20 mJ. The light beam exiting the laser/OPO unit is noncoherent (i.e., divergent), so it is not a true laser beam. However, it does produce very high pulsed energies that present the same hazards as a laser beam and is managed as a laser accordingly. The beam is attenuated to a power in the range of 200 nW/20 nJ through the use of neutral density filters with optical density (OD) ranging from 3.0 to 7.0 before contacting the samples being tested to not *overpower* the samples.

A more powerful than necessary laser/OPO unit is used for this work since it is tunable to the wavelength needed for the sample tests. The power level of the attenuated light beam is unlikely to cause an eye injury; however, laser safety glasses are required whenever the unit is operating.

Other standard laser safety controls are also in effect for this activity, such as an automatic beam shutter that is interlocked with the lab entry doors, beam blocks, and nonreflective surfaces.

The light beam is used to excite carrier pulses in semiconductor systems, solid-state devices, and photovoltaic cells. Instrumentation records the minority carrier lifetime characteristics of the samples being tested, the primary objective of this activity. The samples are mounted in a target box, which is a box providing light and radio frequency shielding of the sample (i.e., the box is not necessarily a safety control, although it does enhance the safety of the activity). The laser/OPO unit and target box are mounted on a standard laser table and are situated about 6 ft. apart. The light beam is open to the lab in between the laser/OPO housing and the target box with the beam shutter and attenuating ODs situated in between the two.

At the time of the incident, a researcher (R1) and his team leader (TL) were testing new sample instrumentation. When there was an apparent problem with the new instrumentation, the TL went to another part of the lab to obtain a different test sample.

Initial investigation findings indicate that while this was occurring, R1 removed the neutral density filters in an attempt to obtain a response from the test sample. Removal of the neutral density filters resulted in full beam power onto the test sample.

(At some point during this process, R1 removed his laser safety glasses.) R1 then manipulated the test sample with a pair of stainless steel tweezers in a further attempt to obtain a response from the test sample. At this point, R1 reports he experienced seeing a flash of light off of the surface of the test sample. He states that he did not feel any pain or sensation when this happened. He then noticed dark floaters in his field of vision (right eye) on looking up at a scope located on an overhead shelf.

The worker informed his team leader of this condition. Both reported to Medical Services where R1 described his symptoms as seeing *floaters* and a yellowish-orange spot in his upper field of vision. The worker was transported by his team leader to Denver Eye Surgeons for medical assessment and treatment. The initial report from Denver Eye Surgeons is that the worker sustained a 2-mm retinal burn and is being referred to a retinal specialist for examination and treatment as necessary. Denver Eye Surgeons administered a steroid injection to reduce inflammation and swelling.

A follow-up examination was conducted on Monday, January 24, 2005. The size of the damaged eye area has reduced from the initial 2-mm burn to a 500-mm wound in the top layer of the retina and a 200-mm wound in the second retinal layer. The swelling and inflammation immediately following the incident may have slightly distorted the initial damage estimate.

The preceding is the official accident report submitted a few days after the event. Now let us take a closer look with a little more informative description and uncover a few additional details.

At approximately 12:00 PM, Wednesday, January 19, 2005, a worker sustained a laser injury to his right eye while working in SERF Lab E-218. Contractor is a grad student, who has been at the laboratory for 3 years. He was working in a two-person lab, principal investigator (PI) and himself. Experiment involved decay elements and exposure to radio frequency. Laser protective eyewear is used in this lab. The use of an oscilloscope is critical to this project. The oscilloscope can be read with laser protective eyewear if one is close to the scope. Unfortunately, if one is near the experimental setup, the oscilloscope is difficult to read. The setup was well established and consisted of a combination of open and closed beam portions. The PI had left the room to obtain another sample. The student was pulling out beam attenuators and reached in to remove the target. He received a reflection off the target onto his right eye; he was not wearing his laser protective eyewear for he was viewing the oscilloscope. The equipment in use at the time of the incident incorporates a Class 4 YAG laser. The wavelength being used was 750 nm and a power of 200 mW/200 nJ attenuated to 200 nW/20 nJ used to illuminate a sample. Standard operating procedures consisted of a number of boilerplates, and do's and don'ts around lasers. Laser safety training at facility is computer based. At this time, no record of the student's training can be found. Also it is uncertain if the laser safety officer (LSO) performed any assessment of the setup. The LSO is primarily an industrial hygiene. The senior researcher and facility Environmental Health and Safety (EH&S) point of contact are required to review the paperwork once a year. In this case, that is the LSO.

The light beam is used to excite carrier pulses in semiconductor systems, solid-state devices, and photovoltaic cells. Instrumentation records the minority carrier lifetime characteristics of the samples being tested, the primary objective of this activity. The samples are mounted in a target box, which is a box providing light and radio frequency shielding of the sample (i.e., the box is not necessarily a safety control, although it does enhance the safety of the activity). The laser/OPO unit and target box are mounted on a standard laser table and are situated about 6 ft. apart. The light beam is open to the lab in between the laser/OPO housing and the target box with the beam shutter and attenuating ODs situated in between the two.

At the time of the incident, a researcher (R1) and his team leader (TL) were testing new sample instrumentation. When there was an apparent problem with the new instrumentation, the TL went to another part of the lab to obtain a different test sample.

The investigation team determined that a reflected light from a laser beam to the right eye was the direct cause of the injury. The root and contributing causal factors and area for improvement are briefly summarized here.

ROOT CAUSES

The grad student removed his laser eye protection (LEP) part way through the activity, in direct violation of an established work procedure.

- The work planning for this activity by the RI and grad student was less than adequate.
- Management and supervision of laser activities was less than adequate, both for this specific activity and for laboratory-wide activities.
- The hazard controls established in the safe operating procedure (SOP) for E-218 were less than adequate, and the processes for reviewing and updating all laser activity SOPs were less than adequate.
- The roles and responsibilities of the LSO may not meet current consensus standards and the LSO resources available to support laboratory activities may not be adequate.
- Laser safety training is less than adequate based on the demonstrated performance of staff, and improvements to the training record keeping processes are needed.
- Reliance on laser protective eyewear is being utilized where additional engineering controls may be reasonable and appropriate.

Figure 1.2 is an image of the laser lab. Figure 1.3 shows the two pairs of laser protective eyewear worn in the lab. You should be able to pick out a number of good practices that are missing. Before any discussion of those and ways to improve the setup, look at the number one problem.

FIGURE 1.2   View of oscilloscope in lab.

FIGURE 1.3   Eyewear used in lab.

PROBLEM NUMBER 1

This experimental setup was close to a year old, but still looked very much like it did on week 4. I am sure even the users who were the developers of the system felt it could be laid out better. To make the necessary changes would mean taking some time away from the actual work. Here is where the critical fault lies. No one thought his or her or anyone else's safety was worth the day it might have taken to reconfigure the experimental setup.

ADDITIONAL ISSUES

No safe means was set to allow the laser to stay on while samples were changed. A possible solution would be a shutter, or even a beam dump could have blocked the beam and allowed safe sample exchanges.

I have used this accident as a teaching exercise (I only wish it was hypothetical, rather than a true case). Following is a list of comments from my students:

- Student should not have been left alone
- Beam was open on the table
- Clutter on the table
- Poor cable management
- Oscilloscope should be relocated
- Oscilloscope fine where it is
- Use of shiny tweezers
- Laser emission light out
- Standard operating procedures too generic
- Poor eyewear choice
- Lack of warning sign
- Poor understanding of hazard

I am sure you might have some additional information that will explain some of the preceding comments.

MY SPIN ON WHY THIS HAPPENED AND HOW SIMPLE STEPS COULD HAVE AVOIDED THE ACCIDENT

During setup of a project or initial proof of principle, safety is not always a high priority. Organizations where safety is an integral step of any project are far ahead of the safety curve. In the accident described earlier, the simple fact is at no time did anyone consider it worth the time of effort to reevaluate the experimental setup or ask can we do this in a safer or more efficient manner.

Why did the beam path have to be completely open? Why not the system be shut off during sample change or a beam block to alternate path been setup during the change process? An interlock or mechanism could have been put into place to prevent one from removing the neutral density filter. For that matter not only could the oscilloscope have been relocated, but better eyewear could have been acquired, allowing visibility.

The answer to all these questions is not things were working so why mess around with it. So let us ask again were things really working or had random chance just not caught up with them.

Making Laser Safety Easier

Now the question is how one makes laser safety easier, an article the author wrote for SPIE Professional magazine helps point us in the right direction. Building on that article, following are five steps to help reduce the odds of a laser injury or incident.

On-the-Job-Training

Today every institution where Class 3B and Class 4 laser use is present has some requirement for fundamental laser safety training. The driver for this training may be from fear of regulatory bodies, compliance with ANSI Z136 series requirements, or just doing the right thing. How much can we expect from this laser safety introduction? It should make the user more aware of potential laser hazards, maybe more inclined to follow laser safety rules. The truly critical laser safety training is on-the-job-training (OJT). Effective OJT and mentoring is the most proven means to prevent or avoid a setup leading to an accident. The trainer needs to have the patience to explain the setup, what the goal is, and what problems they have encountered. The trainer should also explain the sources of known reflections, non-beam hazards, and how to use safety tools such as viewers and eyewear. The trainee needs to be observed doing tasks, even if they were hired as the expert on the equipment, at the very least to evaluate their safety culture. The trainee also needs to feel they can ask questions and receive supportive responses. The mentor's or trainer's body language can unknowingly discourage any further questions. Both parties need to be in agreement prior to letting the new users' work unsupervised. See Chapter 4 for sample forms to document OJT. OJT depending on the task or system is usually more than a 1-day operation. Much also depends on the experience of the one being trained. However, neither should be in a rush to get the job done. For our goal is confidence in getting the task performed by all parties and a safe operation; see Chapter 10 for greater insight and explanation.

Housekeeping

Numerous laser incidents can be traced to poor housekeeping. Laser housekeeping is not the childhood activity of hiding everything under your bed or now optical table, but rather setting time aside to keep work spaces (optical table and work area) clean. By clean I mean free of unused tools, equipment, reflection sources, cleaning solvents, storage boxes, and so on. Granted space is an ever-present problem as well as the time to clean, storage lockers, and shelves over the optical table can help. Setting a dedicated day and time to clean in your schedule is the best proven solution (once a month, once a week is better). Waiting for that open afternoon in your schedule will never happen. The best solution is to try incorporating housekeeping into your daily routine. Address problems as they arise. This will make housekeeping a less daunting task. The more clutter you have from things you might need one day to just excess optics on the table sends a subliminal message to observers and funding sources. The message is if the work is performed in such a disorganized mass how good can their science be? How reliable are their results?

Maybe what the industry needs is an equivalent to a Merry Maids cleaning service, Optics Dusters are Us. Now there is a business opportunity for you.

Look for Stray Reflections

In reviewing laser accident reports in the research and academic environment, the most common scenario is the lack of eyewear usage during laser alignment. A number of these cases involve a failure to look for stray reflections. A 1 mW/cm^2 (continuous wave) in the wavelength range of 400–1400 nm equals 100 W/cm^2 at one's retina. This is about the maximum one's eye can tolerate. It is easy to see (excuse the pun) that the typical 4% reflection off an uncoated optics could easily be above the safe threshold. As a reminder, the macula/fovea portion on one's retina where 400–1400 nm light is focused to is the critical vision portion of your eye. Hence, as alignment is performed, one needs to go optic to optic and be checking for stray reflections at each. Beam splitters have been involved in several accidents when the user forgot to block the unneeded split. This is why the use of remote viewing and motorized mounts should always be encouraged and considered, especially as a system matures from proof of principle to routine operation.

Contain Reflections

Following from preceding step, if stray reflections are identified or expected, containment or blocking those reflections is critical to preventing laser accidents. Sometimes it is as simple as placing a card behind or above an optic in a mount. The author has encountered incidents where users have a known unblocked reflection that seems unlikely anyone will be in front of, because the angle is too steep or the power is too low to cause injury. In each case, some circumstances arose that placed someone in danger or the output was higher than believed. In conclusion, beam blocks, perimeter guards, table enclosures, and beam dumps all add to your safety and those around you. USE THEM!

Eyewear

Finally, I would be banished from the U.S. laser safety community if I did not mention laser protective eyewear. It is fitting that eyewear be mentioned last. They are not your first line of defense but rather the last. It is the four steps mentioned earlier that provide the real safety. Laser eyewear is like a seat belt, if all road rules and conditions are followed it will never be needed but if any fail it might save your life. Your laser eyewear is your seat belt. We have two types of eyewear:

1. *Full protection*: This has a sufficiently high OD to block any direct or stray beams and should always be used with invisible wavelengths.
2. *Alignment eyewear*: This should be used with visible beams when they need to be seen.

Today with new filter and styles of laser eyewear, one can be found for just about all applications.

SUMMARY

This chapter discusses a real-life incident and five areas where laser safety can be addressed. This should give one sufficient food for thought to allow them to revisit their own laser use area and find ways to make it safer. If not, read this chapter again.

2 Who Is Responsible for Laser Safety?

Ken Barat

CONTENTS

INTRODUCTION

If one reads any ANSI Laser Standard, in particular Z136.1, Z136.3, and Z136.8, it would appear the laser safety officer (LSO) has the overwhelming responsibility for laser safety. This fact is firm, and without someone responsible for laser safety it will not happen. As with any activity, there are really many parties that play a role. In the case of laser safety, several groups have a role to play in providing, performing, or evaluating laser safety. A common word today or expression is "Line Management Accountability" pushing down safety responsibility to the user level. Regardless it is the role of the LSO to either see that controls are being carried out or help develop those controls. The controls must be workable or they will be defeated by the individual users. So let us go down the command chain for laser safety, starting with institution management.

LASER SAFETY TEAM

1. Institutional management
2. LSO
3. Laser safety committee
4. Laser supervisor

5. EH&S support
6. Laser user #1

Sitting outside but exerting pressure is regulatory bodies, internal and external.

MANAGEMENT (THE FAMOUS "THEM")

The employer will provide employment and a safe and healthful place of employment for his employees. This statement comes from the California Occupational Safety and Health Administration (OSHA). Similar wording and expectation can be found in many regulatory sources. As part of the responsibility for a safe work place, from a laser safety perspective is the appointment of an LSO. It should be noted that this appointment does not have to be and as a rule of thumb is not a full-time position. Once the appointment is made, the employer through its management chain takes on certain responsibilities. As an editorial comment, the candidate selected makes a strong statement from management to the work force. Technical skills, customer service level, interpersonal skill, and how they see the LSO role (maintenance or proactive) all come across as how committed management is to really having a laser safety culture or just meeting regulatory requirements. Quoting the ANSI Z136.1 2007 version Section 5.4, "the management shall provide for training to the LSO on the potential hazards, control measures, applicable standards, and any other pertinent information pertaining to laser safety and applicable standards or provide the LSO adequate consultative services. The training shall be commensurate to at least the highest-class laser under the jurisdiction of the LSO; Training shall also include consideration for the evaluation of any non-beam hazards associated with the lasers and the laser systems under the jurisdiction of the LSO." This training should not be a one-time course, but ongoing to keep the LSOs.

Let us not let management off the hook too soon.

Institutional management must set the tone that safety is part of every job, not an afterthought. An even harder idea is safety needs to come before schedule. Living these words is much harder than saying them. Many a firm has failed the when the rubber meets the road test.

SUPERVISOR, RESPONSIBLE INDIVIDUAL, PRINCIPAL INVESTIGATOR, OR WORK LEAD RESPONSIBILITIES

The supervisor represents line management and sets the tone for laser safety within a research or local user group. Due to competing responsibilities, the supervisor may spend little time in the actual laser use area. In these instances, the supervisor may designate one or more work leads who act as laser safety supervisors. Work leads take on the responsibilities of the supervisor in terms of laser safety. This does not alleviate the original supervisor from the responsibility that safe work is carried out. The supervisor or work lead must ensure that all laser users receive adequate and appropriate laser safety training on two levels: (1) completion of a fundamental laser safety course and (2) training from the supervisor, work lead, or designee with respect to the hazards and safe operation of the laser system(s), commonly called on-the-job training (OJT). It is at the OJT level that a safety difference can be made. A detailed list of supervisor responsibilities follows:

- Ensure that all personnel complete the laser safety course.
- Develop, provide, and document OJT to personnel in all procedures and techniques required for the safe use of specific lasers and optical systems in their laboratory or use setting.
- Ensure that all personnel obtain a baseline laser eye examination (if required by the institution) and after any suspected case of laser eye exposure.
- Prepare a standard operating procedure (SOP) or laser use authorization document for laser operation and ensure that the provisions of the said document are properly implemented and diligently followed by the laser users.
- Ensure that any laser safety devices (interlocks, etc.) are functioning properly.
- Ensure any visitors receive a site/experimental hazard orientation as part of any laser use area tour when lasers are in use.
- Ensure that personnel are allowed to work only on tasks for which their formal laser safety training and OJT have been completed and documented.
- Visitors and new staff with incomplete training may work only under the direct line-of-sight supervision of an authorized laser user and only for a set maximum number of days (recommendation: 30 days).
- Ensure the LSO is notified of proposed laser acquisitions.
- Ensure the LSO is notified of changes to the laser control area configuration that impact safety.
- Ensure the LSO is notified of changes in laser use or laboratory conditions that impact safety.
- The supervisor shall insure that all laser users have taken the required training.

LASER USER RESPONSIBILITIES

Final responsibility for laser safety rests upon the individual laser user. They must have received proper fundamental laser safety training and OJT that allows them to perform their tasks and be comfortable enough to speak up in times of question or doubt. They must also understand their work goals and be familiar with the institution's "Stop Work Authority" policy. Users have the following responsibilities:

- Attend appropriate training before operating any laser/laser system.
  - This includes not only laser training (but possible non-beam safety), that is, electrical safety if appropriate.
- Receive appropriate OJT.
- Read, understand, and follow all applicable procedures in the SOP and insist that other personnel in the laser lab (coworkers, visitors, repair staff [in-house or vendor]) do the same.
- Receive medical surveillance, where applicable.
- Work in a safe manner following laboratory policy and procedural requirements.

- Promptly report any malfunctions, problems, accidents, or injuries that may have an impact on safety.
- Immediately report any suspected laser eye exposures to the laser supervisor, health services, and the LSO.
- Complete retraining when applicable.
- Bring real-life issues to the attention of group leadership or LSO.
- Have flexibility to get corrections or modification made to work procedures.
- Be aware that complaining to oneself or within peers generally does not produce any positive changes, have a mechanism to raise concerns to a higher level.
- Protection from fear of supervisor or perception on career if accident reported or problems raised.

LASER SAFETY OFFICER RESPONSIBILITIES

Note: A number of ANSI Z136.1 sections are quoted in this chapter and book; while new versions may be issued after the publication of this book, the section numbers may change but the content will remain in the ANIS Laser Standards.

The key to effective laser safety per the ANSI Z136.1 and Z136.8 standards is the role of the LSO. "An individual shall be designated the LSO with the authority and responsibility to monitor and enforce the control of laser hazards and to effect the knowledgeable evaluation and control of laser hazards. The LSO either performs the stated tasks or ensures that the task is performed." This fits within the scope of an integrated safety management plan (see Chapter 4). The LSO has the authority and responsibility to monitor and enforce the control of laser hazards. The LSO has the authority to suspend, restrict, or terminate the operation of a laser or laser system if he/she deems that laser hazard controls are inadequate.

The majority of LSOs are not full-time laser safety officers, rather an individual who will have the responsibility to see that laser safety is achieved in addition to other duties. The ANSI Z136.1 standard Section 1.3 defines LSO as an individual who shall be designated the LSO with the authority and responsibility to monitor and enforce the control of laser hazards and to effect the knowledgeable evaluation and control of laser hazards. The LSO either performs the stated task or ensures that the task is performed. There shall be a designated LSO for all circumstances of operation, maintenance, and service of a Class 3B or Class 4 laser or laser system, and there should be a designated LSO for Class 3R lasers and laser systems.

The standard also states management's responsible in Section 5.3.2.1.

The management shall provide for training to the LSO on the potential hazards, control measures, applicable standards, medical surveillance, and any other pertinent information pertaining to laser safety and applicable standards or provide to the LSO consultative services. The training shall be commensurate to at least the highest class of laser under the jurisdiction of the LSO.

This LSO is the key in obtaining laser safety. The role of the LSO is while not mandated to perform all LSO duties, the LSO is required to see that they are addressed. Remembering that if all you have is Class 1 products, it does not matter if you have

1 or 100. If no access to hazard levels is possible, no LSO is required. If units get open for service with beams on, someone needs to make sure precautions are taken so that users, service staff, or passersby are not injured; generally that person is the LSO.

For positive laser safety, the LSO's duties, responsibilities, and authority include, but are not limited to

- Maintaining the laser safety program.
- Evaluating laser hazards, approving mitigation plans, and providing technical advice for safe laser operations.
- Reviewing and approving all SOPs that reference laser hazards.
- Developing and reviewing alternate control measures than those listed in ANSI Z136.1, under the authority Z136.1 grants to the LSO.
- Performing and documenting, at least annually, a laser safety audit of each Class 3B and Class 4 use area while the laser is operational, if possible.
- Upon request, calculating and verifying nominal hazard zones.
- Ensuring and providing calculations of laser eye hazard parameters (maximum permissible exposure [MPE] and optical density) that will ensure eye protection. Advise laser users and supervisors on appropriate eye protection (full protection of alignment eyewear). This includes frame styles and prescription options.
- Ensuring laser safety training is provided for Class 3B and Class 4 laser users, through lecture, web-based, or other techniques.
- Ensuring the appropriateness of OJT.
- Investigating all instances of suspected laser eye exposure.
- Participating in investigations of beam as well as non-beam-related accidents in laser facilities.
- Additional duties are required or referenced in ANSI Z136.1.
- Resolving conflicts between ANSI Z136.1 and regulations in other standards regarding lasers.
- Maintaining laser inventory system.
- Developing temporary control areas/temporary work authorizations.

KEY LSO TOOLS

SUBSTITUTION OF ALTERNATE CONTROL MEASURES

The ANSI Z136.1 Laser Standard establishes the LSO's authority to modify the control measures required for Class 3B and Class 4 lasers or laser systems. Upon documented review by the LSO, control measures—engineering controls in particular—can be replaced by administrative or other alternate controls that provide equivalent protection. The approval of these controls is incorporated into the SOP for the laser use. An example would be the use of curtain maze or posting in place of an entrance interlock.

TEMPORARY CONTROL AREA/TEMPORARY WORK AUTHORIZATION

The concept of a temporary controlled area (TCA) or temporary work authorization for laser operation comes from the ANSI Z136.1 and Z136.8 standards and is of great practical value in a research setting; its purpose is to allow an authorization of laser work in settings where a formal SOP does not already exist. Examples would be acceptance testing of laser equipment in a lab during initial setup, a short-term change to an existing experiment, short-term laser repairs, and an appropriate laser safety plan to follow while an engineering control is being repaired. The LSO will generate a TCA/TWA memo listing the circumstances and then create control measures to be followed by all parties in the TCA/TWA. All parties will sign the memo indicating they understand the controls and will abide by them. This memo is to be posted at the work site and usually has a 2- to 4-week duration and limited extension capability. Sample controls might be

- No unattended laser work is allowed.
- All users must read and sign the memo.
- Special notice alignment sign must be posted.
- The user will verify that any protective housing by-pass device has been removed before returning the unit to normal operation.
- If any of these conditions are not followed, the laser work must stop.

Some of the duties called out that make up the LSO responsibilities are listed as follows:

- Classification
- Hazard evaluation
- Control measures
- Procedure approval
- Protective equipment
- Signs and labels
- Facility and equipment
- Safety feature audits
- Training
- Recordkeeping
- Accident investigation
- Approval of laser system operations

GUESTS AND VISITORS

Guests, visitors, and new employees without approved laser safety training (or with incomplete laser safety training) may use lasers *only* under the line-of-sight supervision of an authorized laser user.

SERVICE AND REPAIR PROVIDERS

These people need to realize the vital role they play in laser safety. They are looked upon by many to be the experts on systems and therefore the source of laser safety controls and guidance. A casual word or outright disregard for laser safety can have disastrous effects.

Service and repair providers, if they are outside vendors, must be qualified to perform the requested work. They shall receive an orientation to the hazards in the laser use area and be briefed on the area SOP controls. Laser users must be aware that they are responsible for the safety of these individuals and their safety compliance with institutional rules. It is an institution's right to request a safety plan from a vendor. This outlines how they will behave and what rules they will follow while on your site. The following is a sample safety plan.

VENDOR SAMPLE SAFETY PLAN (THIS LONG VERSION CAN BE CULLED DOWN)

POLICY STATEMENT

The policy is to provide a safe and healthful working environment for all personnel through proper inspection, guidance, and adherence to safety codes and standards. All work shall be done in a safe manner. Safety is a management responsibility. To prevent injuries, illnesses, accidental fires, and property damage, all supervisory personnel shall demonstrate the ability to recognize hazards and take necessary steps to eliminate existing and potential hazards. All supervisors and employees shall perform their duties in compliance with the required safety codes and standards.

PURPOSE

This safety, accident prevention, and fire prevention plan has been compiled to make all personnel working on the project thoroughly aware of the need to eliminate all possible causes of accidents.

SAFETY RESPONSIBILITIES

Safety is the responsibility of service management. (Insert company representative's name) shall be the designated on-site safety representative and he shall be responsible for administering the safety program of this laser-servicing project, as required, to carry out this subcontract.

Due to the nature of this laser-servicing contract, there are no construction superintendents nor weekly inspections as this is not applicable to this contract.

Each employee shall be held responsible for performing his/her work in a safe manner in accordance with this safety plan. Any employee who, in the judgment of management, knowingly commits an unsafe act or creates an unsafe condition, disregards this safety policy, or is a repeated safety or health offender will be subject to disciplinary actions up to and including discharge. All employees shall be ready at all times, without fear of reprisal, to correct unsafe conditions or to report hazards at the worksite to their supervisors or management directly or through the anonymous hazard reporting system.

ACCIDENT REPORTING

All individuals who are injured shall report the accident, however minor, to their immediate supervisor. All incidents and accidents shall be reported to the site firm facility point of contact within 2 working days. If a serious accident occurs, the facility point of contact shall be notified immediately. Supervisors shall obtain all pertinent information so that appropriate forms can be completed within the required time period and forwarded to the representative within 5 working days. An injury log (record of injuries) will be kept by the human resource department or a similar group.

EMPLOYEE TRAINING

Chemical Handling

Service engineers are trained on the hazards and control measures for handling any of the required cleaning or surface preparation materials. Copies of the material safety data sheets (MSDSs) will be maintained by the (place company name here) representative.

Laser Safety Training

Service engineers are qualified to perform the required servicing operations and have been trained on the safe handling and operation of lasers and laser systems. Service engineers will follow the required laser safety protocol to ensure safety to themselves and other personnel in the area.

First-Aid Stations

The designated facility point of contact could be ES&H representative or division safety team member shall instruct the (insert company name) service engineer(s) in emergency procedures, first-aid, and the location of the first-aid station.

Safety Eye Washes

The designated facility point of contact could be ES&H representative or division safety team member shall communicate the location of the fire extinguisher nearest to the work area to the (place company name here) service engineer.

MEDICAL SUPPLIES AND ASSISTANCE

First-Aid Kits

First-aid kits shall be available to the service engineer(s) as provided by home institution. First-aid equipment shall be inspected regularly for completeness, and an employee qualified in first aid shall be on-site during the normal work shift.

Emergency Phone Numbers

Emergency phone numbers are to be posted on a bulletin board available to the service engineer(s) or an emergency phone number card will be provided to the service engineer(s). Depending on the nature of the accident and the first aid or medical care required, the necessary assistance shall be requested by phoning the emergency number indicated on the card.

PERSONAL PROTECTIVE EQUIPMENT

It is the responsibility of (insert company name) to provide all employees with training on employer-provided personal protective equipment necessary for each operation. Personal protective equipment shall be worn as required by appropriate codes and standards.

- Hard hats are not required for this contract.
- All employees while performing any operation in which a hazard to the eyes exists shall use eye protection. Examples of such operations are laser operation and surface cleaning/preparation with chemicals.
- Respiratory protective equipment is not required for this project.
- Welding shields or goggles are not required for this project.
- Sturdy work shoes or boots are required in construction areas but are not required for this project.

HAZARDOUS MATERIALS CONTROL

Hazard Communication Program

A written plan for compliance with state and federal worker right-to-know laws is a requirement. This plan includes specific safety responsibilities for all levels of management and for employees. All coherent employees receive training on this program.

Identification

Chemicals may be used in the course of laser servicing. An inventory of all potentially hazardous materials shall be available on-site, along with an MSDS for each material. These shall be available at all times to employees and other persons affected by the materials (e.g., other subcontractors on the same job site).

All chemical containers shall be labeled describing the product's identity and its hazard(s), at a minimum.

Worker Notification

Before using any hazardous material, each worker shall be aware of the requirements of this section and be trained in the proper use, disposal, and special handling procedures to be followed for each material (e.g., when respirators are to be worn).

Handling

While handling hazardous chemicals or solvents, employees shall follow directions and comply with any warnings or cautions affixed to the labels.

Any questions concerning the use of such chemicals and personal protective equipment shall be directed to the supervisor.

Chemical Waste Disposal

The service engineer is responsible for proper disposal of chemical and hazardous waste generated by the service engineer in the performance of this subcontract. Preexisting hazardous material removed and discarded as waste by the subcontractor in the performance of this subcontract shall be disposed of by host site.

TRAFFIC CONTROL

Posted safety, fire, instructional, and traffic signs will be followed.

FIRE PREVENTION

Hot Work Operations

A hazardous work permit for hot work (welding, brazing, etc.) is not applicable for this project. However, small soldering operations may be required to perform repairs. It is the responsibility of host site to advise EH&S if a hazardous work permit is required for small bench top-type soldering.

Smoking

Smoking will be allowed only in areas designated where there are safe receptacles provided for smoking materials.

Nonhazardous Waste Disposal

Accumulations of combustible waste material, dust, and debris shall be removed from the structure and its immediate vicinity at the end of each work shift, or more frequently as necessary for safe operations. Good housekeeping shall be maintained, and access will be kept clear at all times.

Fire Alarm Reporting

A public fire alarm box and telephone service to a responding fire department or equivalent facilities shall be readily available to the service engineer. Instructions on the emergency phone number card shall be used to notify the fire department immediately in case of fire. The local fire department number shall be conspicuously posted near each telephone.

Access for Firefighting

Access routes for firefighting equipment shall be maintained. Fire hydrants and fire department connections shall be kept clear of any obstructions.

Fire Extinguishers

The designated facility point of contact ES&H or division safety team member shall communicate the location of the fire extinguisher nearest to the work area to the (place company name here) service engineer.

Solvents

Minute quantities of flammable or combustible solvents shall be used for cleaning purposes as specified in the task identification process.

Sanitation and industrial hygiene shall comply with the following listed standards:

- Toilet facilities shall be provided at the work site.
- Potable drinking water shall be available at the worksite.
- Proper ventilation shall be maintained to avoid possible harmful buildup in areas where toxic fumes, dust, vapors, or gases may be produced. Respiratory protection shall be supplied when adequate ventilation cannot be provided.

GENERAL AND SPECIAL INSTRUCTIONS

General Instructions

All employees shall comply with the following general instructions:

- Comply with this plan, assist all other employees in doing so, and report all dangerous conditions or practices immediately to their supervisors.
- When injuries occur, the first step is always to provide medical care for the injured and eliminate immediately any apparent cause of the injury. If a cause is not apparent, the work area and equipment shall be secured until qualified authorities determine the cause.
- No one shall be permitted to work while his/her ability or alertness is impaired by illness, fatigue, medication, or other causes.
- Reporting to work under the influence of alcohol, stimulants, tranquilizers, or barbiturates or using them during working hours will be cause for permanent removal from the work site and grounds for disciplinary or legal action if warranted.
- No guard, safety device, or appliance shall be removed from tools, machinery, or equipment except for the purpose of making repairs. Persons qualified to make the repair shall only do such removal, and they shall first disconnect any power source and have the tool, machinery, or equipment in a safe area.
- Employees shall not handle electrical equipment, machinery, vehicles, or air and water lines in a manner outside of the scope of their regular duty except with specific instructions from their supervisors.

- Employees shall not enter trenches, ditches, or any other subsurface area without specific instructions from their supervisors.
- If employees observe sandblasting dust, asbestos fibers, smoke, or other possibly dangerous pollutants in the air of a workspace, they shall contact their supervisors for instructions.

Laser Safety

- Refer to the ANSI Standard Z136.1 Safe Use of Lasers and Z136.8 Safe Use of Lasers in research, development, or testing for additional guidance as well as relevant sections (name company) Laser Safety Policy.
- Follow any additional safety procedures specified by the client site.
- Restore all interlocks (protective housings, etc.) upon completion of work.
- Service staff will be responsible for establishing an exclusion zone around laser equipment when interlocks are bypassed or the potential of laser reflections above MPE are present.
- Wear laser safety glasses (alignment or full protection) with appropriate wavelength and optical density coverage for the laser being worked on.

Electrical Safety

Follow *all* rules of safe practice for work on energized systems if this will be necessary (refer to XXX Company Electrical Safety Policy and Program) or follow procedures specified by the client site, whichever is more protective:

- The step-by-step procedures in the service manual will be followed when working on energized equipment or when de-energizing equipment. The service engineer(s) will not perform repair procedures that are not provided in the service manual without first contacting the appropriate internal service organization or the service manager.
- Work on exposed energized electrical systems with voltages greater than 245 V requires two qualified persons who will work on the equipment and shall be positioned within visible and audible range of each other.
- In the case of outside contractors, the contractor may be one of the qualified workers and must follow appropriate technical manuals for the equipment to be serviced. Adherence to equipment technical manual procedures for work on energized equipment regardless of voltage shall be reviewed and approved by the lead scientist.
- A qualified person is described as one who has been determined by his/her supervisor to have the skills, knowledge, and abilities to safely perform the work to which he/she is assigned.

All lockout and tagout operations shall conform to institutional policies. An energy source included any source of electrical, mechanical, hydraulic, chemical, thermal, ionizing and nonionizing radiation, or other energy. Follow correct lockout and tagout procedures, including:

Identify and isolate all energy sources, use an "active" means of locking out breakers, plugs, or switches not in your direct control, and create a controlled work area to prevent the entry of unauthorized personnel (refer to XXX Company LOTO Plan, or follow procedures specified by the client site, whichever is more protective).

Work in Confined Spaces

- Employees shall not enter or work in confined spaces (such as tanks, vaults, holds, or manholes) without specific instructions from their supervisors. All confined space entries shall be made in accordance with the provisions 29 CFR 1926 and 1910 requirements. The subcontractor shall arrange confined space entries through the construction inspector or program construction coordinator.
- Work in confined spaces is not applicable to this service contract.

Explosives Safety

- Explosives safety is not applicable to this service agreement. Vendor employees will bring no explosives on-site.

Ladders and Stairways

- Employees required to use ladders and stairways shall be trained in their safe use and in the recognition of hazards related to them.
- No employee shall use a ladder that is defective or does not meet OSHA requirements.
- Wooden ladders shall not be painted. They may be treated with linseed oil.
- Splicing of ladders is prohibited.
- Work shall be arranged so that employees are able to face ladders and use both hands while climbing.
- The use of ladders to transport heavy or awkward-shaped items is prohibited.
- Stepladders shall never be used as straight ladders. They shall be fully opened at all times except when in storage. Employees shall not be allowed to stand on the top step or end cap of stepladders.
- Stairways shall meet OSHA requirements.

Scaffolding

- Work with or on scaffolding is not applicable to this service contract.

Machinery and Vehicles

- Unless it is part of their regular duties for which they have had adequate training, no employee shall operate machinery or equipment without specific instructions and guidance. Only licensed operators shall operate vehicles.
- Work with machinery is not applicable to this service contract. Vehicle operation is limited to the service engineer's vehicle.
- The need for servicing or repairs shall be reported to the supervisor.
- Working under suspended loads is forbidden.
- Employees are prohibited from riding booms, loads, slings, hooks, or lift-truck forks or platforms.
- Air hoses shall not be disconnected until they are bled and pressure is securely turned off at its source. All air hoses shall meet the requirements of 29 CFR 1926.302(b)(Federal OSHA Construction Safety Standards) and Title 8 CCR.
- Operators or employees working near units producing noise levels in excess of prescribed standards shall wear adequate devices for protection of hearing.
- No vehicle shall be operated in a reckless or careless manner or at a speed that is not reasonable and proper with regard to weather, traffic, surface condition, visibility condition, load, or type of vehicle.
- All vehicular accidents that occur on the laboratory site, of whatever size and nature, whether injury or noninjury, shall be reported immediately to the (place company name) supervisor and the designated on-site point of contact.

Welding and Cutting: General

- Welding and cutting is not applicable to this laser service contract.

Excavation, Trenching, and Shoring

- Excavation, trenching, and/or shoring are not applicable to this laser service contract.

Fall Protection

- All floor or wall openings and platforms that expose workers to a fall of more than 4 ft. shall be covered or protected by guardrails.

Radiation Safety

- Use of any radioactive material by (vendor name) is prohibited. The host shall be responsible to identify to all vendor personnel any ionizing radiation sources in the vicinity of the laser servicing area.

Hand Tools

- All hand tools, whether self-owned or company-furnished, shall be maintained in safe condition. Unsafe tools shall not be used until repaired.
- Guards required on power tools shall be used at all times. Switch-locking devices shall comply with the requirements of 29 CFR 1926.300 and CCR Title 8. Power grinders shall have protective shields.
- Use of power tools is limited to battery-powered tools.
- All electrical power extension cords shall be used and maintained as specified in 29 CFR 1926.405 and 4 16, and CCR Title 8. Splices shall have insulation equal to that of the cable.

Nonelectrical Work Performed Near Exposed High-Voltage Power Distribution Equipment

- Nonelectrical work performed near exposed high-voltage is not applicable to this laser service contract.

Lockout and Tagout

- All lockout and tagout operations shall conform on-site requirements (your company name here). An energy source includes any source of electrical, mechanical, hydraulic, chemical, thermal, ionizing and nonionizing radiation, or other energy.
- The step-by-step procedures in the service manual will be followed when working on energized equipment or when de-energizing equipment. Our service engineer(s) will not perform repair procedures that are not provided in the service manual without first contacting the appropriate internal service organization to the service manager.
- Information on laser servicing while the laser is in operation is specified in the Subcontract Hazards List.

Service engineers are responsible for establishing a temporary control area to warn and protect others during their activities involving laser radiation. This may include posting of signs or blocking off walkways.

REGULATORY BODIES

OSHA is the national regulatory body that encompasses laser safety. While OSHA has laser safety resources available for the user on its web site, its actual deeds are disappointing. Few OSHA inspectors check for laser safety during routine OSHA audits, but they will respond to worker complaints. Many of these are due to the employees not receiving adequate training.

The few state agencies that have the potential to effect laser safety in the workplace have as a rule (with few exceptions) equal as disappointing.

To end on a positive note, these regulatory agencies at a minimum raise laser safety awareness in their states.

3 Performing a Complete Laser Hazard Evaluation

Ken Barat

CONTENTS

LASER OR LASER SYSTEM'S CAPABILITY OF INJURING PERSONNEL

Here are the components of the actual laser source itself.

Type of laser: Each one is interested to know the details of the wave whether it is pulsed or continuous. The effect of the laser on the nature and the level of control measures and hazards the laser poses to an individual system depend on the nature of the wavelength(s) being generated, ultraviolet, visible, near-infrared, mid-, or far-infrared.

Output of laser: Are we talking about milliwatts, nanojoules, or megawatts, or joules of output? These answers and wavelength will have a dramatic effect on possible laser protective eyewear requirements.

Classification of laser: All lasers are classified into three categories, 3R (3A), 3B, or Class 4.

BEAM PATH OF A LASER SYSTEM

The laser beam should be examined once it leaves the laser source. In a very similar way, one considers one's commute from leaving the security of one's garage to one's work designation. Are we one of the lucky ones who has a short commute of several minutes to a long arduous commute of highways, tunnels, bridges, and drivers of less skill that oneself?

In laser terms, the path of the beam could be open, contained in fiber optics, or enclosed. In addition, just like the driver going down a steep grade, the beam could be amplified or go through nonlinear optics. Therefore, it produces a change of lanes in our driving example but for photons it is a change of wavelength. This could occur several times along with possible chirp stretching or compression. Any of these steps or a combination of them will affect the safety requirements one might apply to a system.

INTERACTION OF LASER BEAM WITH ITS INTENDED TARGET

Once the laser radiation reaches its designation just like our driver reaching work, many options lie ahead, from that great day at work to violent meetings.

A percentage of the beam may be reflected off a target. The interaction of a beam may generate gases as by products requiring ventilation. An intense pulse laser beam may generate ionizing radiation in the form of neutrons; gamma or x-rays even activate products, hence generating additional ionizing radiation.

Maybe the end of the beam path is delivered through a robotic arm, which now introduces new concerns for evaluation.

ENVIRONMENT IN WHICH THE LASER IS USED

Now we have to consider factors from the work place and how they contribute to our hazard evaluation. Do they make our job easier or harder? A place such as a clean room may do both. While adding to access control and therefore helping keep the unauthorized out, cleanness requirements may make it harder to implement other controls.

Some other common laser-use environments are the operating room, manufacturering floor, fabrication area, and our chief interest research laboratory.

PERSONNEL WHO MAY BE EXPOSED TO THE LASER RADIATION

When you think of these people, who should come to mind? I would say authorized laser users, ancillary staff, visitors, and in some case consumers. As each of these groups requires evaluation, items such as training requirements, personnel protective equipment (PPE), and even ergonomic factors require consideration (Figure 3.1).

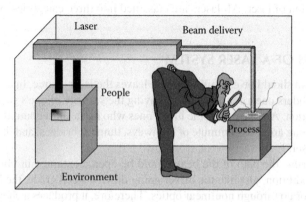

FIGURE 3.1   Cartoon representation of hazard evaluation process.

Step 1 Laser and ability to cause injury: Following are the points to be considered:
- Type of laser(s)
  - Excimer, commercial, original equipment manufacturer, homemade
- Output/class
- Power supply
- Fixed position
- Moveable (fiber on breadboard, diode)

Step 2 Beam path(s), what risks are associated with the setup once the laser beam leaves the laser radiation source?
- Open beam
- Enclosed
- Fiber
- Combination
- Wavelengths present
  - 800 to 400, 1064 to 532 to 266, tunable
- Amplification
- Pick off beams
- Multibeam paths
- Vertical beams (periscopes)
- Beam expansion, beam focusing
- Reflections
  - Expected and unintended

Step 3 Process interactions: What happens when the beam reaches its destination? Is everything enclosed such as in a reaction or vacuum chamber, or is it open to air and human contact?
- Reflections
  - Expected, back reflections, unintended directions
- Fumes
  - Ventilation requirements
- Robotics
- All activity in a chamber
- Open interaction area
  - Blue light
  - Plasma
- Equipment safety items
  - Interlocks, viewing windows

Step 4 Environment: Where is the action happening, all in one room or is the beam passing from one room to another? Is the laser source far from its termination point?
- Clean room
- Factory/job shop
- Operating room
- Fabrication area
- Research lab
- User facility
- Each setting has its own special items and solutions

- Additional environmental concerns
  - Grounding
  - Dye pumps
  - Access controls
    - Access interlocks or other means
  - General work space
  - Windows

Step 5 People: Who is in the laser-use area(s)?

- Consumer
- Users
- Ancillary
- Visitors
  - Tours
  - Collaborators or Guests
  - Service vendors
  - Summer students, interns/all year

These individuals may need you to consider the following:

- Training
  - Basic and on the job
- Ergonomic factors
- PPE
- Safety plan
- Procedures
- Organizational authorization

SUMMARY

If the five-step hazard evaluation process is followed with open eyes and a new perspective, a safe work setting can be generated for all.

ROLE OF A USER IN HAZARD EVALUATION

For a laser safety officer (LSO) to perform the hazard evaluation cited earlier, they need an explanation of the work process by the user. The best way would be a demonstration of how they operate. From this conversation or demonstration, the LSO can see opportunities for safety improvements that might not be apparent to the user. As suggestions are made, they need to be made as "What if we try this or can this be done." Why this approach, well there might be valid reasons why the suggestion will not work on the flow of the experimental procedure. Of course, the suggestion might be doable and the user had not considered them or was even aware of them.

ANSI Z136.8 SAFE USE OF LASERS IN RESEARCH, DEVELOPMENT, AND TESTING

The ANSI Z136.8 standard contains a number of sample audit forms. One is for the laser safety program itself and the others are for laser use areas. They are certainly worth a look for ideas of areas to look at and confirm functionality.

LASER AUDIT FORMS (SEE CHAPTER 4 FOR SAMPLE AUDIT FORMS)

There are many versions of laser audit forms available on the web, some are more compete than others. One thing many have in common is questions or check boxes on Center for Devices and Radiological Health (CDRH) requirements for the laser such as aperture label, logo label, protective housing label, and emission indicator. In the author's opinion, these questions can be dropped. If the laser in question is a certified commercial product by law they must have all these items. It should not be the responsibility of the user or the LSO to replace or place such labels and if the engineering product items are missing (remote interlock connector, emission indicator), trying to add these could void a warranty or be an unnecessary burden to the user. The ANSI Z136.8 standard states that such items are preferred but not required. They add little to a laser safety audit.

Some useful audit items usually overlooked are the following (many more suggested questions can be generated, but these are good food for thought):

What is the level of lighting during normal use?
Do potential reach in hazards exist?
Are remote viewing systems in use?
Are fiber ends labeled?
Is a fiber scrapes box needed and presented?
Where is laser protective eyewear stored?
Are users sitting at workstations protected?
Do beams cross a walkway, if so what precautions are taken?
Is there a written procedure for entryway interlock checks?
Is there a log of entryway interlock checks and frequency?
Is unattended open beam work performed?
Are contact lists up to date?

4 Documentation of Laser Safety

Ken Barat

CONTENTS

In today's world safety has two faces: real safety and documentation of that safety. First, "is one safe?" and second, "can we prove through documentation that we are safe?"

Going without an incident is not proof that one is safe; it might be because the users and institution have the correct polices and follow them. It could be no one has fallen into the pit that is in front of them, just walking around it. This is where safety documentation becomes very important. Documentation alone will not provide safety, but it can help staff focus on safety elements. A safety tool that has found great success within many institutions is Integrated Safety Management (ISM). This chapter explains ISM and provides examples of laser safety documentation; consider them templates for your manipulation and adoption.

ISM has several steps that build and reenforce safety. As with any philosophy or technique take it with a grain of salt and remember flexibility not a rigid approach will allow and produce the greatest adherence and safety. ISM is the corporate philosophy that sets the official safety culture.

INTRODUCTION TO THE INTEGRATED SAFETY MANAGEMENT

ISM is the means by which EH&S requirements are integrated into the planning and execution of work. It consists of two related components: underlying principles and operations (functions or processes). Institutions must systematically integrate

ES&H into management and work practices at all levels so that missions are accomplished through effective integration of ES&H management into all facets of work planning and execution. In summary, the overall management of ES&H functions and activities becomes an integral part of mission accomplishment. Another way of stating this is when putting together a grant proposal, the cost of safety needs to be incorporated into the proposal, not an afterthought. ISM has seven defined guiding principles that are the fundamental policies to use in the management of ES&H. They are as follows:

1. Line management responsibility for environment, health, and safety
   - Line management is responsible for the protection of the public, the workers, and the environment. More specifically, all employees, contractors, and visitors who function as managers or represent management are responsible for integrating EH&S into work and for ensuring active communication up and down the management line and with the workforce.
2. Clear roles and responsibilities
   - Clear and unambiguous lines of authority and responsibility for ensuring EH&S are established and maintained at all organizational levels within the laboratory and for work performed by its contractors. At the institutional level, this principle is manifested in contract language, position descriptions, and work authorization documents.
3. Competence commensurate with responsibilities
   - All personnel must possess the experience, knowledge, skills, and abilities necessary to discharge their responsibilities. Competence includes training, experience, and fitness for duty.
4. Balanced priorities
   - Resources (financial and human) are effectively allocated by staff to address EH&S, programmatic, and operational considerations. Protecting the public, workers, and the environment is a priority whenever activities are planned and performed.
5. Identification of safety standards and requirements
   - Identification of EH&S standards and requirements. Before work is performed, the associated hazards are evaluated and an appropriate set of standards and requirements is known by employees, contractors, and visitors. These standards and requirements, if properly implemented, provide adequate assurance that the public, workers, and the environment are protected from adverse consequences. The appropriateness of the current standards set will be established at least annually.
6. Hazard controls (including adverse environmental impact) tailored to work being performed. Administrative and engineering controls must exist to prevent and mitigate hazards. They are tailored to the work and associated hazards being performed. This tailoring or flexibility must be recognized by the institution.
   - This requires judgment to be exercised at the appropriate decision level and technical expertise.

7. Operations authorization
   • The conditions and requirements to be satisfied for operations to be initiated and conducted are clearly established and agreed upon.

ISM has five defined core functions for integrated ES&H management that comprise the underlying process for any work activity that could potentially affect the public, the workers, and the environment.

1. Define the scope of work
   • Missions are translated into work, expectations are set, tasks are identified and prioritized, and resources are allocated. A clear definition of the tasks that are to be accomplished as part of any given activity is critical.
2. Analyze the hazards
   • Hazards and environmental aspects associated with the work are identified, analyzed, and categorized. Analysis and determination of the hazards and risks associated with any activity; in particular, risks to employees, the public, and the environment.
3. Develop and implement hazard controls
   • Applicable standards and requirements are identified and agreed upon, controls are established to prevent and/or mitigate hazards, environmental aspects are identified and evaluated for reduction, the ES&H envelope is established, and controls are implemented. Controls that are sufficient to reduce the risks associated with any activity to acceptable levels. Acceptable levels are determined by responsible line management, but are always in conformance with all applicable laws and standards.
4. Perform work within controls
   • Readiness is confirmed and work is performed within the ES&H envelope established. Conduct of the tasks to accomplish the activity in accordance with the established controls. Staff needs to recognize when work is starting to exceed defined limits. This should trigger reauthorization.
5. Provide feedback and continuous improvement
   • Feedback information on the adequacy of controls is gathered, the efficiency of reducing environmental impacts is researched, opportunities for improving the definition and planning of work are identified and implemented, line and independent oversight is conducted, and, if necessary, regulatory enforcement actions occur. These five core functions are applied as a continuous cycle with the degree of rigor appropriate to address the type of work activity and the hazards and/ or environmental aspects involved. Implementation of a continuous improvement cycle for the activity, including incorporation of employee suggestions, lessons learned, and employee and community outreach, as appropriate.

ISM SUMMARY

The five functions of ISM continuously feed back into each other. Management and staff can make ISM a valuable tool by always looking for areas and processes to improve. The guiding principles and core EH&S functions are closely related.

DOCUMENTATION OF SAFETY

What follows next are, as stated in the beginning of this chapter, examples of laser safety documentation; consider them templates for your manipulation and adoption.

SAMPLE LASER CHAPTER

Editorial comment, laser safety chapters seem to fall into two styles, one the most common contains a great deal of tutorial information, as if it is a training tool. The other is more of what one needs to do to be authorized for laser work at the facility. This author has written both, but over the years has found users tended to prefer the latter than the former.

X.1 LASER SAFETY POLICY

The objective of the laser safety program is to ensure that there is no unintentional exposure of laser radiation in excess of the maximum permissible exposure (MPE) limit to the human eye or skin. This policy on laser safety requires that all lasers and laser systems be operated in a manner comparable to the American National Standards Institute (ANSI) Z136.1, Standard for the Safe Use of Lasers, and ANSI Z136.8 Safe Use of Lasers in Research, Testing or Development (one may wish to add others is required such as the Z136.6 for Outdoor laser use). Additionally, the program is designed to ensure that adequate protection against collateral hazards is provided. These collateral hazards include the risk of electrical shock, fire hazard from a beam or from use of dyes and solvents, and chemical exposures from use of chemicals and vaporization of targets.

To implement the policy properly while giving the greatest possible latitude to the researcher in consideration of the needs in a research setting, all laser operations at (fill in institutional name) must be reviewed and approved by the laser safety officer (LSO) (see LSO responsibilities). The requirements for laser safety are multilevel and include engineering controls, administrative controls, and training.

This chapter does not describe the theory behind the laser or a number of laser safety–related topics such as bio-effects and how to select laser eyewear. Laser safety background and instructional material can be found (you can list in an appendix or separate document or just delete). This includes material on

- Laser classification
- Biological effects or wavelengths
- Laser eyewear selection parameters
- Commercial laser product requirements

- Explanation of nonbeam hazards
- Sign making
- Use of laser pointers on-site
- Laser safety tools and resources

X.2 OVERVIEW CHART

Training Requirements

Class 3B and Class 4 users	EH&S course no. required
Retraining	3-year frequency
Classes 1–3A	EH&S no. recommended
Medical surveillance	Contact health services, extension
If exposure suspected	

Standard Operating Procedures (SOP)

Classes 1–3A/R work	Not required
Class 3B and Class 4 laser users	Required

Control Measures

Interlocks	Contact LSO first
Class 3B and Class 4 requirements	See this chapter section

Laser Supplies

Warning signs	Contact LSO
Laser labels	Contact LSO
Eyewear literature	Contact LSO
Eyewear holders	Contact LSO
Curtain material	Contact LSO

X.3 RESPONSIBLE PARTIES FOR IMPLEMENTING THE LASER SAFETY PROGRAM

X.3.1 PRINCIPAL INVESTIGATOR

The principal investigator (PI) represents line management and sets the tone for laser safety within a research group. PI must ensure that all laser users receive adequate and appropriate laser safety training on two levels. Fundamental laser safety course and training from the PI or designee with respect to the hazards and safe operation of laser system(s), commonly called on-the-job training (OJT). A detailed list of PI responsibilities follows:

- Ensure that all personnel complete the required laser safety course.
- Ensure documented OJT to operate lasers safely.
- Ensure that all personnel report to health services for laser eye examinations, as outlined in the section on "Note from ANSI Z136/1-2007 on, 'Baseline eye examinations are no longer required'," and after any suspected case of laser eye exposure.

- Prepare a SOP for laser operation and ensure that the provisions of the SOP are implemented.
- Ensure any visitor receives a site/experimental hazard orientation as part of any laser use area tour.
- Ensure that visitors and new employees are allowed to work unsupervised only on tasks for which their formal training and OJT are completed.
- Ensure the LSO is notified of proposed laser acquisitions.
- Ensure the LSO is notified of changes to the laser control area configuration that impact safety.
- Ensure the LSO is notified of changes in laser use.

X.3.2 Work Team Leader

Many times due to competing responsibilities, the PI may spend little time in the actual laser use area. In these cases, the PI will select a work team leader. This person acts as a laser safety supervisor (LSS).

- LSS shall ensure that required and appropriate safety procedures are followed.
- Ensure that any laser safety devices (interlocks, etc.) are functioning properly.
- See that users receive required training, institutional and on the job.
- Check that laser eyewear is in good condition and is appropriate for the laser applications in use.
- Keep inventory system current (this includes tracking inventory additions, transfers out of lab, lasers going into storage, and eventual disposals).
- Notify LSO of any changes that might impact laser safety.

In short, they take on all the safety responsibilities of a PI as listed in the section on "Principal Investigator."

X.3.3 Laser User Responsibilities

Final responsibility for laser safety rests upon the individual laser user. They should have received proper fundamental laser safety training, OJT that allows them to perform their tasks and be comfortable enough to speak up in times of question or doubt. Understand their work goals and be familiar with the institution's "Stop Work Authority" policy. Without user buy-in, no safety program can be successful. User responsibilities include the following:

- Attending appropriate training before operating the laser or laser system.
- Receive and document appropriate OJT.
- Receive applicable medical surveillance.
- Have read applicable SOP.
- Comply with established policy, SOP, and other procedural requirements.
- Work in a safe manner following laboratory policy and procedural requirements.
- Promptly report any malfunctions, problems, accidents, or injuries, which may have an impact on safety.
- Immediately report any suspected laser eye exposures to the laser supervisor, health services, and the LSO.
- Complete retraining when applicable.

X.3.4 LASER SAFETY OFFICER

The key to effective laser safety per the ANSI Z136.1 and Z136.8 standard is the role of the LSO. "An individual shall be designated the LSO with the authority and responsibility to monitor and enforce the control of laser hazards and to effect the knowledgeable evaluation and control of laser hazards. The LSO either performs the stated tasks or ensures that the task is performed." The LSO has the authority and responsibility to monitor and enforce the control of laser hazards. The LSO has the authority to suspend, restrict, or terminate the operation of a laser or laser system if he/she deems that laser hazard controls are inadequate. The LSO's duties and responsibilities include, but are not limited to the following:

- Maintaining the institutional laser safety program
- Providing technical advice for laser operations
- Reviewing and approving SOPs that reference laser hazards
- Developing and reviewing alternate control measures than those listed in ANSI Z136.1 and Z136.8 under the authority Z136.1 and Z136.8 grants to the LSO
- Performing an annual laser safety audit of each laser SOP
- Performing observational visits to laser use areas
- Approving specific safety plans for laser diode and fiber use when applicable
- Upon request, calculating nominal hazard zones
- Ensuring and providing calculations of laser eye hazards parameters (MPE and optical density) that will ensure eye protection. Advise laser users and supervisors on appropriate eye protection (full protection of alignment eyewear). This includes frame styles and prescription options
- Ensuring laser safety training is provided for Class 3B and Class 4 laser users, through lecture, web-based, or other techniques
- Ensuring the appropriateness of OJT
- Investigating all instances of suspected laser eye exposure
- Participating in investigations of beam- as well as nonbeam-related
- Additional duties are required or referenced in ANSI Z136.1 and Z136.8

X.3.4.1 Substitution of Alternate Control Measures

The ANSI Z136.1 and Z136.8 Laser Standard establishes the LSO's authority to modify the control measures required for Class 3B and Class 4 lasers or laser systems. Upon documented review by the LSO, control measures in particular engineering controls can be replaced by administrative or other alternate controls that provide equivalent protection. The approval of these controls is incorporated into the SOP for the laser experiment. An example would be a use of curtain maze or posting in place of an entrance interlock.

X.3.4.2 Temporary Control Area/Temporary Work Authorization

The concept of a temporary laser-controlled area comes straight out of the ANSI Z136.1 standard (greatly expanded in Z136.8). In the research setting, this is of great value. Its purpose is to allow an authorization of laser work in settings where a formal authorization (SOP) does not already exist. Examples would be acceptance testing of laser equipment in a lab during initial setup, a short-term change to an existing experiment, or a plan to follow while an engineering control is being repaired. The LSO will generate a temporary control area (TCA) memo listing the circumstances and then control measures to be followed by all parties. All parties will sign the memo indicating they understand the controls and will abide by them. This memo is to be posted at the work site and usually has a 2–4-week duration and limited extension capability.

X.3.4.3 Laser Safety Protocols and Interpretations

The LSO shall develop and issue laser protocols (interpretations) of ANSI Z136.1 and/or Z136.8 requirements. These maybe necessary when elements of one's program meet the spirit of ANSI Z136.1 or Z136.8 while not the letter. Examples could be use of Master Key. These protocols will be reviewed by the Laser Safety Committee to ensure they fit into research environment and needs. Such protocols will be reviewed by the Laser Safety Committee to ensure they fit into research environment and needs.

X.3.5 LASER SAFETY COMMITTEE

The institutional safety committee has established a subcommittee to advise the laser safety program on laser safety issues that arise. This subcommittee provides a check and balance approach to the EHS Laser Safety Program. The responsibilities of the LSC are to

- Recommend the establishment or modifications of laser safety policies.
- Review laser-related accidents.
- Review protocols and interpretations by the LSO.
- Review cases, which involve repeated infractions of laser safety rules and recommend actions.
- Meet on a regular basis (minimum quarterly).
- Review appeals from users in disputes between themselves and the LSO determinations.

X.3.6 EH&S HEALTH SERVICES GROUP

- Arrange or perform laser eye examinations.
- Advises laser users and LSO of any ocular abnormalities that could be attributed to laser exposure or that could be relevant to laser use.
- Arrange laser eye examinations when an injury is suspected.

X.3.7 EH&S PROFESSIONALS

Provide guidance in handling laser-associated hazards such as

Handling of laser dyes and other toxic materials
Ventilation requirements for laser targets and toxic materials
Hazardous gas controls
Electrical hazards
Seismic controls

X.3.8 PURCHASING

- Advises LSO of all laser purchases

X.4 REQUIREMENTS FOR LASER USE

X.4.1 REQUIREMENTS FOR CLASS 1 PRODUCT: LASER SCANNING CONFOCAL MICROSCOPE, CELL SORTER, AND LASER INTERFEROMETER

The Center for Devices and Radiological Health, the Food and Drug Administration body, tasked with developing laser light performance product safety standards defines a Class 1 product as any laser product that does not permit access during the operation to levels of laser radiation harmful to eyes or skin. Class 1 levels of laser radiation are not considered to be hazardous.

Many times these Class 1 laser products are used in work areas where no attention or training on laser safety has taken place. Therefore when open for service, these Class 1 laser products may introduce a potential laser hazard in work area and staff unaccustomed to the laser safety considerations.

During routine operation of a Class 1 product where no laser radiation exposure is possible, no laser safety training, medical surveillance, or user restrictions from a laser safety perspective are required. Each unit should have an individual responsible for its proper use.

Service or maintenance: Yielding accessible laser radiation, the following steps will be taken:

1. Notification of accessible laser radiation will be made by posting access doors with a "Notice Laser Service/Alignment in Progress" signs, or setting up an exclusion zone near the unit with the same posting
   a. If necessary doors, windows, and so on will be blocked
2. The user/service person will wear the appropriate laser protective eyewear for the wavelength and exposure output expected
3. The individual performing the work is responsible for the safety of anyone entering the use area
4. A staff member shall be present whenever a non-company person (vendor service person) is working on energized laser equipment
5. Unattended open laser work is not allowed
6. The LSO will be notified of the work (extension)

X.4.2 OPEN BEAM CLASSES 1, 2, 3A, 3R, 1M, 2M LASERS OR LASER SYSTEMS

By their classification definition when used as intended, these laser or laser systems will not present a hazard to the user or those around them. When personnel not familiar with the low-hazard nature of laser operations are present, a sign advising of the low-hazard nature of the operation may be appropriate. In use areas where there is a likely chance of the beams being viewed through a telescope, microscope, binocular, eye loop, or collecting optics, a review by the LSO or posting of caution signs will be required.

X.4.3 GENERAL REQUIREMENTS FOR CLASS 3B AND CLASS 4 LASER SYSTEMS

Authorization to use Class 3B or Class 4 lasers or laser systems is granted to laser user through a formal authorization process. The authorizing document is referenced to as a SOP. Only the PI can submit a SOP for review. There are situations where the LSO may approve use of Class 3B and Class 4 lasers or laser systems without a SOP. These are in selective predetermined situations which will be described in the section on "Temporary Work Authorization," or the ANSI term Temporary Control Area. Class 3B and Class 4 lasers or laser system use in communications is exempted from the SOP requirement, but training requirements do exist for service staff. The requirements for Class 3B or Class 4 laser or laser system are as follows:

1. An approved and current standard operating procedure (SOP)
   a. This entails completion of required training
      i. Completion of laser safety training
      ii. Completion of tri-annual refresher training
2. Obtaining a baseline laser eye examination
3. Approval of laser controls by the LSO
4. Having all lasers listed in the Laser Inventory System
5. Use of approved laser protective eyewear when required

X.4.3.1 Medical Surveillance, Note No Longer a Shall Item

Laser eye examinations will be performed as follow-up to suspected eye exposure or at the request of the LSO or user. It is strongly recommended that the laser user receives an additional laser eye examination on termination of employment.

For those who routinely receive ultraviolet exposure as part of their experimental work, an annual skin evaluation is recommended.

X.4.3.2 Training

All employees and guests who are authorized to work unsupervised with Class 3B or Class 4 laser must complete two laser safety elements.

Laser training course: This a fundamental laser safety course. In addition, it is highly recommended that anyone who regularly works in a Class 3B or a Class 4 laser area attends this course. The laser safety course is offered regularly and can also be customized for specific laser user groups. All laser operating personnel should complete the approved laser safety training before working unsupervised with Class 3B or Class 4 lasers.

OJT: The second and most important element of laser safety is OJT. The PI or designee shall train/orientate staff on the hazards of the specific experimental work to be done. Laser users are responsible for knowing the safety requirements that apply to their specific laser or laser system and for knowing the contents of the applicable activity hazard document. OJT shall include location and mitigation of potential hazardous beams/reflections, hazards associated with the work, and the use of all required personal protective equipment (PPE). The depth of OJT must be tied to the responsibilities and degree of hazard. It can apply to the entire experiment or elements of the research work (specific laser systems or experimental operations). This OJT must be documented. The format for documenting OJT can be as part of the SOP or paper documents, which must be available for review. The PI is ultimately responsible to ensure that all personnel, visitors, and students with access to the laser use area have a clear understanding of the controls associated with laser operation.

X.4.3.2.1 Refresher Training

Laser users must participate in refresher training at least once every 3 years.

X.4.3.2.2 Class 3B and Class 4 Use Area Control Measures

The goal of Class 3B and Class 4 laser or laser system controls is to prevent exposure to hazardous levels of laser radiation by authorized and nonauthorized individuals who may enter the control area. Each particular set of controls will be site-dependent on a laser hazard review by the LSO (see Hard evaluation method below for guidance). For clarification,

Access control, can be nondefeatable interlock, defeatable interlock, and card key access and administrative control (posting). See the LSO for the correct access control approach

Beam control, unless planned all laser beams shall be contained onto the optical table, this can be through beam blocks, perimeter guards, and complete table enclosures

Exiting control area, no beam shall unintentionally leave the laser use area, this can be accomplished by curtains, barriers, or beam blocks

Posting, entrances to these laser user areas will be posted with approved laser warning signs

Eyewear, as required approved laser eyewear will be provided and worn

Training, supervised or conducted by an individual knowledgeable in laser safety.

Authorized, only staff who have met Class 3B and Class 4 use requirements can use those class laser unsupervised

X.4.3.3.3 Hazard Evaluation Method

Regardless if it is the users' self-assessment of their laser setup or a laser safety review by the LSO, the following method of safety analysis is recommended. Consider the laser use area as five discreet components. Evaluate each for hazards and control measures.

The laser source(s):

What class laser(s) is being used?
Does it have any inherent risks?
Toxic gases (i.e., excimer laser)
Power supply
Cooling lines

Beam path:

Is the beam path entirely enclosed?
Is the beam path entirely open?
Can sections of the beam path be enclosed?
Do wavelengths change along the path?
Does the beam intensity change along the path?
Does the beam diameter change along the path?
Does the laser path leave its original plane (i.e., become vertical)?
Does the beam path leave the optical table?

Environment:

What type of access control does the laser use area have or need?
    Interlock
    Posting alone
    Card access

Does the beam have a possible direct route out of the control area?
Where should protective eyewear be stationed?
Do nonbeam hazards exist in the area and are they being mitigated?
    Electrical
    Seismic
    Housekeeping
    Chemical, and so on

People:

Will all who work in area be authorized laser users?
Do all users have required training?

X.5 SPECIAL TOPICS

X.5.1 Guests and Visitors

Guests, visitor, and new employees may use lasers under the direct supervision of an authorized laser user (for new users a maximum of 30 days before completing the training and medical surveillance requirements).

X.5.2 Laser Alignment

The majority of laser accidents in research activities occur while aligning the laser or similar beam manipulation activities. All possible steps should be taken to prevent any such accidents. These activities should receive prolonged and detailed OJT as a precautionary measure. Only laser users who have received OJT and are authorized by the PI or designee shall perform laser alignment/beam manipulation activities.

The following are laser alignment protocols that should be in every laser alignment procedure:

- The laser beam is never to be viewed directly.
- As a precaution, reflective jewelry, ID badges, and so on will be taken off by those handling the laser.
- Whoever manipulates or moves optics shall be responsible to check for stray reflections. When found, those reflections shall be contained to the optical or experimental table(s), even if they are below an eye hazard level.
- Alignment should be conducted optic-to-optic and constant checking for stray reflections.
- Alignment procedures are performed at the lowest possible laser output.
- Remote viewing approaches shall always be considered.
- Use of coaxial visible beam shall be considered.
- The use of alignment eyewear for visible beam is allowable with the approval of the LSO.

Alignment Eyewear Approval
Alignment eyewear is an option for use with visible beams and gives partial visibility for beam observation from diffuse or attenuated reflections, but not full protection from the direct beam. Use of alignment eyewear will be limited to situations where exposure to a direct beam is prevented, and thorough justification for the use of alignment eyewear must be provided by the laser user.

This must be specifically stated as part of an AHD and approved by the LSO. Approval for the use of alignment eyewear must be renewed annually, based on an evaluation of the experimental conditions by the LSO. The following criteria will be used as guidance in determining whether alignment eyewear is appropriate for particular circumstances and for selecting the appropriate eyewear optical density (OD)*:

1. Protection from exposure to the direct beam provided by beam blocks and/ or enclosures
2. Alignment eyewear providing protection from stray reflections from uncoated surfaces, typically ~4% of the direct beam; OD reduction ~1.4
3. Alignment eyewear providing protection from diffuse scatter at >0.5 m observation distance; OD reduction ~4–5
4. Attenuation of direct beam *below* Class 3B level

X.5.3 SUSPECTED LASER INJURY

Laser users must report all laser accidents no matter how minimal to the laser safety supervisor (LLS) and LSO. Accidental or suspected laser beam exposure is a serious event. Key item is to keep the individual calm.

In case of suspected laser injury operations:

- Seek medical attention for the individual exposed.
- Notify others in the area.
- Stop and safe out work area.
- The laser setup will remain unchanged to allow for analysis of the cause of the accident.
- Call (insert phone number) during normal work hours.
- Off-hours call (insert phone number if different from above).
- By cell phone, call 911.
- Exposed employees will be transported to health services for evaluation. Notify laser use supervisor and LSO.

The LSO will perform a follow-up investigation on all reported suspected injury incidents. The goal is to determine the factors that lead to the incident and develop corrective actions to prevent its reoccurrence at any LBNL use site. When appropriate, a lessons learned will be generated.

X.5.4 LASER PROTECTIVE EYEWEAR

The energy emitted from laser or the reflection of a laser beam can present a highly concentrated energy source, sufficient to cause permanent eye injury. Although engineering controls are preferred to reduce this hazard, it may be necessary or prudent to use laser protective eyewear. The eyewear should be matched to the wavelength(s) emitted and for the laser intensity. Laser protective eyewear must be marked with the optical density of the lens for which protection is provided (OD and wavelength). How well the eyewear fits along with visual light transmittance

---

* Determination of the appropriate OD (and justification) must be made for specific circumstances by the laser user in consultation with the LSO.

round out the remaining leading factors in selecting laser eyewear. Laser protective eyewear can either be full protection or alignment style.

Full protection will block a direct or reflective strike from the laser source in question for a minimum duration of 10 seconds. Full protection shall be worn whenever one is dealing with solely invisible beams.

Alignment eyewear is an option for use with visible beams and gives partial visibility, but not full protection from a direct or reflected strike.

In the research setting, there may be times when no commercial laser protective eyewear is available to meet experimental needs. At these times, the user and LSO must reexamine beam controls and if necessary select a less than optimal protective eyewear.

X.5.5 Requirements for Laser Use by Staff Off-Site from [Name Institution]

The safety of laser users on site is extremely important. No employee shall work in a facility off-site where they feel their safety is at risk. If such an occasion arises, either stop work or discuss with local management or contact the LSO to see what assistance the LSO can grant the host site. Off-site locations fall into two categories, where (name institution) has management responsibility and where (name institution) does not have management responsibilities.

STANDARD OPERATING PROCEDURE

The *noun* "standard operating procedure" has two meanings.

1. The procedure that would normally be followed
2. A prescribed procedure to be followed routinely

If any document is overused in laser safety, it is the SOP, or better yet is any document that does not live up to its potential or intended purpose it is the SOP. Just to beat this one more time, it is like the first round draft choice that does not make it out of training camp.

Why am I being so hard on this cornerstone of laser safety? Just for that very reason, so many emphasize the SOP but in reality once signed most users do not think about it for at least a year. Think about yourself: once you pass your driver's license exam how often have you thought about the meaning of a green curb, and how to signal a turn with our hands, or the stopping distance when one is going 55 mph? Not to say this is universal, this author has met users and institutions where the SOP is used as intended, it lives up to its potential.

So getting to the SOP, it is really a contract between the user and institution on how laser work will be conducted. It outlines hazards, most importantly the controls, training requirements, maybe eyewear choices, laser inventory, what to do in an emergency, how alignment is conducted, and maybe a few other things.

The SOP as most commonly used is not an SOP. It really does not tell one how to do the work, what knob to turn, and how far. Rather it serves as a safety document, explaining the hazards involved in the work and how those hazards will be mitigated. Over years of experience, I have found two types. One that is very generic and full of "boilerplate" material and the other that is very operation- or site-specific. A certain amount of boilerplate is good; it does make sure essential safety requirements are incorporated into the document. The SOP should be considered a contract between the user and safety department. Not a license to run wild, but a performance contract with expectations for all. Therefore what one wants is a marriage of both. This SOP template allows that.

SAMPLE SOP

SOP Template (Term PI = Principal Investigator, which some institution use RI = Responsible Individual, or Supervisor or Big Shoot, Work Team leader, or the Man).

MITIGATION OF LASER HAZARD

Laser radiation from Class 3B and Class 4 laser sources present a real potential eye and skin hazard. Within an instant, a life-changing accident can occur. The purpose of this laser SOP is to outline general and specific controls that will allow the user to work with laser in a safe manner. It is critical that users complete the user-specific sections of this schedule. The general guidance presents what is accepted practice for safe work, but it is the specific controls and user compliance for this SOP that will make all the difference.

GENERAL LASER SAFETY POLICY

1. All laser operations shall be conducted in accordance with the EHS Chapter on Lasers.
2. All operating personnel with possible exposure to laser radiation shall have had laser safety training (Course no., a laser eye exam [if applicable] and OJT).
3. Laser areas shall be kept as clean as possible. Reflective objects such as tools, optics, and screws shall be kept away from laser beams.
4. An authorized laser operator shall be in the lab whenever a laser is operated with noninstitutional personnel present. Visitors accompanied by an authorized individual who shall assure that all are wearing required laser protective eyewear and protective clothing.
5. The institutional LSO shall be notified if any lasers are added or any experimental changes are made to an approved laser setup that impact safety. Upon such notification, a hazard evaluation will be performed.
6. Before turning on the laser system, the operator is responsible to inform all in the laser lab that laser operation is pending. If protective eyewear is required, the operator is responsible to see that all staff in the room is wearing the proper PPE. This extends to
   a. Anyone entering the room once the laser is on, if they are at risk of exposure.
   b. Communicate your intentions to others present at all times (e.g., before opening/closing shutters, removing beam blocks, or actions that might put others unintentionally at risk).

7. Modifications determined to have a minor impact on safety per the PI/work team lead and LSO will be amended to this SOP without formal EH&S review. These modifications will be reviewed at the time of the SOPs' annual renewal.

LASER

All Class 3B and Class 4 lasers authorized under this SOP must be listed in the laser inventory, this includes homemade units.

Commercial lasers in use   [ ]
Commercial OEM lasers in use   [ ]
Homemade laser in use   [ ]

BEAM PATH

The beam path is considered to be the space between the laser source and intended termination of the beam. This may include beams transported through other lasers. Within the beam path, wavelengths may change and output may increase or decrease or both. The beam could be open or enclosed or a combination of both.

Beam Controls

Whenever possible (within experimental considerations), appropriate enclosures, barriers, beam blocks, or beam tubes shall be applied to contain laser radiation below the threshold that could cause eye or skin damage.

When working with Class 4 open laser beams, all beams shall be terminated with nonflammable beam dumps at the end of their required beam paths to prevent collimated or focused beams from leaving the laser tabletops or other experimental areas.

No lasers beams, direct or scatter, are allowed to unintentionally exit the laser use area.

Total beam path enclosure provides the greatest level of laser safety.

Describe how beam containment to the optical table is achieved (include the use of beam blocks and perimeter guards, curtains, etc.). In some cases, you may wish to describe how any beams are blocked from exiting the laser use area.

Open Walkways

Any beam path between tables or between laser tables and targets where an open walkway exists shall have a control in place to warn or block those in the room from crossing the pathway (e.g., hinged tubes, beam tubes, retractable tapes, or chains).

Describe control measures for open walkways.

Beam Alignment

The majority of laser accidents in research activities occur while aligning the laser. All possible steps will be taken to prevent any such accidents. Alignment for the particular systems in this SOP shall be covered in required OJT of fully authorized users.

The laser beam is never to be viewed directly.

As a precaution, reflective jewelry, ID badges, and so on will be taken off by those handling the laser.

Only laser operators authorized by the PI may perform laser alignment activities.

Whoever manipulates or moves optics shall be responsible to check for stray reflections. When found, those reflections shall be contained to the optical or experimental table(s), even if they are below an eye hazard level.

Alignment should be conducted optic-to-optic and constant checking for stray reflections.

Alignment procedures are performed at the lowest possible laser output.

Describe any specific laser alignment procedures.

Beam Interaction

Once the laser beam reaches its goal, a consequence is realized. It might be raising an energy level or cutting a material. Laser radiation interaction with a variety of target or sample materials can generate beam target interaction hazards. The purpose of this section is to cause the user to think about what these interactions might be including reflections from a reaction chamber. Beam interactions could include ultraviolet exposure, ionizing radiation, reflective material, and fumes.

Describe any safety precautions that will to be taken to mitigate beam interaction.

ENVIRONMENT

This section responds to the laser use area, which could be an isolated room or a multiuser facility.

Posting

The laser use area must be posted with an approved warning sign that indicates the nature of the hazard. Wording on the sign will be specified by the LSO.

An approved laser warning sign has to be posted in all entrances.

Indicate if you will be using a "Notice Alignment in Progress" sign, during alignment activities.

ACCESS CONTROL

In accordance with the ANSI Z136.1 standard, access control for Class 3B or Class 4 laser use areas can be an engineering control (i.e., interlock) or administrative/ procedural control (posting or locking of doors). Not every Class 4 laser use area requires an interlocked room. After consultation with the PI/work team leader, the LSO shall determine room access controls.

Room Access Interlock

In laser use areas with room access interlocks, the interlocks shall be tested every 6 months. The PI or designee will note the results on the interlock check form. The interlock check is conducted following a written procedure (it is suggested the procedure be placed into the upload schedule). Any access code to the laboratories shall be issued to authorize personnel only.

Noninterlocked Access
Access control may be achieved by key card control, locking of doors, or posting. Describe access control for this laser use area.

Lighting Conditions

Room illumination can have a dramatic effect on eyewear selection and general safety within a laser use area, indicate level of illumination:

Normal room light [ ] low light illumination [ ] Complete darkness [ ]

Workstations

A common error is to have a workstation in a laser use area, where the users are in line with the laser beam path. In such cases, a partition between the user and optical table is required.
If this condition is present describe user protection steps.

Unattended Laser Work (Only to Be Added When Needed)
In the event of unattended operation of nonenclosed lasers and noninterlocked room, the following controls shall be in place:

A sign shall be posted outside the laboratory, which states that an unattended laser operation is under way (the sign can be obtained from the LSO).
Emergency shutdown procedures and emergency contacts shall be posted at the door.
The LSO can review and allow exemptions for access restricted areas.

Nonbeam Hazards
List any hazards associated with this SOP and the appropriate level of controls (i.e., electrical, chemical, and mechanical).

Personnel

This section is to emphasize training requirements. Actual training requirements and level of compliance can be found in the section on "Training Schedule," excluding OJT.

Training
All laser operating personnel should complete LBNL-approved laser safety training before working with Class 3B or Class 4 lasers. The PI shall be responsible to ensure that all personnel, visitors, and students with access to the laser use area have a clear understanding of the controls associated with laser operation.

OJT Requirement

The PI or designee shall train/orientate staff on the hazards of the specific experimental work to be done. This will include location and mitigation of potential hazardous beams/reflections, hazards associated with the work, and the use of all required PPE. *This OJT will be documented*, including a signature by the trainer and trainee.

Training Schedule

List here all courses required by this SOP. This may include electrical safety, pressure safety, hazardous waste handling, and so on.

Eyewear

- Laser operators shall use the appropriate laser protective eyewear (have the appropriate optical density and/or reflective properties based on the wavelengths of the beams encountered and the expected exposure conditions).
- This eyewear shall be stored in such a manner as to protect its physical integrity.
- There shall be sufficient laser protective eyewear in hand for users and expected visitors.
- Keep in mind that the need for laser eye protection must be balanced by the need for adequate visible light transmission.
- Laser eye protection shall be inspected periodically to ensure that it is in good condition.
- The eyewear will be reviewed by the LSO or other safety professional during periodic audits.

Alignment eyewear is for laser adjustment work where hazardous laser radiation occurs in the visible portion of the spectrum (400–700 nm). This eyewear reduces but does not completely block the visible spectrum. It does allow the user to see the beam and therefore perform laser alignment (adjustment) activities. The alignment filters should attenuate the radiation level to Classes 2–3A. The use of alignment eyewear can only be authorized by the LSO. That authorization is only valid for a year and for a set of specific circumstances, demonstrated to and evaluated by the LSO.

MAINTENANCE

Service vendors are responsible for providing technicians who are properly trained and qualified to work with laser systems.

Service vendors must follow safety guidance from the SOP and the laser chapter of the EHS manual; this includes the use of laser protective eyewear.

Room access interlocks must receive a functional operational check every 6 months. This check needs to be documented and should follow a written procedure.

EMERGENCY RESPONSE

Authorized laser users will be familiar with the building emergency plan, location of emergency equipment, and emergency procedures for fires, earthquakes, and evacuations. Emergency shutoff for lasers is done at the electrical panel circuit breakers or by electrical shut off switches.

SUSPECTED LASER INJURY

Accidental laser beam exposure is a serious event. In the case of suspected laser injury, operations will be ceased and the laser setup will remain unchanged to allow for analysis of the cause of the accident. Exposed employees will be transported to (insert location, i.e., fire house, medical) for evaluation. Call (insert number for assistance). Make sure to notify any staff in the area, laser use supervisor, and LSO. Key item is to keep the individual calm.

ANNUAL REVIEW

Annual review will be 1 year from approval date. If new hazards have been introduced, a full EH&S review will be required. If no changes other than users have been made, renewal can be granted by user's division safety coordinator. Every 3 years a full EH&S review is required.

ALIGNMENT EYEWEAR APPROVAL FORM

A critical component of laser activities is laser beam alignment. This activity is the central point of a large percentage of unintended laser eye exposures. One of the major contributing factors is not wearing protective eyewear. This problem centers on the issue of beam visibility and eyewear blocking that beam. To combat this, the Laser Safety Program promotes the use of remote viewing (cameras, etc.) and engineering controls. When that is not feasible, the use of alignment eyewear is preferred over no eyewear being worn.

Alignment eyewear is an option for use with visible beams and gives partial protection and visibility for beam observation from diffuse or attenuated reflections, but not full protection from the direct beam. Thorough justification for the use of alignment eyewear must be provided by the laser user. This must be specifically stated as part of an SOP or a memo approved by the LSO.* LSO approval for the use of alignment eyewear must be renewed annually, based on an evaluation of the experimental conditions by the LSO. The following criteria will be used as guidance in determining whether alignment eyewear is appropriate for particular circumstances and for selecting the appropriate eyewear OD:

1. Protection from stray reflections from uncoated surfaces, typically ~4% of the direct beam; OD reduction ~1.4
2. Protection from diffuse scatter at >0.5 m observation distance; OD reduction ~4–5
3. Attenuation of direct beam *below* Class 3B level
4. Protection provided by beam blocks and enclosures

---

* Determination of the appropriate OD (and justification) must be made for specific circumstances by the laser user in consultation with the LSO.

List OD and wavelength coverage of alignment eyewear being used if applicable.
Alignment-specific controls (to be completed by user):
(Examples: reduce output, use of ND filters, use of Iris, use of beam enclosures,
remote viewing)

EXAMPLE

*The alignment eyewear (specified in the preceding discussion) will provide ade-
quate protection against diffuse reflections of laser beams (from matte surfaces).
However, it will not protect users against direct viewing of laser beams or some
specular reflections. The following alignment conditions must be carefully followed to avoid the potential for eye injury:*

   Alignment eyewear is to be used only during "fine" beam adjustments, with as
      much of the laser beam path enclosed as possible to eliminate the risk of
      specular reflections.
   With alignment eyewear, the beam may be viewed only as a diffuse reflec-
      tion (from a matte surface, at a distance >0.5 m), or with remote viewing
      instruments.
   Alignment eyewear will not be used when inserting optics into the beam path,
      since the OD is not sufficient to protect against specular reflections (even
      from uncoated surfaces).
   Beam blocks and enclosures must be used to terminate all stray reflections if
      alignment eyewear is to be used.

ALIGNMENT EYEWEAR APPROVAL

As LSO I approve the use of laser alignment eyewear within the context of the
above SOP. This approval is based on my evaluation of laser alignment controls, present beam set up & outputs as demonstrated to me by:

_____,
   and the assurance that this technique and controls will be followed by
   all authorized to perform laser alignment.
   Prior to the use of alignment eyewear the user must check for stray
   beams and take all practical steps to reduce the intensity of the laser
   beam(s).

Approval granted on _____,
expires one year from this date.

_____
LSO

TEMPORARY CONTROL AREA AUTHORIZATION FORM 1

To: User (list specific name)
From: LSO
Re: Temporary Control Area Authorization
Date:
Expires:

I authorize the use of the _____ laser systems for testing and initial set up activities. Be clear this is not permission for routine operation. This is based on my evaluation of the work area, in consideration that your:

- SOP has been submitted for modification
- Acceptance testing is required for vendor payment
- SOP is being written
- Service work is required

and the following of the safety controls listed below. I grant this approval as LSO per ANSI Z136.1 Safe Use of Laser, the intuitional EHS manual, chapter 16. The engineering and temporary administrative controls listed below are sufficient to mitigate the laser hazard. All workers (staff) in the laser use area need to be informed of the controls and sign this memo indication their understanding and compliance of the controls.

TEMPORARY CONTROL AREA CONTROLS

Sample control measures

1. The beam path will be contained within the optical table.
2. During periods of laser use, a warning sign and flashing LED light will be posted on the lab entrance.
3. The secondary entrance will be posted with a warning sign (supplied by the LSO).
4. In cases of unattended open beam use, the panels on the optical table will be closed and an unattended operation sign will be posted at the front table panel (visible from the entrance door).
5. All laser users should have taken laser safety training.
6. The use of alignment eyewear for the visible beams is authorized.
7. The dye laser solution will be mixed in a hood.
8. The dye pump will be situated in secondary containment.

9. The laser operator is responsible for the safety of anyone entering or working in the room, segregation of the room (non-laser zone) is allowed upon review by the LSO.
10. If any of these conditions are not followed, the laser work must stop.

Signatures

_____

_____

Upon completion of the TCA, fax signed form to LSO @####

TEMPORARY CONTROL AREA AUTHORIZATION FORM 2

Temporary Work Authorization

Work as described below may be performed during the stated period after all required concurrences and authorizations have been obtained.

Effective date: **Expiration Date:**
Work Location: Building   Room
                                             Maximum duration: two weeks
Work Scope (describe work including permitted and prohibited activities, boundaries and "stop points" as appropriate):

Controls required I am issuing TWA as Laser Safety Officer for////////per ANSI Z136.1 Safe Use of Laser, EH&S Manual laser chapter. My hazard evaluation conclusion is that engineering and temporary administrative controls listed below are sufficient to mitigate the laser hazard:

Personnel included in this authorization (signature denotes verification that training in the provisions of this Temporary Work Authorization has been provided)

Supervisor

Name	Signature	Date
Name	Signature	Date
Name	Signature	Date
Name	Signature	Date
Name	Signature	Date
Name	Signature	Date

Concurrences and Work Authorization

Principal Investigator Concurrence

Name   Signature   Date

EH&S Concurrence

LSO   Signature   Date

LASER INVENTORY

Note: *A laser inventory is not a requirement of Z136.8*
The *noun* "inventory" has five meanings.

1. *Inventory*, stocklist—a detailed list of all the items in stock.
2. Stock, inventory—the merchandise that a shop has on hand; "they carried a vast inventory of hardware."
3. *Inventory*—(accounting) the value of a firm's current assets including raw materials and work in progress and finished goods.
4. Armory, armoury, *inventory*—a collection of resources; "he dipped into his intellectual armory to find an answer."
5. *Inventory*, inventorying, stocktaking—making an itemized list of merchandise or supplies on hand; "the inventory took two days."

The *verb* "inventory" has one meaning.

1. *Inventory*—make or include in an itemized record or report; "Inventory all books on the shelf."

A laser inventory has the elements of all the preceding definitions. It is also one of the elements of a laser safety program most overlooked. In particular when one is talking about a comprehensive institutional inventory. While your property management group may keep an inventory of equipment that will contain lasers, why should there be a laser inventory? Well here are some reasons to support keeping an inventory database.

Having an inventory allows one to direct users of similar lasers to help each other when problems arise.
Can show you where to apply safety efforts as well as shifts in technology direction.
It might be a regulatory requirement.
The New York State Department of Labor requires that institutions possessing Class 3B and Class 4 lasers must maintain an inventory of such lasers and update the inventory at least annually. The State of Arizona has a similar regulation.

It might be an internal institutional requirement.

The more lasers you have the easier it may be to document.

If one decides to keep a laser inventory, be aware not to populate the inventory form or documentation with too much data. When is too much too much? First let's consider the minimum needed for a solid inventory.

- Owner data: Division, individual responsible for equipment
- Location data: Building, room, and so on
- Identification of the laser: Manufacturer, model, serial number (is nice)
- Laser specifications: Class, continuous wave or pulse, wavelengths

Now we are crossing the line toward too much.

Maximum output, output actually used, wavelengths, pulse duration, pulse reparation rate, energy per pulse. Why is this too much, for these parameters can change and will most likely not be updated on the inventory form? While they are essential for calculation of protective eyewear they really have no place on an inventory form or listing.

The question that comes to mind now is how do I get this inventory information. Here are several approaches:

1. Annual physical inventory (interns and summer students are great for this).
2. Send inventory form to users to confirm and update annually. This will still require some sort of auditing of a percentage of responses for accuracy.
3. Have a mechanism where the purchasing department notifies you of all approved laser purchases (you may miss loaners and trial uses).
4. Make inventory check part of audits.
5. Make an inventory part of SOP documentation.

Some combine the laser inventory with the laser use authorization, here is such a form from that approach:

LASER REGISTRY

I. PI Information

Principal Investigator _____ **Phone** _____

Department _____ **School** _____

Date _____

II. Personnel who use laser system

Name	ID#	Status (student or staff)
_____	_____	_____
_____	_____	_____
_____	_____	_____

III. Laser System Information

1. **System location (Building/Room#)** _____
2. **Laser warning sign on door (Y/N)** _____
   Wording on sign _____
3. **Do users wear safety goggles?** _____
   Type/Manufacturer _____
4. **Are goggles available for visitors?** _____
   Type/Manufacturer _____
5. **Service for laser: in-house (Y/N)** _____
   Contract service company's name _____
6. **Is there a written SOP available?** _____

Complete table below:

Manufacturer	Laser 1	Laser 2	Laser 3
Model #			
Serial #			
Class (1,2,3a,3b,4)			
Type (CW,Pulsed)			
Description (i.e., He-Ne,ND: YAG)			
Wavelength(s)			
Maximum Power/Peak Power (Watts or Joules)			
Pulse Duration (repetition rate)			
Emerging Beam Dimensions (mm)			
Use (holography, alignment, etc.)			

LASER SAFETY AUDITS

The *verb* "audit" has two meanings.

1. *Audit*, scrutinize, scrutinise, inspect—of accounts and tax returns; with the intent to verify
2. *Audit*—attend academic courses without getting credit
   1. Checklist—a list of items (names or tasks, etc.) to be checked or consulted

Audits do have a useful place in laser safety. Also several layers of audits exist. The overall goal is to keep the user's focus on laser safety. They all too easily can become items to find documentation gaps. They should help the laser user stay safe by pointing out items the user may have missed or underestimated its potential for injury.

Even a casual visit to a laser setup area can be considered an informal audit. A comprehensive look around is not required, just a good observation can pick up on the overall safety culture of an area. Focusing the user on housekeeping items can be of great value to all parties.

The audit form can take many forms. A clever idea today is to have the form on a electronic tablet and automatically transfer the data into an electronic searchable database.

Here are some sample forms of varying complexity; you should notice a great deal of similarity for they are all auditing the same topic laser use:

Form 1

Name: ....   Dept. & Section: ....

Building and Room No: ....   Ext. No: ....

Building and room(s) to which this assessment applies: ....

1. Do all the laser sources have the appropriate classification and warning label(s)?   **YES/NO**
    Points to consider:
    Laser products need to bear signs conforming to CDRH or EN 60825-1
    If commercial laser products have been modified their classification should be
    checked.

2. Is the use of optical viewing permitted within the laser area?   **YES/NO**
    If YES, please summarize the precautions taken to preventing hazardous levels of
    laser exposure.
    State if Class 1M and 2M lasers are in use.

3. Are dazzle-susceptible activities (e.g. vehicle driving, working at heights)   **YES/NO**
    permitted within the laser area?
    If YES, please summarize the precautions taken to control these activities
    and/or visible laser beams
    (If no visible-beam lasers, state 'None')

4. Are there normally Class 3B and Class 4 open beam paths in the laser area?   **YES/NO**
    If YES, please indicate which of the following control measures below are
    in place:
    a. All beam paths are enclosed as much as is reasonably practicable.   YES/NO/NA
    b. All beam path components that generate errant beams are locally enclosed.   YES/NO/NA
    c. All beam paths are properly terminated.   YES/NO/NA
    d. All unprotected open horizontal laser beams lie above or below normal eye   YES/NO/NA
        level.
    e. All lasers and optical components on the beam line are securely mounted.   YES/NO/NA
    f. Shiny surfaces (including jewelry) are not permitted around laser beam   YES/NO/NA
        paths.

g.	Laser beam paths do not cross walkways.	YES/NO/NA
h.	All upwardly directed beams are shielded to prevent human exposure.	YES/NO/NA
i.	Laser sources and beam paths are kept under the control of competent persons.	YES/NO/NA
j.	Information of the current laser hazard is clearly displayed at each and every point of access to the laser area.	YES/NO/NA
k.	Low level lighting is provided for 'lights-out' operations.	YES/NO/NA
l.	Persons at risk of exposure to the laser radiation have received adequate laser safety training and instruction.	YES/NO/NA
m.	A visible or audible warning of the potential laser hazard is provided.	YES/NO/NA
n.	Unauthorized persons are prevented from gaining access to the laser area.	YES/NO/NA
o.	Precautions are in place to safeguard visitors entering the laser area.	YES/NO/NA
p.	Multiple wavelengths.	YES/NO/NA
q.	Laser safety eyewear is provided.	YES/NO/NA
r.	Those at risk of exposure have received an eye check.	YES/NO/NA

If NO to any of the above, please summarize precautions that are taken to control these activities:

5. Do all 3B and 4 laser operations take place within a Laser Use Area (LUA)?   **YES/NO**

   If YES, please indicate which of the following control measures below are in place:

a.	The LUA presents a robust physical boundary that isolates laser radiation from personnel outside the area.	YES/NO/NA
b.	The LUA boundary (including windows) is opaque at the laser wavelengths and without gaps.	YES/NO/NA
c.	Points of entry from hazard-free to laser hazard areas within the LUA (e.g. a door or opening from a room for changing or data collection) carry current laser hazard information.	YES/NO/NA
d.	All hazards are clearly identified at all access points to the LUA.	YES/NO/NA
e.	Where different laser wavelengths are accessible in the LUA at different times accurate status information is displayed at all access points.	YES/NO/NA
f.	The laser hazard cannot extend beyond the LUA if a door into the LUA is opened.	YES/NO/NA
g.	The laser hazard is automatically terminated if an unauthorized person enters the LUA.	YES/NO/NA
h.	The laser hazards from separate laser experiments within the LUA are isolated and information of the current laser hazard within a sub-divided area is clearly displayed at points of access.	YES/NO/NA
i.	Independent non-laser activities are prohibited within the LUA.	YES/NO/NA
j.	Prior warning is provided if laser hazards are introduced from outside the LUA.	YES/NO/NA
k.	Laser beams entering the LUA from other (adjacent) areas are under sole and overriding control from within the LUA.	YES/NO/NA
l.	Temporary restrictions are imposed for servicing and other non-routine activities within the LUA.	YES/NO/NA

If NO to any of the above, please summarize precautions taken to control these activities:

6. Is laser safety eyewear provided?   **YES/NO**

    If YES, please which of the following control measures below are in place:

    a. Laser safety eyewear provides sufficient protection for each accessible   YES/NO/NA
        hazardous laser wavelength (including wavelengths that could be generated
        by non-linear effects).

    b. Laser eyewear is properly stored and maintained in good condition.   YES/NO/NA

    c. The eyewear clearly identifies the laser/area within the LUA it is   YES/NO/NA
        suitable for.

    d. Lighting levels are appropriate for the visual transmission of the eyewear.   YES/NO/NA

    e. The colors of warning signs and lights are effective when viewed through   YES/NO/NA
        the eyewear.

    If NO to any of the above, please provide a brief justification:

7. Are there non-beam hazards associated with laser use (including during servicing   **YES/NO**
    and maintenance)?

    If YES, are control measures in place to address the following hazards:

    a. Fire hazard (with **Class 4** laser beams)   YES/NO/NA

    b. Laser generated fume hazard   YES/NO/NA

    c. Electrical hazards   YES/NO/NA

    d. Explosion hazard   YES/NO/NA

    e. Secondary and collateral radiation   YES/NO/NA

    f. High-pressure gas hazard   YES/NO/NA

    g. Trip hazards and sharp corners at head height   YES/NO/NA

    h. Other non-beam hazards   YES/NO/NA

Form 2

Laser Owner Name   Organization   Responsible Line Manager

Responsible Laser Safety Officer   Location of Laser Use

LASER INFORMATION

Manufacturer Date of Manufacture Inventory Number

Model Number Serial Number

Laser Type Wavelength(s) (nm)

CW Average Power (W)

Pulsed Pulse Rate (Hz)

Pulse Duration (s)

Pulse Energy (J)

Beam Diameter or Dimensions at Aperture (mm)

Beam Divergence, Full Angle (mrad)

Laser Classification (circle) Class 1 Class 2 Class 3A Class 3B Class 4

Commercial Laser Product   Yes No

Protective Housing Interlocks Functional   Yes No

Key Switch (or Computer Code)   Yes No

Emission Indicator	Yes No
Beam Attenuator	Yes No
Remote Interlock Connector	Yes No
Equipment Manual Available	Yes No
Class Warning Label	Yes No
Protective Housing Label	Yes No
Aperture Label	Yes No
Manufacturer's Label	Yes No
Certification Label	Yes No

Options/Modifications/Comments

Laser Controlled Area Report

CONTROLLED AREA PERSONNEL INFORMATION

Laser Controlled Area Location

Controlled Area Supervisor Organization

Responsible Line Manager

Approved Laser Personnel

LASER SAFETY EYEWEAR INFORMATION

Manufacturer Model	OD @ nm OD @ nm OD @ nm
Manufacturer Model	OD @ nm OD @ nm OD @ nm

CONTROL MEASURES

Environment: Indoors Outdoors Lasers In Use Single Laser Multiple Lasers

Area Warning Sign; Acceptable Revision Required Access Warning Lights Manual Automatic Functional

Entryway Interlocks: Manual Automatic Defeatable Bypass Functional Not Required (Explain)

Barrier at Entryway	Yes No
Laboratory Windows Covered	Yes, No, NA

Eyewear Available At Entryway In Controlled Area

Panic Button Labeled At Entryway Functional

Unattended Operation	Yes No

Beam Path Enclosure Total None (Open) Partial

Stray Beams/Reflections Limited By Beam Blocks

Path Covers Table Curbs Barriers Curtains

Hazard Warning Labels On Beam Blocks Path Covers Table Curbs Barriers Curtains

Standard Operating Procedure	Yes No
Alignment Procedure	Yes No
Trained Personnel	Yes No

Other Beam Controls

Approved Alternate Controls

Electrical Hazard During Operation	Yes No
Adequate Controls of Electrical Hazards	Yes No
Electrical Hazard During Alignment or Maintenance	Yes No

Other Non-Beam Hazards

Lasers in Controlled Area Comments

Conditions of Approval

Responsible Laser Safety Officer

Form 3

Name: **Organization:**

Laboratory #:   **Principal Supervisor:**

Laser Classification:

Classification	Number present	Wavelength(s)
Class 1	_____	*Not required*
Class 2	_____	*Not required*
Class 1M/2M	_____	_____
Class 3R		
Class 3B		
Class 4		

The following recommendations apply mainly to class 3B and class 4 laser laboratories.

Is the laboratory a *single laser facility, multi-user*, or with *multiple lasers*.
    Comment:

Laboratory System Controls:

	Present	Comment
Door interlocks or alarm	o	
Good room illumination (to reduce pupil size)	o	
Light colored; diffuse wall surfaces	o	
Nearby equipment should have matt or low reflectivity surfaces	o	
Light colored; diffuse wall surfaces	o	

    Comment:

Engineering controls:

		Applicable to class		Comment
Visible ON indicator				
	located	o Inside	3B, 4	o
	o Outside	3B, 4	o	
Ventilation	o General dilution	3B, 4	o	
	o Local exhaust	3B, 4	o	
Unintentional specular reflections				
	minimized/removed	3B, 4	o	
Beam terminated at end of useful path		2, 3, 4	o	_____

Comment:

Form 4

Administration Information

SOP Title

Auditor	Date	PI	Division	Laser Supervisor (POC)	Building	Room

Reason for audit

Present during audit

_____

Documentation

SOP current (circle)	Yes		No	
All Lasers Listed in SOP (circle)	Yes		No	

Laser Information

Highest Class Laser	3B		4	
User Manuals present	For All	Some		None

Lasers in storage

Alignment Lasers in use	Y		N	Type/Quantity	HeNe	Diode	IR

New laser(s), list with specifications on last page

Environment   YES   NO   NA   NOTE

Main entrance Door Posted

Posting accurate (wavelength)

Contact information

Readably visible

Ancillary doors

Entry through curtain

Windows and doors coverings

Illuminated sign

    Functional

Access control

Administrative means

    Explain

Interlocked

    By pass available

    E-stop present

Last interlock checks date:

    Interlock functioning

    Written Interlock check procedure

    Interlock to shutters

    Interlock to power supply

Housekeeping

On optical table

In laser use area

Space at beamline

Unattended operation

Post unattended sign

Beam Path

Totally open beam path

Completely enclosed beam path tubes perimeter _____ Panels class 1 product fiber

Combination path, _____ % open _____ % enclosed

Lasers & optics secured to table

Beam properly contained

Beam blocks _____ Secured _____ Loose

Perimeter guards

Table top enclosures

Other means (describe)

Beam in line with workstations

Evidence of laser burns or cross hairs on walls

Reflections contained, specular as well as diffuse

Beams blocked from directly exiting open door or window

Beams required to leave table?

Crosses walk way (controls in place)

Describe Passes into adjacent room/chamber

Describe means and controls

YES   NO   NA   NOTE

Non-essential materials out of beam path

Upward directed beams

Blocked

Vertical Labels used

Collecting optics used in room

Fiber optics in use _____

Fiber ends labeled _____

Container for sharps _____

Fiber conduit labeled _____

Personnel Factors

Laser eye exam by all laser personnel

Laser safety training current

Has staff read SOP?

Correct eyewear available (OD & wavelength)   Proper storage, where   outside use area   inside use area   Sufficient quantity on hand   Condition of eyewear: Very Good   Good   Fair Damaged

Labeling problems

Skin protection needed, if yes available

Process Interaction
Are gases/vapors/fumes controlled?
Electrical items

Optical tables grounded?

Commercial equipment?   All   Some home made

Seismic concerns

Table(s)

Work area

Associated non beam concerns related to this work

List

Chemicals   Optic cleaning solvents

2nd containment

Additional comments and notes

Note

LASER BINDER

A laser safety management system as presented here is a tool that is designed to demonstrate the current status of one's laser safety and to act as a focus for where you need to be and how you are going to get there. While the format presented here is paper-based, there is no reason why all of the documents should not be stored electronically. However, from experience, most people seem to find the paper version easier to use. In an electronic version, certain sections could be linked and updated from larger databases, a good example would be a training database. In this way, as a user completes a course, it would be entered in the programs training matrix.

The approach we have generally used is to use a ring-binder file with 13 sections. Consider the binder as a central point for laser safety documentation. However, with use it may be found that your laser application needs more or less than 13 sections. They may not even have the same titles as those suggested here.

First, one needs to decide on the level of the laser safety management system. At the highest level, we could produce a program for the whole organization. The advantage of this is that a standard could be set across the organization. We could have a program for the site, department, section, laboratory, or individual laser application. As we get closer to the laser application, more details are likely to be included. We also need to exercise some caution. If we have a laser safety management system for each laser application, and there are a number of applications in the laboratory, they cannot be treated in isolation.

The following are institutional binder sections:

1. Safety structure
2. Safety committee
3. Standards and regulatory
4. Training program
5. Audits and reports
6. Risk assessments
7. Control measures
8. Incident investigation
9. External liaison
10. Miscellaneous

I suggest starting with the following 13 sections for user binders:

User binder
    1. SOP
    2. Training/authorized worker list
    3. Interlock checks
    4. Alignment procedures
    5. Audits and reports
    6. MSDS/URL
    7. Checklists
    8. Engineering safety notes
    9. Contact list
    10. ES&H chapters
    11. Lessons learned
    12. Emergency procedures
    13. Miscellaneous
        Optional sections
    14. Safety structure
    15. Specific procedures

16. Safety committee minutes
17. Standards and regulatory items
18. Risk assessments

Some of the sections remain fairly static throughout the life cycle of the laser application; others will need to be updated regularly. Therefore, it is important that the file represents the current status of laser safety. It could be audited by external organizations, including regulatory officers. The impression they get of your organization could be colored by what they find in the file. Much of real-life inspections hinges on perception. Some sections of the binder are reference and therefore appearance. A well-kept program binder provides a very good perception of laser and safety awareness and concern.

A benefit that comes with time is that the regulatory officers get to recognize the format of the laser safety management program files. Even the existence of the file produces a positive impression.

The following presents guidance on what to include in each section of the file. The actual contents will depend on your development of the laser safety management program and are likely to change with time, as you find out what works best for you.

STANDARD OPERATION PROCEDURES

The first section provides an opportunity to describe the laser application, hazards, and control measures. The level of detail will depend on the level of the laser safety management program: corporate, application, or somewhere in between.

TRAINING/AUTHORIZED WORKER LIST

This section should contain a list of all required training courses required by this SOP. A way to list the courses is a spreadsheet matrix. All staff working under the SOP is required to sign the matrix. This indicates they have read the SOP and have completed the training requirements. The PI may add or delete names and course requirements for individuals as needed. This is a simple way for the RI to keep track of training and demonstrate that all staff have read the SOP.

INTERLOCK CHECKS

The laser use area may require access controls such as door interlocks. For that matter, there may be other safety devices that require periodic inspection or verification of operation. This section of the binder is the place for such documentation to be kept. Not just the log showing a history of dates but any procedure that needs to be followed to perform the inspection. A copy of the periodic (set within SOP) interlock checklist is found here. Make entries on this form and continue on the same page until the form is filled.

ALIGNMENT PROCEDURES

When one considers that over 60% of laser accidents that occur in the research setting are during the alignment process, it is easy to see why a section in this binder should be dedicated to alignment. In this section, either generic alignment guidance should be found or specific alignment procedures for the laser operation. While specific is desired, the general alignment is adequate (if followed).

AUDITS AND SELF-ASSESSMENT REPORTS

The auditing of laser use area is critical to good laser safety and a required task of the LSO or designee. This section will contain a copy of such audits. If management or the users perform any self-assessments or other audits, those records should be here (this does not include and regulatory inspections). Program management or self-inspection should be encouraged. Specifying status or required corrective actions may be found here. Audit is a very powerful tool to demonstrate that you are doing what you say you are doing. This section should include some kind of audit plan. It may be that you decide to dedicate a specific time to audit the complete laser safety management program, or you break the audit down into a number of smaller tasks that are carried out on a regular basis throughout the year.

There will be results from the audits—they may say that everything is still completely up to date and the operational use of the laser safety management program correlates with what actually happens. If so, this is still a valuable result that needs to be recorded. If there are either noncompliances or something needs to be improved, then this needs to be recorded and an action plan developed.

The frequency of auditing-specific aspects of the laser safety management program will depend on the risk if that part of the program fails. For example, a safety critical control measure may need to be audited frequently—some even each time the laser is used.

Should you be visited by an enforcement officer who is assessing your laser safety then this part of the file can be extremely useful. If, for example, you have an interlocked enclosure and someone manages to get exposed to the laser beam because the interlock did not work, can you demonstrate that the interlock had only recently failed? If the interlock checks were part of a weekly safety audit, then you may be able to demonstrate that it was working at the last check. This will generally only convince the enforcing officer if there is a written record of the test, signed and dated. If you have not been carrying out the safety audit, then the interlock may have failed shortly after installation.

MSDS/URL

Material safety data sheets (MSDSs) are for chemicals. The most current MSDSs are accessible on the web, and it is recommended that a web address be listed here. For some research labs, it would be easy to fill a binder just with MSDSs. Workers may wish to keep hardcopies here for particular chemical used under this SOP.

CHECKLISTS

This section is available for additional checklists that may be needed under the SOP.

ENGINEERING SAFETY NOTES

If specialized equipment such as vacuum vessels is used for this project, this section allows one to keep a copy or reference where an engineering or safety documentation exists to verify the manufacturer of the equipment.

CONTACT LIST

This section will contain useful contacts on various EH&S concerns. It should be updated annually, otherwise the information may no longer be accurate, that is, change of safety personnel. The PI may add extra names applicable for the SOP or facility.

Specific names and contact details should be included for everyone involved in laser safety, appropriate to the level of the laser safety management program.

ES&H CHAPTERS

Most institutions have a Health & Safety Manual. Such a manual will list the hazard and requirements for working with such. Examples would be lasers, electrical, toxic gas, and radioactive materials. This section would contain hardcopies of the most relevant chapters. One could just list web address, but having a hardcopy for staff to reference generally means a stronger chance of actually being used.

LESSONS LEARNED

Information and possible copies of lessons learned applicable to this SOP activity. Knowing that lessons learned is a strong laser safety reenforcement tool, providing lessons learned for present and future users can only help the laser safety program.

EMERGENCY PROCEDURES

This section would include the institution's emergency response information, earthquake, fire, chemical spill, and so on. In addition, the PI may add SOP-specific emergency instruction and his own emergency response contacts here.

MISCELLANEOUS

There will invariably be things that do not fit into one of the other sections. They could be filed in here. However, as part of the audit of the file, it may be obvious that some of the items that have been built up in this section could form a new part of the file. If one of the other sections is not being used, then you could switch the information from here to there.

OJT Form

While the performance of OJT is critical, its documentation can be just as valuable in seeing that the activity is performed.

One can have a form that covers OJT of work tasks, such as how to operate a certain laser system and/or OJT that shows the person has learned "core" laser safety values. Review the following two forms.

Rationale

All laser users at LBNL are required to take a basic laser safety course (EHS280) or equivalent web course.

While this class is important to the general understanding of hazards, it is the site-specific training that will prevent most laboratory injuries. For example, laser alignment/beam manipulation has proven to be the activity related to the majority of laser accidents in the research setting. It presents the greatest opportunity for unplanned/unintentional reflections onto the eye or skin. In recognition of this, only individuals with the skills and system/hazard awareness are allowed to perform "beam on" laser manipulation and alignment work. The individuals listed in the following have demonstrated core laser safety skills for their work site by OJT.

OJT is carried out by

- Mentoring the individual through instruction on the work process and safety steps
- Demonstrating hands-on skills
- Observing the individual perform the activity
- Receiving feedback from the trainee

SOP No   Building   Room(s)

CORE LASER SAFETY PRINCIPLES

1. Selects proper eyewear
2. Checks condition of eyewear
3. Alerts others before turning on laser and of open beams
4. Checks for stray reflections thoroughly
5. Blocks stray reflections
6. Demonstrates beam detection methods
7. Understands controls for different intensity levels
8. Read and familiar with controls per SOP
9. Familiar with equipment safety features
10. Communicates with others

Your signature indicates confirmation that the individual satisfactory demonstrates the 10 core Laser Safety competencies.

Note: One might wish to add a "Task" column to the form below, for step by step OJT

Name OJT Provider   Signature Recipient of OJT   Signature   Date completed

_____

_____

_____

_____

_____

_____

_____

_____

_____

5 ANSI Safe Use of Lasers and Laser Safety

Ken Barat

CONTENTS

INTRODUCTION

The American National Standards Institute (ANSI) through the Z136 series has a number of laser safety standards, all geared toward users. Starting out with Z136.1 Safe Use of Lasers first issued in 1973 (at the request of the federal government), a number of additional Z136 application standards have been developed and issued. These include medical use of lasers and outdoor use. The ANSI standard Z136.8 entitled Safe Use of Lasers in Research, Development, or Testing is the most significant application standard for any facility where laser and product development is taking place.

The two most important items to remember are the following: first, these standards are suggested user guidance and all call on the individual laser safety officer (LSO) to apply their controls as needed and modify per use application. Second, the majority of regulatory bodies have adopted them as gospel. These two applications are not always compatible.

Any text that tries to explain all the ANSI controls will date itself as new standards and modifications of the existing standards come about. Technology is always advancing. Control measures that once stood tall for the argon ion laser fall short for the diode array. This is why the standard puts so much on the LSO to evaluate the setup and select and approve the proper controls. All of this supports Chapter 3 on hazard evaluations.

The laser user and laser safety community owes a great debt to those who worked on and developed the early ANSI standards and editions, particularly when one considers all this has been accomplished on a volunteer basis. Over time, the ANSI standards and international rules for safe use of laser may/will meld together. No matter when that may happen, the LSO will always serve a critical function.

I have been active on several ANSI standard and technical committees, and I can easily say I have met some of the finest people in the laser field working to make things better for all of us on those committees.

ANSI Z136.8 Safe Use of Lasers in Research, Development, or Testing, as mentioned in Chapter 1, is a great improvement in guidance to laser users and management. Some of the additions to laser safety guidance are given in this chapter.

IS Z136.8 ACCEPTABLE TO REGULATORY AUTHORITIES?

Z136.1 allows guidance from specialized/application standards, such as Z136.8, to take precedence within the scope of that standard. "Other special application standards within Z136 series may deviate from the requirements of this standard. Each deviation is valid only for applications within the scope of the standard in which it appears."

Therefore, guidance from Z136.8 can be used for research and development (R&D)-specific applications that may not have been fully addressed or may be in conflict with Z136.1. It is up to the discretion of the LSO to determine which specialized standards apply to their particular environment, just as an outdoor range officer will look toward Z136.6 the outdoor laser standard rather than Z136.1. Remember Z136.1 is a horizontal standard and was designed to encompass a wide range of laser applications, but it always expected the generation of vertical/application standards to provide more specific and germane guidance. The following sections are some highlights from Z136.8.

HAZARD EVALUATION

Since the first laser, ANSI standard laser hazard evaluation has been broken down into three factors: the capability of the laser to injure people; the environment in which the laser is used; and, finally, the personnel who may be exposed to laser radiation. These three factors have served the laser user community well for 40 years. Today, advances in lasers and their application in R&D have called for an expansion of the hazard evaluation parameters. Therefore, Z136.8 added the beam path and process interaction. Today's beam path can be complex and made up of many elements. It can be in fiber, with nonlinear optics wavelengths can change, with tunable laser techniques and chirped pulse amplification the pulse duration and peak owner can be manipulated over a wide range. Because of these and other beam path options, the standard committee felt it was important to call out the beam path of hazard evaluation. As to process interaction, the use of laser beams to generate high harmonics, and ionizing radiation made it clear that this element needed to be called out to the LSO and user for evaluation.

LASER ALIGNMENT EYEWEAR

The use of alignment eyewear has been ongoing in the laser use community in one form or another, which includes looking over full protection. The goal is to provide visibility of beams during alignment and still provide some eye protection to the user. This is preferable to one removing their eyewear to see the beam and having no protection. The challenge for the LSO is what optical density to suggest that allows visibility and reduces the transited beam to a Class 3R level or less. Section 4.5.2.10, Alignment Eyewear, of Z136.8 addresses the issues with suggestions for both continuous-wave and pulse lasers.

LASER USE LOCATIONS

Defining the following laser use locations assists the LSO in applying their professional judgment and aiding in justification validation:

1. Unrestricted location—unlimited access (hallway in a public building)
2. Restricted location—access granted to authorized people and limited for the general public (a research laboratory)
3. Controlled location—access, occupancy, and activities subject to strict control (R&D area with positive access control and video surveillance)
4. Exclusion location—occupancy possible but denied during the operation of the laser (free electron laser machine room or beam path)
5. Inaccessible location—occupancy not possible due to dimensions (enclosed beam path on an optical table)

ADDITIONAL TOPICS

Use of noncertified lasers and the appropriate administrative, procedural, and engineering controls are discussed.

The Special Considerations section discusses visitors and spectators, laser user facilities, and export controls, as well as useful information on fiber optics and laser robotics.

Education and Training is expanded to include the importance of on-the-job training (OJT) for users in the dynamic environment of R&D.

Updates information on standard operating procedure requirements.

Discusses the management of multiple types of laser use areas, including indoor Class 3 and 4 lasers, laser exclusion areas, and airspace (includes references to related Federal Aviation Administration and ANSI sources).

Information on non-beam hazards has been condensed to R&D-relevant information. Z136.8 includes full-color signage and diagrams.

Included in the figures section is current ANSI Z136.1 and Z535.2 compliant signage for

- Class 2 and 2M lasers
- Class 3R, 3B, and 4 lasers
- Temporary laser-controlled areas
- Unattended laser operations
- Warning signal word

There are diagrams and explanations for exterior entryway controls, multiple entry point controls, and Class 4 entryway controls for laboratories with and without entryway interlocks.

INTRODUCTION OF NEW APPENDIXES

The Z136.8 standard includes three new appendixes as well as all R&D-relevant appendixes from Z136.1. New appendixes contain useful information not addressed in the main standard, whereas appendixes pulled from Z136.1 are modified to focus on information that is relevant to the R&D.

Appendix B: Sample Forms

Two sample laboratory audit forms
   Sample laser program self-evaluation form

Appendix C: Frequently Reported Incidents

Discusses the most common causes of accidental eye and skin exposures to laser radiation and accidents related to non-beam hazards.

Appendix H: Laser Product Safety Rules

New engineering controls: many engineering controls present in Z136.1 are not found in this standard. To maintain proper emphasis on their importance, "CDRH Sec. 1040.10 Laser Products" is included as guidance for Appendix A: Supplement to Section 1:

   LSO requirements and responsibilities
   Guidance on laser safety committees
   Personnel responsibilities (supervisors and employees)

Appendix D: Education and Training

Class 1M, 2, 2M, and 3R awareness training
   OJT
   Laser pointer awareness
   Certifying laser products
   Going back to Z136.1, the following are some of the critical items.

SECTION 1: ESTABLISHMENT OF THE LASER SAFETY OFFICER

The laser safety program established by the employer shall include provisions for the following:

1. Designation of an individual as the LSO with the authority and responsibility to effect the knowledgeable evaluation and control of laser hazards and the implementation of appropriate control measures, as well as to monitor and enforce compliance with required standards. The specific duties and responsibilities of the LSO are designated in the Normative Appendix A. Throughout the body of this standard, it shall be understood that wherever duties or responsibilities of the LSO are specified the LSO either performs the stated task or assures that the task is performed by qualified individual(s).
2. Education of authorized personnel (LSOs, operators, service personnel, and others) in the safe use of lasers and laser systems and, as applicable, the assessment and control of laser hazards. This may be accomplished through training programs. Employers should consider the benefits of initiating awareness training for employees working with and around lasers and laser systems greater than Class 1. If training is warranted for embedded lasers, it shall extend to those routinely around the systems, who will be present when service and maintenance occurs.
3. Application of adequate protective measures for the control of laser hazards as required in Section 4 of the standard.
4. Incident investigation, including reporting of alleged accidents to the LSO, and preparation of action plans for the prevention of future accidents following a known or suspected incident.
5. Formation of a laser safety committee when the number, hazards, complexity, and/or diversity of laser activities warrant it.

SECTION 4: CONTROL MEASURES

ESTABLISHING CONTROL MEASURES

Control measures shall be devised to reduce the possibility of exposure of the eye and skin to hazardous levels of laser radiation and other hazards associated with laser devices during operation and maintenance.

The LSO shall have the authority to monitor and enforce the control of laser hazards and effect the knowledgeable evaluation and control of laser hazards and conduct surveillance of appropriate control measures. The LSO may, at times, delegate specific responsibilities to a deputy LSO or some other responsible person.

APPLICABILITY OF CONTROL MEASURES

The purpose of control measures is to reduce the possibility of human exposure to hazardous laser radiation and to associated hazards.

"In some cases, more than one control measure may be specified. In such cases, more than one control measure that accomplishes the same purpose shall not be required."

Substitution of Alternate Control Measures (Class 3B or Class 4)

The engineering control measures recommended for Class 3b and Class 4 lasers or laser systems, upon review and approval by the LSO, may be replaced by procedural, administrative, or other alternate engineering controls, which provide equivalent protection. This situation could occur, for example, in medical or R&D environments. Accordingly, if alternate control measures are instituted, then those personnel directly affected by the measures shall be provided the appropriate laser safety and operational training.

These empower and allow the LSO to do the job at hand. Working safely is compatible with high-quality work and innovative work.

6 Understanding Optics

Bill Molander

CONTENTS

INTRODUCTION

Optics as a science and a technology covers a tremendous range of material. Fortunately, only a limited subset of this material is essential to the field of laser safety. With a few exceptions, this chapter concentrates on the narrower field of geometrical optics, which is the branch of optics that deals with lenses, mirrors, and optical instruments that use lenses and mirrors. After a brief discussion of the nature of light and the relation of geometrical optics to the more comprehensive fields of physical optics and quantum optics, the fundamentals of the behavior of rays within the geometrical optics formalism are presented. Next, how the behavior of rays can be used to make lenses that focus light and produce images is addressed. The eye is one optical system that is important for laser safety personnel to understand, and it is discussed next. Finally, some optical devices that are commonly encountered in laser research and development laboratories are described.

NATURE OF LIGHT

To set the stage for the following treatment of geometrical optics, it is helpful to provide an outline of the more comprehensive modern understanding of the nature of light. Concepts such as the relation between wavelength and color are frequently referred to in geometrical optics but are best introduced as part of the wave theory. Furthermore, in some situations, such as the behavior of very-high-quality optical systems or the propagation of laser beams, the geometrical theory is not completely adequate, and some concepts from the wave theory are required. Finally, some of the common optical components that are discussed in the final section can only be understood by referring to the quantum theory of light.

Because vision is so central to our perception of the world, it is not surprising that people since the beginning of recorded history have tried to understand the nature of light and vision. The concept of a *line of sight* and the observation of shadows have from ancient times suggested that light is something that travels in straight lines. We can introduce the somewhat abstract concept of a ray of light based on this as shown in Figure 6.1. A ray is a single geometric line of light that originates at some light source, like a point on the candle in the figure, and proceeds until it is stopped at some object. The candle emits rays of light in all directions and from each part of the flame. The rays can be different colors. Some rays from the candle are yellow; others are blue or orange. In this view of the nature of light, the reason we can see the candle is that some of the rays from the candle travel to our eyes. This seemingly simple concept that we perceive visual objects only when rays of light enter our eyes can be very helpful in understanding optical phenomena.

This concept of light as some form of energy traveling in straight lines originating at a source explains many observations regarding the behavior of light, but early in the eighteenth century scientists began to discover cases in which the behavior of light was more complicated. In the phenomenon called *interference*, when two light rays strike the same point the amount of light energy at the point can be more or less than the total energy in the two rays. Another phenomenon called *diffraction* is observed in which light energy from a source can reach locations where no straight-line path from the source exists.

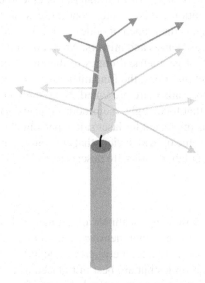

FIGURE 6.1 The light emitted by a candle can be thought of as a collection of rays that travel outward from each point on the flame in straight lines in all directions.

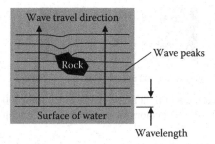

FIGURE 6.2   Water waves bend (diffract) around an obstacle. The distance between successive peaks is the wavelength.

These effects were explained by the wave theory of light. In this view, light is a wave similar to water waves or sound waves, which are familiar to everyone. The process of interference occurs with waves because when two waves come together they can add up to a larger wave if the crests and troughs of the waves occur in the same location or cancel each other if the crest of one wave coincides with the trough of the other. Diffraction of waves is also readily observed in water waves. If a barrier such as a rock is placed in the path of a wave, the wave can spread around it, as shown in Figure 6.2. As we all know, sound waves commonly go around barriers, so we do not need a straight line of sight from a speaker to our ear to hear a person speaking. An important parameter of a wave is the distance from one wave crest to another, called the *wavelength*.

Another important parameter is the frequency of the wave. Imagine a cork floating on top of the water. As waves go past it, the cork bobs up and down. The rate at which the cork goes up and down is the *frequency*. For example, if the cork completes two up-and-down cycles in 1 second, the frequency would be 2 inverse seconds. The unit of an inverse second is also called a *hertz*. There is a very close connection between wavelength and frequency for any wave. If the wavelength increases, the frequency decreases. In fact, the product of wavelength and frequency is always equal to the speed of the wave.

Of course, light is not a water wave. Rather, it is an electromagnetic wave as are radio and television waves and x-rays. Because other familiar waves travel in some medium (e.g., water waves in water or sound waves in air), scientists initially believed that there must be some medium through which light waves travel. Eventually, however, it was shown that light can travel in a medium consisting of only free space.

The wave theory of light clarified several other mysteries about the nature of light. First, the color of light was found to depend on the wavelength, the distance from one wave crest to the next. The wavelength of light is very small. The shortest wavelength of visible light is about 0.0004 mm or 400 nm. This corresponds to deep purple light. The longest wavelength is about 0.0007 mm or 700 nm. This long wavelength is a deep red. Each wavelength corresponds to a single primary (rainbow) color, as seen in Figure 6.3. Any color that is not a primary color consists of a mixture of waves with various wavelengths. Figure 6.3 only shows the range of visible light. Although the human eye is not sensitive to wavelengths outside this range,

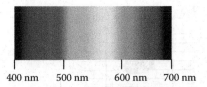

400 nm   500 nm   600 nm   700 nm

FIGURE 6.3   Each wavelength of light in the visible range from 400 to 700 nm is perceived as a single primary color. All other colors are combinations of multiple wavelengths.

electromagnetic waves with shorter or longer wavelengths do exist. Fundamentally, electromagnetic waves of any wavelength behave just like light in the visible range. When the wavelength is not too far outside the visible range, the technology for generating, propagating, and detecting the radiation is very similar to the technology used in the visible range. Therefore, radiations with wavelengths a little shorter than visible (ultraviolet) or a little longer (infrared) are sometimes referred to as light somewhat loosely. Many lasers emit wavelengths outside the visible range, and this has important consequences in laser safety. The hazard is more severe because it cannot be seen. As with all waves, as the wavelength increases the frequency decreases. The frequency for light at 400 nm is about 750 terahertz (TH; tera = million million, i.e., 10^{12}), and the frequency for light at 700 nm is about 430 TH.

Radio waves are also electromagnetic waves with a much longer wavelength than visible light. Gamma rays are electromagnetic waves with a much shorter wavelength. These types of radiations propagate through free space just like visible light; but they interact with matter in very different ways, and their associated technologies are very different.

In all these electromagnetic waves, the things that are going up and down in analogy to the surface of water in a water wave are really abstract objects, the electric field and the magnetic field. These are both forces. The electric field is usually the more important. It is oriented to push an electrically charged particle in a direction perpendicular to the direction that the wave is traveling. This is analogous to a cork bobbing up and down in a water wave. The motion of the cork is perpendicular to the direction of travel of the wave. The electromagnetic wave in three-dimensional space can exert a force in any direction that is perpendicular to the direction of travel. If the direction is always the same throughout the wave, it is called *linear polarization*. Circular and elliptical polarizations are also possible. Any polarization can be described in terms of two special linear polarizations. Often, these two special polarizations correspond to the electric field as horizontal or vertical.

The wave theory can be used to determine the speed at which light travels. In free space, it travels about 300 million m/s. Light can travel through other media as well, for example, water or glass. When it goes through other media, it travels at a slower rate.

Wave theory at first glance seems so different from the geometrical ray point of view that it may seem impossible that both theories are describing the same phenomenon. However, some similarities can be seen by thinking again about water waves. A wave in the open ocean travels in a straight line. The wave peaks are perpendicular to the direction in which the wave travels. So, we could talk about rays for water waves as the direction of travel. A wave encountering a large island, for example, does not bend around it so there is a *shadow* of the island. It is only when there are multiple waves that can interfere or smaller objects for which the bending of waves is observed that describing waves in terms of rays breaks down. This is exactly the situation encountered in the wave theory of light. All of geometrical optics can be derived from the wave theory as an approximation that is perfectly adequate in many situations. In this chapter, with a few exceptions, we need only this approximation.

The wave theory of light has been very successful in predicting the behavior of light with high accuracy. Nevertheless, there is still more to the story of light. Near the beginning of the twentieth century, a number of observations of the interaction of light with atoms and electrons were inconsistent with the wave theory. This led to the quantum theory of light. In this theory, light is composed of tiny particles of energy called *photons*. Although it might seem that photons and waves are completely different concepts, the quantum theory shows that they are very closely related. Just like the ray theory can be shown to be an approximation of the wave theory, the wave theory itself is an approximation of the quantum theory. The quantum theory of light is beyond the scope or needs of this book. A few concepts from this theory are discussed in the final section of this chapter, regarding typical optical devices in laser laboratories.

SCATTERING, ABSORPTION, REFLECTION, AND REFRACTION OF RAYS

With this background of how scientists today understand light, let us return to the ray picture, which is also called *geometrical optics*, and study the behavior of rays after they leave the source that generates them. Other than simply traveling in a straight line forever, there are only a few other fates that can befall a ray within the geometrical optics model. The principle of conservation of energy dictates that all of the energy carried by a ray must be accounted for by one of the processes discussed next.

Consider again the candle discussed when the concept of rays was introduced. If the candle is in a dark room, we notice that when it is lit other objects in the room become visible. This is due to a phenomenon known as *scattering*, which is illustrated in Figure 6.4. One of the rays from the candle strikes the picture on the wall. The ray from the candle stops, but it causes new rays to be emitted from the point where it struck the picture. Rays traveling in all directions are generated. If some of these new rays travel to our eyes, we can see the picture by the light of the candle.

Certain parts of the scattering object are more efficient at scattering blue light, and other parts are more efficient at scattering green. If the illumination contains many colors, then the areas that scatter the most blue light will appear blue. However, the scattering object does not change the color of the rays. That is why in a room illuminated by a red light, everything looks red.

For rays of colors that are not strongly scattered by the object, some of the rays simply cease to exist. They are *absorbed* by the object. How much is absorbed and how much is scattered depend on the nature of the object and on the color of the ray.

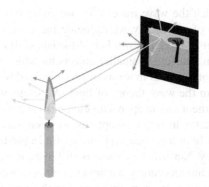

FIGURE 6.4 Light originating at the candle can scatter from objects. This scattered light can then be detected by the eye. This is the scientific explanation of how we see something by the light of the candle. A scattered ray has the same wavelength as the ray that induced the scattering.

Because energy is conserved, the energy that disappears from the light rays must turn into energy in some other form. In this case, it turns into heat in the illuminated object. With ordinary room lights this heat may not even be perceptible, although with bright floodlights the heating due to absorbed light may be very noticeable. If the light source is a powerful laser, the absorbed energy can generate enough heat to melt or burn the object. The scattering from a surface discussed in this section is sometimes called *diffuse reflection*.

Notice that the scattered light from the picture is in many ways similar to the light originally coming from a source like the candle that actually generates light. In particular, light rays emerge from the picture in all directions with various colors and energies. When some reach our eyes, we can see the picture just as we see the candle. So, for many purposes we can consider the picture to be a light source just like the candle even though it is only scattering light rather than generating it. We refer to either of these as an object, a word used frequently in geometrical optics to refer to something that provides the rays that propagate through an optical system.

In addition to the picture in the room, suppose there is a smooth shiny surface, say, a mirror. In this case, another phenomenon in addition to scattering and absorption can be observed. We see in the mirror what appears to be a second candle. This is due to the process of *reflection*. When a ray strikes the mirror, the ray stops, but a new ray is generated. In scattering rays are generated in all directions, but in reflection each ray generates only a single new ray. The new ray has the same color as the original ray but goes in a direction determined by the law of reflection as illustrated in Figure 6.5. First, draw a line perpendicular to the mirror surface at the point where the ray strikes the mirror. This is called the *surface normal*. The reflected ray leaves the mirror at the same angle from the surface normal that the incident ray comes in but on the opposite side of the normal.

How does this law of reflection lead to the appearance of an image? This is illustrated in Figure 6.6. In this figure, light rays are generated by something called a *point source*. Like the light rays, a point source is an idealized, abstract concept, but it is used so frequently in geometrical optics that it is important to understand.

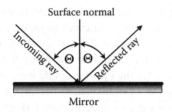

FIGURE 6.5 The law of reflection states that the angle of reflection equals the angle of incidence.

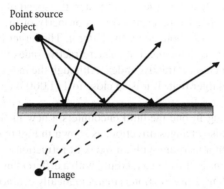

FIGURE 6.6 By applying the law of reflection to all the rays from the point source, which strike the mirror, it can be seen that an image of the point source appears behind the mirror.

A point source is a single point from which light rays emerge in all directions. It can be thought of as a single point on the candle, the picture, or some other object. It is used because it can illustrate the concept of imaging much more easily than trying to explain the behavior of all the rays from a more realistic object.

The rays from the point source that do not strike the mirror behave just as discussed. They might travel in a straight line forever, or they might strike some other object and scatter or be absorbed. The rays of interest here are the ones that strike the mirror; a few are illustrated in Figure 6.6. Each of these rays obeys the law of reflection. So the ones that strike the surface at a small angle generate a reflected ray at a small angle, and the ones that strike the surface at a larger angle generate reflected rays at larger angles. The reflected rays begin at the surface of the mirror. However, if these rays were to be continued backward, as shown by the dotted lines in the figure, they would intersect in a single point that is directly behind the mirror from the point source at the same distance behind as the point source is in front. This point is called the *image point*.

How does this explain what we see when we look at a mirror? Remember that what we see depends only on the light that gets into our eyes. If we are looking at the mirror, the rays entering our eyes are exactly the same as if there is no mirror but instead there is a point source at the location of the image. For interpreting the meaning of the light falling on our eyes, our brain takes no account of the fact that these rays do not originate behind the mirror. Thus, we *see* the image.

To complete the description of imaging by a mirror, note that for some extended sources such as the candle the rays from every point on the object behave just like those from the one point source discussed. Each forms an image directly behind the mirror surface. Therefore, when looking into the mirror the rays reaching the eye are just the same as that would reach the eye if a candle were located behind the mirror. The brain interprets the rays as a candle behind the mirror.

The fourth process that a ray can undergo, *refraction*, is a little more complex. It occurs when a ray traveling in one medium enters a different medium. As mentioned, a light ray travels at a slower speed when it is traveling in a transparent medium other than free space. The speed that it travels with depends on what the medium is and the wavelength of the ray. In water, the speed of a green ray is about 225 million m/s. The speed is determined by a property of the medium called the *refractive index*, usually symbolized by the letter n. The larger the refractive index, the slower the light travels in the medium. The refractive index of free space is 1. All other materials have a larger refractive index. Although the index of refraction of air, as of any material, is larger than 1, it is so close to 1 (1.0003) that in most cases the difference is ignored.

When a ray traveling in one medium encounters a new medium, in addition to changing its speed it also changes direction, as shown in Figure 6.7. In this figure, a light ray is traveling in a medium with an index of refraction n and crosses a plane interface where it begins to travel in a medium with a different index of refraction, n'. As in reflection, the change in direction is mathematically characterized by a change in the angle the ray makes with the surface normal. The change in direction follows a law of refraction similar to the law of reflection. If the second medium has a larger refractive index than the first, the angle gets smaller, as shown in the figure. If the second medium has a lower index, the angle gets larger. Also, the larger the incident

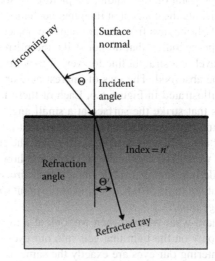

FIGURE 6.7 When a ray goes from one medium to another with a different refractive index, the ray's direction changes. If, as shown here, the second medium has a larger refractive index, the ray bends toward the surface normal.

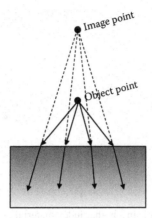

FIGURE 6.8   Rays from an object in a low-index medium refract at the plane interface with a higher index material. An image is formed that appears farther from the interface than the object.

angle, the more the ray bends. If the incidence angle is zero, normal incidence, the ray does not bend at all, although its speed does slow down. The mathematical form of this law, usually called *Snell's law*, is not needed in this presentation. The interested reader can find it in the bibliography. A commonly encountered case is when one of the two media is air, which for most practical purposes has the same refractive index as the vacuum of free space.

As in the case of reflection, instead of a single ray we consider a point source. Figure 6.8 shows a point source in a low-index medium for which some of the rays travel into a higher index medium. All the rays bend toward the normal to make the refracted angle smaller than the incident angle, but the ones with a higher incidence angle bend more. The result is that to an observer below, in the high-index medium the rays reaching the observer's eye appear to come from a point farther away than the object point. Again, if the source is an extended source composed of many, many point sources, each forms an image in the same way. The observer will perceive an image of the extended source at a distance farther from him or her than the actual source. Nothing will be perceived as existing at the actual object location because no rays that seem to originate from this location reach the eye.

When the object is in the high-index medium, the opposite happens. That is, the image appears closer than the object. This can be seen in the photograph shown in Figure 6.9. The tip of the pencil is in water, and the camera is above in air. By following the outline of the part of the pencil above water, you can see that the apparent position of the pencil tip is closer to the surface of the water than the actual tip. The rays from the parts of the pencil that are under water refract when they leave the water. Since these rays are the only way in which the camera, or an eye, can detect the pencil, it appears as if the pencil is located where the refracted rays diverge. If you put your finger in the water and try to touch the tip of the pencil where it appears to be, you will find that it is not there at all. It is, of course, located where you would expect it by continuing the lines of the pencil part that is not in water.

FIGURE 6.9 When the object is in a higher index material as the tip of the pencil is in this figure, the image appears to be closer than the object.

I have discussed all the processes that can occur to light rays but so far only mentioned scattering and absorption on surfaces. These two processes can occur in *bulk*, meaning that they occur within the medium that the light travels. Before concluding this section on the behavior of light rays, I discuss these subjects.

When a light ray travels through a medium other than free space, its energy may decrease as it travels. This can be caused either by bulk scattering or by bulk absorption. In bulk scattering, new light rays going in all directions are generated at each point of the ray's path through the medium. A simple experiment can illustrate bulk scattering. If you shine a flashlight or a laser pointer into pure water, the beam is difficult to see. Pure water does not scatter light very well. If a small amount of powdered milk is dissolved in the water, the beam becomes easily visible. It is important to understand that the light that enters your eye allowing you to see the beam is from the scattered light and not from the beam directly. If you shine the same flashlight through the air, which has negligible scattering, again you do not see the beam. Bulk absorption differs from bulk scattering in that the energy from the ray decreases without generating any new rays. According to the law of conservation of energy, this lost energy must turn into energy in another form. The lost energy in ordinary absorption turns into heat in the medium, which then increases the temperature. Figure 6.10 shows how an incoming ray can undergo all these processes in a material. Usually, all the light rays that are generated have the same wavelength as the incident ray. This is called the *regime of linear optics*. In nonlinear optics, discussed separately, new wavelengths can be generated.

One special reflection mechanism should be mentioned as it is important for both optical devices and laser safety. This is the phenomenon of Fresnel reflection, named after the French scientist who studied it in the eighteenth century. If light is incident on an interface between two materials with different refractive indexes, some of the light is reflected and some is transmitted. The amount reflected depends on the two refractive indexes, angle of incidence, and direction of the electric field (polarization). This Fresnel reflection is why you see a reflection

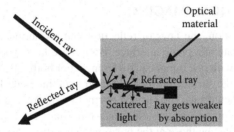

FIGURE 6.10 By conservation of energy, all the energy of an incident light ray must be reflected, transmitted (refracted), scattered, or absorbed. The process of absorption converts the light into heat. The other processes change the incident ray into other light rays traveling in different directions.

in a pane of glass. For ordinary glass, in the visible range of wavelengths at normal incidence the amount reflected is about 4%. This is important in laser safety because Fresnel reflection often leads to stray beams. Sometimes, it is important to reduce Fresnel reflection. This is done using a thin antireflection coating on the interface surface.

Again, I stray from the geometrical view of light to discuss a topic that is becoming important in many situations in laser safety. The topic of saturated absorption requires knowledge of the quantum theory of light to fully explain, but a qualitative sketch can be given here. Usually, absorption is just a property of the material and the wavelength of the ray. On an atomic scale, what is happening is that atoms (or molecules in some cases, but I just say atoms to avoid repeatedly saying atoms and molecules) in certain states can transition to states with higher energies when exposed to light with the right wavelength. Next, an atom or molecule can convert that absorbed energy into heat. Only once has it done that can it absorb more energy. Conversion of the absorbed energy into heat can happen very quickly. However, some lasers today emit very short pulses, as short as a few femtoseconds (one-billionth of a millionth of a second). These pulses, although short, can be extremely intense during the time that they are on. It can then happen that during the time that the pulse is on the intensity is so high that all the available atoms make the transition to the higher energy state and the pulse is so short that they do not have time to convert that energy into heat. No atoms will be available to absorb any more of the energy in the pulse. If this occurs, then any additional energy in the pulse will not be absorbed at all. The absorption is said to be *saturated*. The amount of energy absorbed by saturated absorption may be several orders of magnitude less than that which would be absorbed normally.

Why is this important for laser safety? Most laser-protective eyewear uses absorption to reduce the amount of laser light to which the eye could potentially be exposed. If an ultrashort pulse laser is used, the absorption in the laser eyewear could become saturated. The degree of protection afforded by the eyewear may then be insufficient to mitigate the hazard. When specifying laser eyewear for use with short-pulse lasers, it is necessary to ensure that the eyewear has been tested to verify its suitability at the pulse length used.

LENSES AND OPTICAL IMAGING

In the preceding section, the laws of reflection and refraction were introduced to show how they operate at a plane interface. Both these laws work exactly the same for curved surfaces, with one important difference. For both reflection and refraction at a plane interface, the surface normal was used to define both the incident ray and the reflected or refracted ray. The surface normal is different at each position on a curved surface. For curved surfaces, the surface normals used in the laws of reflection and refraction must be those at the position where the ray strikes the surface.

To illustrate how refraction works at a curved surface, the operation of a simple plano-convex lens is described. The object in the center of Figure 6.11 is the lens. It is a piece of glass with one surface ground into the shape of a sphere and the other surface flat. This is an example of a positive lens. Negative lenses are introduced later.

The rays entering the lens are all parallel to the symmetry axis of the lens. This is called a *plane wave*. Recall that in wave optics the wave peaks are perpendicular to the rays, so the wave peaks here form planes that are perpendicular to the rays. The waves in Figure 6.2 are plane waves before bending around the rock. Another way to think about this is as rays coming from a point source at a large distance to the left of the lens. The rays from the point source fan out in all directions, but the only rays that reach a distant location must be the ones that are traveling in the right direction to reach that location. A familiar example of a point source at a large distance is starlight. The plane wave is important in laser safety because, in many cases, a laser beam can be considered to be a plane wave.

Even though the rays are traveling in the same direction, they strike the first surface at different incidence angles because that surface is curved. The ray in the center enters the lens at normal incidence, and therefore it does not bend at all. The rays farther from the center enter at larger angles the farther they are from the center and therefore bend more. At the flat surface of the lens, the rays bend again when they go from the glass back into air. As a result of these two refractions, the rays that come from the point source traveling in the same direction converge after passing through the lens. In fact, they converge in just such a way that they intersect at a single point called the *focal point* of the lens. The distance from the lens to the focal point is called the *focal length* of the lens.

Consider the ray paths after they pass through the focal point. They fan out from the focal point just as the rays fan out from a point source. They are not traveling in

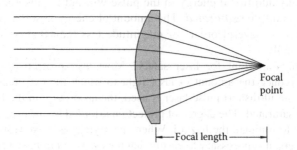

FIGURE 6.11   Refraction at a curved surface can focus a plane wave to a point.

all directions, though, because the lens only collected a limited number of the rays from the source. However, for an observer in the path of those rays the rays are just as if the point source were located at the focal point. Recall that the brain interprets vision based only on the light that enters the eye, so the perception is exactly as if the point source were located at the focal point. Therefore, we call the focal point the image of the point source at infinity (i.e., at a very large distance). The ability to essentially take an object at one location and transform it into an image at another location is very useful and is the basis for many optical instruments, such as telescopes, microscopes, and cameras.

It is at this point that a traditional optics tutorial begins to get quite complex with numerous cases derived from different curvatures, different object locations, and multielement optical systems. The specialized terminology alone can be daunting to the uninitiated. Fortunately for the readers of this chapter, there is not enough space to go into all of this. However, a few more concepts are needed to treat issues arising in the context of laser safety. The first is that of imaging of an extended object rather than a point source and not necessarily at a large distance. Then, I briefly return to the wave nature of light to discuss the role of diffraction in imaging. Finally, I mention negative lenses and virtual images.

In Figure 6.12, the same lens discussed for Figure 6.11 is shown. The object is an extended object. The object can be anything; for example, it could be the candle discussed in the preceding section. For purposes of illustration, the object is just represented as an arrow. The object is not at a large distance from the lens. The effect of the lens on the rays is found by considering each point on the object as a point source. A full picture of what happens can be obtained by considering only two of these points.

First, consider the rays from the base of the arrow. Just as in the case of the distant point source, all the rays come together after passing through the lens to form an image of the base of the arrow. The image, however, is not at the focal point but at a point farther from the lens than the focal point. This is because the rays incident on the lens are not parallel but diverging. The lens bends the rays by just about the same amount as in the previous case; but because they start out diverging, this amount of bending does not give them as much convergence as previously, so they intersect farther from the lens.

Now, consider the rays from the tip of the arrow shown in the diagram. The symmetry axis of the lens is called the *optical axis*. Therefore, the tip of the arrow is called

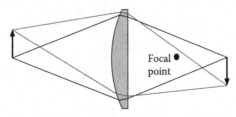

FIGURE 6.12 Imaging of an extended object at a finite distance from a lens can be understood by tracing rays from two points on the object.

an *off-axis object point*. After passing through the lens, these rays also converge, but they converge to an off-axis image point in the same plane as the base of the arrow. The rays from intermediate points on the object also converge to the corresponding points in the image. Imagine that there is a screen located at the plane of the image. A screen is a material that does a good job of scattering light of all colors equally. A person looking at the screen will have light scattered into his or her eye from each location where rays from the object strike the screen. The only light coming from the screen will be from where the light that passed through the lens strikes it. This pattern of light will be in the shape of the object, so an image of the object will appear on the screen. Any image that can be projected onto a screen like this is called a *real image*. Notice that the image is upside down, or *inverted*, from the original object.

The image can be larger or smaller than the object. The ratio of the size of the image to the size of the object is the magnification of the optical system. As the object is brought closer to the lens, the magnification becomes larger and the image moves farther away from the lens. When the object is exactly one focal length in front of the lens, the image moves to an infinite distance.

Sometimes, the wave theory of light cannot be ignored when discussing imaging. In fact, the effect of diffraction has important consequences for laser safety. So far, it has been stated that all the rays from a point source converge to a single point. This suggests that a point source would have an image on a screen that is just a point. It turns out that, due to the wave nature of light, the image of a point source is not quite a point but is smeared out into a *diffraction spot*. The size of this spot increases if the diameter of the lens gets smaller, if the focal length gets larger, or if the wavelength gets larger. There is also the case that the lens may not perfectly converge the rays to a spot. In this case, we say the lens has *aberrations*. In some cases, the aberrations cause more smearing than the diffraction, in which case the diffraction can usually be ignored. On the other hand, if there are aberrations but they cause much less smearing than diffraction, the aberrations can be ignored and the lens is referred to as *diffraction limited*.

Another type of lens and another type of image are illustrated in Figure 6.13. Notice that one surface of the lens is concave. The object is again a point source at a large distance from the left edge of the page. The first surface that the rays

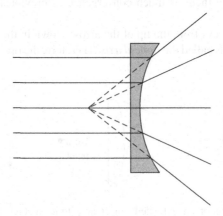

FIGURE 6.13   A negative lens forms a virtual focus.

encounter is flat, and all the rays strike it at normal incidence. Therefore, the rays do not bend. At the second surface, the rays strike at different angles. Because they are going from a higher index to a lower index, their angle of refraction is greater than their angle of incidence. The result is that rays farther from the lens' center diverge more and, to an observer of the light passing through the lens, they appear to come from a point on the other side of the lens. This point from which the rays diverge is called the focal point of the lens. Because the focal point is on the opposite side of the lens compared to the positive lens, this is called a *negative lens*.

Notice that the rays never actually pass through the focal point; they only appear to come from it. Therefore, this image of the point source cannot be projected onto a screen. For this reason, this type of image is called a *virtual image*. Virtual images can also occur with positive lenses when the object is closer than the focal length. A magnifying glass is a positive lens used in this way. When looking through a magnifying glass, you are seeing a virtual image of the object that is larger than the object.

Mirrors with curved surfaces are also common in laser technology. Just as with lenses, because the rays from a plane wave strike various points on a curved mirror at different incidence angles, the rays can be brought to a focus for a concave mirror or diverge from a virtual focus for a convex mirror.

Complex optical systems such as camera lenses or the setups in laser research laboratories consist of combinations of the lenses and mirrors discussed. When more than one optical element is used, the system is analyzed by considering each element in turn in the order that light strikes it. The image produced by the first element, whether it is real or virtual, becomes the object for the second element, and so on until the last element produces the final image.

EYE

Once the basics of imaging are understood, a wide array of optical instruments can be demonstrated: the telescope, the microscope, the camera, and so on. Details of the construction and operation of these optical instruments can be found in the bibliography at the end of this chapter. For purposes of laser safety, the most important optical instrument is the human eye; therefore, this one instrument is discussed here.

A highly simplified model of the eye is shown in Figure 6.14. The eye is a nearly round globe about 1 in. in diameter. Most of the eyeball is filled with a fluid called the *vitreous humor*, which has an index of refraction close to that of water. There is in addition a *lens*, which has material with various indices somewhat higher than that of water. The outermost part of the eye is the *cornea*. The cornea is the interface between the air and the fluid inside the eye. It is one of the few places on the body where living cells are exposed to the environment. There is also an *iris*, which controls how much light enters the eye. At the back of the eye is an area called the *retina*, which consists of light-sensitive cells that convert light to electrical impulses that are transmitted to various parts of the brain via the optic nerve. There are many other features of the eye: muscles, aspheric surfaces, structures to reduce light scattering, different types of light-sensitive cells, and other fluids. The ones pointed out here are all that are required to understand the functioning of the eye.

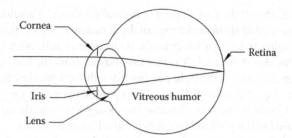

FIGURE 6.14 The eye is a very complex optical instrument. The drawing shows only a basic summary of the important parts of the eye.

Light arriving at the eye from a distant point source is shown in Figure 6.14. The rays are bent due to several curved surfaces at which there is a change in index of refraction. This is a *multielement lens*. Most of the bending is done at the cornea, but the small amount of bending at the lens is also very important. In the figure, the distant source is imaged on the retina. If the source were closer, we have learned that the image would move farther away from the lens and would not fall on the retina. However, muscles in the eye are able to bend the lens and change its curvatures. This has the effect of shortening the focal length and thus moving the image back to the retina.

The lens can only be deformed by a limited amount. If the object is brought in so close that the lens cannot be deformed enough to bring the image back to the retina, then the object cannot be clearly imaged. The closest location that an object can be placed and still be clearly imaged is called the *near point*. In young people, this is approximately 25 cm. As the eye ages, the lens becomes less flexible and the near point moves farther away. Reading glasses are positive lenses that make up for the reduction in focusing power.

The light-sensitive cells on the retina are remarkably sensitive. However, if the amount of light reaching them is too small, they will not be activated. The amount of light can be increased by increasing the size of the *pupil*, which is the opening in the iris. On the other hand, in a very brightly lit room there may be too much light. Then, the pupil will get smaller to reduce the amount of light entering the eye.

The figure indicates that the distant point forms a point image on the retina, but as mentioned, due to the wave nature of light, the image is not a point but a diffraction spot. The eye is very close to a diffraction-limited optical system.

The complicated operation of the eye, including the change in the lens' shape, opening and closing of the iris, and processing of the information sent to the brain, is mostly controlled by the unconscious mind. This highly sophisticated optical system can function for as much as a century. It is very remarkable.

EYE AND LASER SAFETY

The laser today is ubiquitous. There is one in every CD and DVD drive. They are used in communications, in store checkout counters, as laser pointers in presentations, and in land surveying. In all of these applications in which the general public makes use of lasers, lasers are either configured in such a way that they are not hazardous or completely enclosed within an inaccessible beam. In industrial and research and development applications, however, this is not always the case. High-power and even relatively low-power lasers can potentially cause severe, irreversible eye damage.

There are a number of different mechanisms by which lasers can cause damage to the eye. The most common one can be understood simply by looking again at Figure 6.14. Suppose that the plane wave incident on the eye is not from a distant point source but from a laser emitting a visible wavelength (400–700 nm). The eye will focus the light onto the retina just as if the laser beam is starlight, but the amount of optical energy involved is many orders of magnitude larger. As with the point source, this energy is absorbed by the light-sensitive cells. However, as discussed, absorption changes the light energy into heat energy. If the heat generated causes temperature to rise too much, the light-sensitive cells will be destroyed. The area over which the temperature rises can be larger than the spot itself due to heat conduction. If this occurs on the part of the retina associated with peripheral vision and if the laser is not too powerful, this could cause a blind spot in the field of vision. However, if it occurs in the small region of the retina that is used for detailed viewing, the result could be partial or complete blindness. With short laser pulses, the damage may be caused not by heat but by ablation, in which parts of the cells are essentially exploded. In the blue and green regions of the spectrum, laser light can also break chemical bonds in the cells and cause damage even if there is not enough increase in temperature to destroy the cells.

The light-sensitive cells in the retina are not sensitive to light with a wavelength longer than about 700 nm. However, the lens, the vitreous humor, and other parts of the eye are still transparent to rays with wavelengths as long as 1400 nm. This means that lasers with wavelengths in this range can cause the same injuries described. This presents a very dangerous situation because the light from these lasers cannot be seen. In the visible range, the hazard due to laser eye exposure is mitigated to some extent by the natural aversion response (blink), which limits the amount of time that the light illuminates the retina. Of course, there is no such aversion response with invisible light. The wavelength range from 400 to 1400 nm is called the *retinal hazard zone*.

For wavelengths longer than 1400 nm, little energy is transmitted to the retina. The radiation is absorbed in the lens or other parts of the eye depending on the wavelength. Although the absorbed energy will still cause heating, the temperature rise will be smaller because the energy is absorbed over a large volume rather than in a small spot on the retina. Also, the delicate light-sensitive cells are not heated. The spectral range with wavelengths greater than 1400 nm is sometimes called the *eye-safe region*. It is only eye safe in the sense that the maximum permissible exposures (MPEs) in this region are orders of magnitude higher than those in the retinal hazard zone. There are certainly lasers with wavelengths in this range that can cause serious eye injury.

For lasers with wavelengths shorter than 400 nm (ultraviolet), a similar situation occurs. Again, the eye is not transparent to the radiations. However, there are two important differences between the far-infrared and the ultraviolet wavelengths. First, the

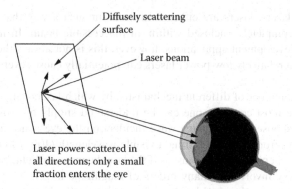

FIGURE 6.15 Diffusely scattered laser light is less hazardous than direct exposure to a laser beam because only a small fraction of the light enters the eye. Under some circumstances, the spot size on the retina can also be larger, which further reduces the hazard.

eye absorbs most of the radiation very close to the front of the cornea. Second, unless the laser intensity is very high any damage to the cornea is due to chemical changes such as those caused by sunburn. This is the same injury that can happen when viewing nonlaser ultraviolet sources. This injury, called *welder's flash*, is very painful but usually causes no permanent injury because the cells on the cornea naturally replace themselves quickly. Of course, this exposure can also increase the risk of cancer. For these reasons, the ultraviolet region of the spectrum is not considered eye safe even though there is no retinal hazard. In fact, for certain wavelengths and exposure durations the MPE for ultraviolet can be lower than the MPE in the retinal hazard region of the infrared range.

Next the difference between direct laser exposure and exposure to diffusely scattered laser light is discussed. Figure 6.15 shows a laser beam impinging on a diffusely scattering surface. As discussed, all the laser energy is carried off by the scattered rays (unless there is some absorption also). Therefore, most of the radiation is harmlessly directed away from the eye. Only a few rays get into the eye. For this reason, diffusely scattered light is much less hazardous than direct laser light. In fact, except for class 4 lasers diffusely scattered laser light is not a hazard. Notice that the farther the eye from the diffuse reflection, the fewer the rays that enter the eye. Therefore, the diffuse hazard decreases with distance in contrast to the hazard of direct laser exposure, which is nearly independent of distance.

There is another way that diffusely scattered light may be less hazardous than direct laser light in some cases, but not always. Sometimes, the laser spot is large enough that when it is imaged by the eye onto the retina the image is large enough to spread out the heat deposited on the retina. This reduces the rise in temperature. This is the reason for the correction factor C_E in American National Standards Institute (ANSI) Z136.1.

OPTICS IN A LASER RESEARCH LABORATORY

An industrial laser safety officer (LSO) may encounter lasers in a wide variety of settings. The evaluation of all the hazards may be very simple or very complex. In some modern laser research laboratories, the task may be daunting just due to the variety of components and systems that may be encountered. This section presents a sampling of the types of equipment that may be present in a laser laboratory along with some comments about the associated laser safety issues. The list is not necessarily exhaustive.

LASERS AND LASER AMPLIFIERS

The laser is a device that is difficult to explain without good knowledge of the quantum theory. Fortunately, the principles of laser operation are not very important for laser safety considerations. The central part of a laser oscillator is some sort of a gain medium, with a crystalline or glass rod, tube of ionized gas, jet of dye, and p–n semiconductor junction being the most common. The gain medium is excited by an energy source called a *pump*. The pump may be an electrical current, a flash lamp, or even another laser. After being excited by the pump, the gain medium emits light, which is recirculated through the gain medium by a resonator until it reaches the desired energy, after which it leaves the resonator through either an output coupler or an electro-optical switch. This then is the output laser beam.

Most lasers emit a narrow beam of light. High-power lasers may emit a beam that is not narrow. Commercial solid-state lasers generally have a beam diameter less than a centimeter, but custom-built, high-power lasers in research facilities may have dimensions of tens of centimeters. Some laser diodes and laser diode arrays emit light that has such a large divergence that it is not really a beam. Many lasers emit rays with exactly one wavelength; others emit several wavelengths, sometimes simultaneously. Still others can be tuned to emit at any desired wavelength within a range.

There are, of course, a great variety of lasers. Commercial lasers exist with wavelengths from 157 nm to 12 million nm. Powers of commercial lasers can be as high as 30 kW. Pulse lengths can be as short as 10 femtoseconds. Custom-built lasers can stretch the parameter range even further.

This large variety of specifications means that the LSO must carefully read the specifications of any potentially hazardous laser. For example, if a laser is capable of emitting several different wavelengths, is the selected wavelength the only wavelength that will be emitted? If even a small amount of another wavelength leaks out, it may be sufficient to pose a hazard. For tunable lasers, does the laser eyewear cover the entire tuning range of the laser?

Some laser systems include laser amplifiers to increase the energy in the beam. The amplifiers may be embedded in a commercial laser along with the oscillator or may be mounted outside the oscillator. A laser amplifier is similar to an oscillator in that it consists of a gain medium and a pump, but usually there is no resonator; the light generated by the oscillator (or a previous amplifier stage) is passed through the gain medium. In some cases, the beam is sent through an amplifier several times. Many times, the beam size is increased as it is sent from one amplifier stage to the next.

LASER BEAM TRANSPORT OPTICS

Mirrors and lenses are used to convey a laser beam from where it is generated to where it will do something useful. Flat mirrors are the most commonly used elements in laser systems. They are nearly always made reflective using a thin-film, multilayer dielectric coating. Dielectric coatings use interference of light to enhance the reflectivity of a surface. They can be tailored to provide nearly any desired reflection characteristic, such as very high reflectivity at the laser wavelength. Unlike silvered mirrors, though, the reflectivity at wavelengths other than the laser wavelength may be small. For a laser operating in the infrared or ultraviolet range, the reflectivity throughout the visible range may be small. To someone unfamiliar with these mirrors, it can be hard to realize that what appears to be just a transparent piece of glass actually reflects 99.9% of the laser light.

Also unlike silvered mirrors, the portion of the laser beam that is not reflected from the mirror is not absorbed; rather, it is transmitted. If, as is sometimes the case, the back of the mirror is polished, the leakage beam will be transmitted out through the back of the mirror. This is one of the many sources of stray beams in laser laboratories. Even if this is only 0.1% of the energy of the original beam, it can easily be a hazardous beam. As the coating ages or as the humidity in the room changes, the amount of transmitted light can increase. If the reflectivity falls to 99%, the loss in the reflected beam might not even be noticed, but the stray beam transmission would increase by an order of magnitude to 1%. Also, the amount of light transmitted can dramatically increase if the coating is scratched or otherwise damaged. For these reasons, it is best practice to locate an opaque beam block behind the mirror and never to count on the reflectivity of the mirror for protection against laser hazards.

One particular hazard of mirrors is that they can create upwardly directed beams. Most of the time, laser beams are directed parallel to the surface of an optical table that is kept below eye level. However, occasionally it is necessary to change the level of the beam. This is done with a pair of mirrors in a periscope configuration in which one mirror directs the beam up and the other then directs it again parallel to the table but at a higher level. If the system becomes misaligned so that the beam misses the upper mirror, the beam travels upward until it encounters something that stops it. In several unfortunate incidents, the upward-traveling beam was stopped by a worker's eye. Extra caution is required whenever upward-directed beams or beams at or near eye level are employed.

Lenses are also used to transport laser beams. One might think that a laser beam propagates a long distance without changing, but diffraction can greatly change the profile of a laser beam. A beam that is created to have a very uniform intensity over its cross section can develop hot spots after propagating some distance. This is undesirable if a flat intensity is needed for the application and if the hot spots are intense enough to cause damage to some of the optical elements in the system. A telescope consisting of two (or occasionally more) lenses can project an image of the flat-intensity beam to another location. The telescope can also be used to change the size of the beam, perhaps to increase the beam size for a following amplifier stage. In most telescopes, the laser beam is brought to a small focal spot between the two lenses. With high-power lasers, the intensity in this small spot can be large enough to cause air to ionize, creating something like a small ball of lightning. When the beam passes through this ionized air, the beam is distorted. For this reason, the beam transport telescopes for high-power lasers are enclosed in a vacuum tube. These long, skinny tubes with lenses as vacuum seals and a vacuum hose coming out of them are ubiquitous in laser research laboratories.

Lenses also frequently have their surfaces coated with a thin-film dielectric coating. The purpose of this coating is to reduce reflectivity. The glass that is the lens material will reflect several percent of the laser beam. This may not seem like very much, but a complex laser system can have dozens of lenses, and laser light can be very expensive. The dielectric coating can reduce the reflectivity to a small fraction of a percent.

Even if dielectric coatings reduce reflectivity, there is still some reflection from each surface of the lens. These reflections from the lens are usually called *ghosts*, and they are another source of stray beams in a laser laboratory. They can be more problematic than those from mirrors for a number of reasons. There are more of them. In addition to one from each surface, there can be secondary ghosts in which the light reflected from one surface reflects again after hitting another surface. Often, these reflections cannot be blocked because they are in the beam path. Because the reflections can be from curved surfaces or may pass through curved surfaces, they can be divergent or convergent. The convergent beams, in addition to being an eye safety concern, can cause damage to the laser if they cause a small focus to form on the laser optics. Like the focus in the beam transport telescope, they can sometimes cause air to ionize. The divergent beams (and the convergent beams after passing through focus) can become large and difficult to block. As they get larger, of course, they become less hazardous, but determining the nominal hazard zone of these beams can be very difficult. Often, prudence requires that they all be considered hazardous. This is a major reason why laser eyewear should always be required in a laser laboratory even if the workers are sure that they cannot be exposed to the main beam.

Setting up or realigning a laser beam transport system can be one of the most hazardous operations involving lasers. When first locating a mirror or lens on a table, the laser beam may initially stray far away from the path it is supposed to take. Low-power alignment lasers should be used as much as possible. Part of setting up a laser laboratory should always be locating all the stray beams and blocking them if possible. In a complex system with lasers outside the visible range, this can be a difficult task. The stray beams may take some unexpected paths. Sometimes, building an opaque enclosure around the laser system may be the best way to control these hazards.

Laser beam transport systems frequently include beam splitters. Beam splitters can be of several types. The simplest is a mirror that reflects only a portion of the light and allows the rest to be transmitted. In some cases, such as for interferometry or holography, the transmitted and reflected beams have about the same intensity. In other cases, the majority of the beam is reflected and a small portion is transmitted, often for use in a diagnostic measurement device. Alternatively, the transmitted beam may be the small-intensity beam used for diagnostics, whereas the reflected beam contains the majority of the energy. Uncoated beam splitters fall into this last category. These use Fresnel reflection to generate the reflected beam. This has several advantages. It does not require any coating of the surface, the amount reflected can be very accurately predicted, and the uncoated surface can be much more resistant to laser damage than a coated surface.

Other types of beam splitters are used as well. A dichroic beam splitter transmits some wavelengths and reflects others. This is useful in frequency conversion, discussed separately, since there usually is some unconverted light that must be removed. Finally, polarizing beam splitters transmit one linear polarization and reflect the other.

NONLINEAR OPTICAL DEVICES

All the events that were previously said to be the potential fate of a ray involved either the generation of new rays with the same wavelength or the disappearance of the ray. In the quantum theory of light, it is possible to generate new rays with different wavelengths. Components that do this are frequency conversion devices. Another device that works on the same principle can increase the energy of a beam at one wavelength at the expense of a beam at another wavelength. This is called *optical parametric amplification*. Components that always produce rays of the same wavelength as the incident beam belong to the realm of linear optics. The devices in this section are called *nonlinear optics*.

One very common example of the first process is a harmonic generator. It consists of a special crystal that a laser beam is directed through in a certain direction. Part of the laser beam is converted to a wavelength exactly half of the original wavelength. In other words, a laser beam with a wavelength of 1000 nm would be converted to a wavelength of 500 nm. It is possible but less common to generate other harmonics, such as one-third or one-fourth of the original wavelength, in a single crystal. Instead, these harmonics are generated using the closely related process of sum-frequency generation. To obtain the third harmonic (wavelength one-third of the original), a frequency doubler first converts some of the light to half the wavelength. Next, the unconverted light and the second harmonic are combined in another crystal to produce the third harmonic.

Usually in harmonic generation, the harmonic beam travels along the same path as the original beam. Even if the harmonic generation is very efficient, there will be some residual unconverted light. Unless the residual is removed completely, the hazards of both wavelengths need to be taken into account for safety considerations. This is why it was mentioned that for lasers emitting multiple wavelengths the possibility of more than just the desired wavelength being present must be considered. Lasers emitting multiple wavelengths usually have harmonic generators inside them.

Another nonlinear optical device is the optical parametric amplifier. It is also based on a crystal illuminated by two laser beams of different wavelengths. Of the two, the longer wavelength is called the *pump* and the shorter is called the *signal*. In this process, some of the energy of the pump is extracted and goes into producing more energy in the signal beam. In the process, a third beam is also produced, called the *idler*. Because the energy taken from the pump goes into both the signal and the idler, the signal can at most be increased by half the energy in the pump. Sometimes the idler beam itself is desired, in which case this same process is another frequency conversion process called *difference frequency generation*.

There are many other nonlinear optical processes as well, with Raman scattering and Brillouin scattering being two of the more common ones. In Raman scattering, light is scattered from atoms and molecules. I have discussed scattering of light from atoms and molecules, but in that case the scattered light had the same wavelength as the incoming ray and the atoms were left in the same state as before the scattering. If the atoms change to another energy state during a scattering, the wavelength of the scattered light changes. In Brillouin scattering, light scatters from a sound wave. The change in wavelength in Brillouin scattering is very small. Brillouin scattering is the basis for a device sometimes found in high-power laser laboratories called a *phase conjugator*. A phase conjugator acts like a mirror with a very unusual law of reflection. The reflected ray retraces the path of the incident ray. If there are any aberrations on the beam, the phase-conjugated reflection has exactly the opposite aberration. When the reflected beam retraces its path through the system that produced the aberrations, the aberrations on the beam are cancelled, resulting in a high-quality beam with no aberrations.

Whenever nonlinear optical devices are used, it is important to ensure that all potential wavelengths are included in the safety analysis. Most nonlinear devices must be carefully aligned to work properly. Therefore, the energies of the various wavelengths to be used for safety calculations must always be the worst-case numbers to allow for misalignments or damage to the nonlinear devices.

Diagnostics

Much of the optical and nonoptical equipment in a laser research laboratory consists of devices to measure and record characteristics of lasers or of the experiments for which lasers are used. Some typical measurement equipment includes power and energy meters, photodiodes for pulse width measurements, cameras, interferometers, and spectrometers.

To diagnose the laser beam itself, a sample of the beam must be obtained. There are two ways of doing this. A mirror in the beam transport system can be given a coating that is not highly reflective but has a transmission of a few percent. Then, the beam transmitted through the mirror can be used for diagnostics. Alternately, a diagnostic beam splitter can be used. This is a piece of glass with two flat surfaces that reflects a portion of the beam for diagnostics. The reflectivity of an uncoated piece of glass is 3% to 4%, depending on the type of glass and the wavelength. A dielectric coating can be used to increase or decrease this value, depending on the need. Of course, both surfaces of the beam splitter reflect light. Often, the beam splitter will be wedged (the two surfaces are not exactly parallel to each other) so that the two reflections can be separated. Both of the beams may separately be used for different diagnostics. If only one of the reflections is used, the second becomes another stray beam that must be blocked.

Good diagnostics can be an important aspect of laser hazard management. If the important parameters of a laser can be monitored with remote diagnostics and displayed on oscilloscopes, monitors, or other display devices and, if needed, adjustments can also be performed remotely, the need to access the laser beam can be minimized. Occasionally, diagnostics can also introduce optical hazards. The absorbing surfaces of energy meters are often smooth and can generate specular reflections, which can be hazardous. Similar to many laboratory hazards, this one tends to cause problems primarily during the initial alignment procedure.

CONCLUSIONS

Some of the most basic principles of optics have been discussed in this chapter. By mastering these basics, a better appreciation of the functions of various devices and setups likely to be found in laser research and development laboratories can be obtained. It is important for an LSO to know the right questions to ask to get a realistic assessment of all the optical hazards in a laboratory. Some of these can be very subtle and may not be appreciated by the workers in the laboratory. Being knowledgeable about the equipment in a laboratory and its common uses also enhances the LSO's appreciation of the work that is being done in the laboratory. Solutions to laser safety issues that are developed with an understanding of the scientific purposes for which the equipment is used are more likely to not hamper the work in the laboratory while still mitigating hazards.

Of course, the discussion here only involved the basics of this vast field. To learn more, there are several references given to some of the standard books on several aspects of geometrical and wave optics and to laser technology.

BIBLIOGRAPHY

Fowles, Grant R., *Introduction to Modern Optics, Second Edition*, 1989. Dover: New York.
Hecht, Eugene., *Optics, Fourth Edition*, 2001. Addison-Wesley: New York.
Jenkins, Francis A. and White, Harvey E., *Fundamentals of Optics, Fourth Edition*, 1976. McGraw-Hill: New York.
Johnson, B.K., *Optics and Optical Instruments*, 1960. Dover: New York.

7 Diffraction Gratings for High-Intensity Laser Applications

Jerald A. Britten

CONTENTS

INTRODUCTION

The scattering of light into wavelength-dependent discrete directions (orders) by a device exhibiting a periodic modulation of a physical attribute on a spatial scale similar to the wavelength of light has been the subject of study for over 200 years. Such a device is called a diffraction grating. Practical applications of diffraction gratings, mainly for spectroscopy, have been around for over 100 years. The importance of diffraction gratings in spectroscopy for the measurement of myriad properties of matter can hardly be overestimated. Since the advent of coherent light sources (lasers) in the 1960s, applications of diffraction gratings in spectroscopy have been further exploded. Lasers have opened a vast application space for gratings, and apace, gratings have enabled entirely new classes of laser systems.

Excellent reviews of the history, fundamental properties, applications, and manufacturing techniques of diffraction gratings up to the time of their publication can be found in the books by Hutley (1982) and more recently by Loewen and Popov (1997). The limited scope of this chapter can hardly do justice to such a comprehensive subject, so the focus here is narrowly limited to characteristics required for gratings suitable for high-power laser applications and methods to fabricate them. A particular area of emphasis is maximally efficient large-aperture gratings for short-pulse laser generation.

Light of a wavelength λ incident at angle θ_i on a diffraction grating that has modulation with a period p is diffracted in a direction θ_m according to the grating equation:

$$\sin \theta_m = \sin \theta_i + m\lambda/p \qquad (7.1)$$

where m is the grating order. Figure 7.1 illustrates the geometry of diffraction of light of a single wavelength into various orders from a grating.

Gratings can operate in transmission, reflection, or both. The modulation can be in the form of periodic variations in the density or refractive index throughout the bulk of a material. These are volume gratings. Or, the modulation can be physical corrugations of the surface of a material; these are surface relief gratings.

The use of gratings for high-power laser applications requires that, other than redirecting the incident light according to the grating equation, the grating does nothing else to distort the beam, and that the beam does nothing to the grating, such as heating or damage. These restrictions turn out to be quite serious and severely limit potential candidates for the intended application, as discussed in this chapter.

Of particular note appropriate for this book, dealing mainly with laser safety issues, is the fact that diffraction gratings can direct intense beams in several directions at once, directions that are counterintuitive to researchers used to dealing with normal reflective and refractive optics. This makes the beams more dangerous in general than normal optics in high-power laser applications, and so they must be treated with the appropriate attention to safety, with beam-blocking setups in both reflection and transmission to safely block unused orders.

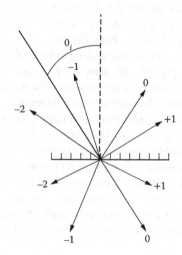

FIGURE 7.1 Orders diffracted from a grating.

TYPES OF GRATINGS AND THEIR USES

VOLUME GRATINGS

Dichromated gelatin (DCG) gratings (cf. Arns et al., 1999) are perhaps the best-known variety of volume grating. These are made by casting a many-micrometer-thick film of gelatin sensitized with a photoreactive dichromate salt onto a glass plate, exposing this plate to a holographically generated interference pattern, and using wet chemistry to change the refractive index of the exposed regions in the gelatin. The resulting film is hygroscopic and must be hermetically encapsulated by a cover plate sealed at the edges to retain the index modulation. DCG gratings are typically used in transmission but can be used in reflection by double passing the light through the film with a reflective surface on the backside plate. They can have very high efficiency and low polarization-dependent losses and can be made quite large. They are used in spectroscopic and display applications but in general have poor diffracted wave front flatness that must be corrected by postpolishing or adaptive optics in a laser application. Also, the organic gelatin material has a very low laser-induced damage threshold in comparison with some transparent inorganic dielectric materials.

Photothermal refractive (PTR) gratings (Efimov et al., 1999) are a relatively recent technology finding increasing use in demanding laser applications. PTR glass is a multicomponent glass doped with small amounts of photosensitive constituents that undergo a very small reduction in refractive index on exposure to ultraviolet (UV) light below about 330 nm followed by a heat treatment. Interference lithography can be used to generate the grating pattern in this glass. PTR glass has laser damage resistance to infrared (IR) light quite similar to BK7 or other multicomponent glasses and is therefore quite useful for many high-power laser applications (e.g., Chung et al., 2006; Liao et al., 2007). At present, the glass and the interference lithography platforms capable of recording into it are limited to a few millimeters in aperture. These gratings also will not be useful for kilowatt-class high average-power laser systems as the bulk material will heat and distort due to low-level absorption by some of the glass constituents.

Fiber Bragg gratings (FBGs) (Kashyap, 1999) are a related type of grating by which an index modulation is induced longitudinally in the core of an optical fiber by exposing the fiber to interference from UV light. The fiber can be a germanium-doped fused silica exposed to 244-nm light or even pure fused silica exposed with 193 nm at high temperature in the presence of hydrogen. FBG structures were an area of intense research for optical telecommunications applications several years ago and are currently under development for short-pulse lasers that require dispersion control.

SURFACE RELIEF TRANSMISSION GRATINGS

Low-cost transmission gratings made by embossing surface relief patterns from master gratings onto plastic substrates, or polymer films on glass substrates, are available from several commercial suppliers. These can exhibit very good efficiency and optical quality but in general are not suitable for intense laser applications due to damage to the films or thermal distortion of the beam by low-level absorption of the laser light. Gratings etched into bulk fused silica (Néauport et al., 2005; Nguyen et al., 1997; Yu et al., 2003) have demonstrated the ability to withstand very high peak power laser pulses at both IR and UV light. High-efficiency transmission gratings used for beam steering, focusing, and wavelength discrimination (Néauport et al., 2005; Nguyen et al., 1997) require periods on the order of the laser wavelength and rather large incidence angles, so that only the 0 and −1 orders exist, and require deep grating structures with height-to-width aspect ratios (H/W) of about 4. These features require ion beam milling to transfer-etch the pattern from the photoresist mask to the bulk material. Gratings of this type (470 × 420 mm in aperture) have been demonstrated for a high-energy beam-steering and -focusing application (Néauport et al., 2005). Low-efficiency beam-sampling transmission gratings that diffract and focus a very small but precisely known fraction of the total beam energy to an energy diagnostic have been fabricated at 400 × 400 mm aperture by wet etching the mask pattern into the bulk fused silica using a weak hydrofluoric acid solution (Yu et al., 2003). This is possible due to the very low aspect ratios of the features required for low-efficiency gratings, typically H/W of about 0.01. It is possible to create multi-function, multiscale, low-aspect ratio grating structures by wet etching that simultaneously performs wavelength discrimination and beam-sampling functions (Britten and Summers, 1998).

GENERAL LIMITATIONS OF TRANSMISSION GRATINGS IN HIGH-POWER LASER SYSTEMS

High average-power lasers: Parts per million absorption in multicomponent-glass transmissive optics can very quickly cause thermal distortion and lack of focusability of the beam in multikilowatt lasers (Yamamoto et al., 2006). A probable exception would be high-purity fused silica surface relief gratings.

Intense nanosecond pulsed lasers: Again, high-purity fused silica is acceptable in many cases, but coherent addition of the electric fields generated by more than one propagating order in the bulk of the glass can cause laser damage. This is true even if the grating is on the output face of the optic as it is typically not possible to totally suppress backreflections of 0, −1, or possibly higher orders.

Intense short-pulse lasers: Very intense laser beams operating at tens of picoseconds or shorter can cause dielectric breakdown even in atmospheric gases and so are propagated in vacuum. Needless to say, all-reflective optics are required for beam manipulation in this case.

SURFACE RELIEF REFLECTION GRATINGS

Surface relief reflection gratings are probably the largest class of gratings in terms of numbers and application space and certainly the most important for intense short-pulse lasers. As with transmission gratings, relatively low-cost, high-quality gratings replicated in epoxy or other polymeric films from holographically written or mechanically ruled master gratings and overcoated with a reflective metal layer are readily available from commercial sources. However, to obtain the highest possible diffraction efficiency, wave front quality, and thermal loading performance, the large-aperture short-pulse laser community for years has demanded high line-density gratings produced by laser interference lithography (holographic gratings). The most intense short-pulse lasers require very large aperture gratings to reduce the power density on the grating surface to below the optical damage level. The first petawatt-class laser system, built at Lawrence Livermore National Laboratory (LLNL) in the mid-1990s (Perry et al., 1999), used internally produced gold-overcoated holographic master gratings of 94 cm in diameter. These are still the largest monolithic gratings made, and now several laser systems around the world use gratings of the same design to do basic high energy–density and nuclear fusion–related research (Danson et al., 2005; Kitagawa et al., 2004).

At the same time the Nova Petawatt was being built, a new class of reflective diffraction grating was being developed, also at LLNL. Multilayer dielectric (MLD) gratings (Perry et al., 1995; Shore et al., 1997) combine an MLD high-reflector stack with a grating etched into the top layer or top several layers to create a grating having theoretically 99.9% diffraction efficiency into the −1 order with no absorption of the light and therefore much higher laser damage threshold. Consequently, just as MLD high reflectors have supplanted metallic mirrors for high-power laser systems, MLD gratings have supplanted gold-overcoated gratings for most, but not all, short-pulse laser applications. The remainder of this chapter briefly reviews short-pulse generation, then describes in detail the design and manufacture of large-aperture holographic gratings.

PLANE GRATINGS FOR LASER PULSE COMPRESSION

The technique of chirped-pulse amplification (CPA) (Strickland and Mourou, 1985) uses one or more gratings with other optics in a "stretcher" to temporally disperse a low-energy, broadband, short-pulse beam by a factor of about 10^3. The stretched beam is then amplified by conventional gain media without undergoing nonlinear self-focusing. Gains of about 10^6 are possible. The amplified, stretched pulse is then sent through a "compressor," typically containing two to four gratings that undo the temporal dispersion of the stretcher to create an intense pulse of nearly the initial pulse duration. A schematic of CPA is shown in Figure 7.2. The requirements of the compressor gratings in particular are quite demanding. Typical compressor designs use four grating bounces.[4] The beam size and intensity are in large part limited by the size and damage threshold of the compressor gratings. Therefore, maximizing the efficiency and damage threshold of the gratings has an enormous impact on energy delivered to the target as well as the cost of the laser system.

DESIGN CONSIDERATIONS

High-energy petawatt-class lasers being built today operate at 1053 nm, use Nd:glass as the laser medium, and are designed to produce pulses from a few hundred femtoseconds to several picoseconds in duration. These pulse durations require a high diffraction-efficiency bandpass less than 20 nm, a condition easily met by MLD gratings. To maximize diffraction efficiency and peak power-handling ability, MLD

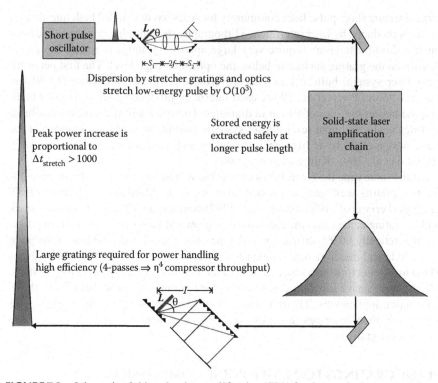

FIGURE 7.2   Schematic of chirped-pulse amplification (CPA) for short-pulse laser generation.

grating designs have evolved to high-line densities (1700–1800 lines/mm) operating at high near-Littrow incidence angles, comprising hafnia (HfO_2)/silica (SiO_2) layers with the top grating layer made of silica. The Littrow angle is defined as that in which the -1 order diffraction angle is the same as the incident angle. HfO_2 and SiO_2 are the materials of choice based on extensive experience for high-intensity nanosecond-pulse MLD high reflectors (Wu et al., 2001). Because it is a high-bandgap material with a high intrinsic laser damage threshold (Stuart et al., 1996) and is also amenable to deposition and subsequent processing incurred during grating manufacture, SiO_2 is the material of choice for the grating layer.

We use an in-house grating design code based on the work of Li (1993) in conjunction with commercially available thin-film codes to design MLD gratings. Commercial grating design software (Grating Solver Development Company, Allen, TX; www.gsolver.com) is available as well and offers good agreement with results of our in-house code, although the e-fields generated in the stack and grating are not readily viewable by it.

Typically, the stack is composed of 20–60 layers. The main criteria are that it should be a high reflector at the use wavelength and angle range, and that it should have a relatively thick SiO_2 capping layer. The large number of layers gives the designer a great deal of flexibility to incorporate other features into the stack, such as antireflection properties at the holographic exposure wavelength and angle (Shore et al., 1997) and etch-stop layers for precise grating depth control (Britten et al.,

1996a). Grating design codes are used iteratively with the thin-film design codes to optimize the etched depth and duty cycle (ratio of grating line width to period), shape, and layer thicknesses for the application of interest. Of particular importance for large-aperture gratings is a design that is robust to small duty cycle and depth errors inevitable in large-scale manufacturing. Lamellar (vertical sidewall) grating profiles are relatively easy to produce with high-contrast photoresist masks and collimated ion beam etching and exhibit high laser damage thresholds. The following discussions pertain to this type of structure.

Figure 7.3 shows an MLD grating design with a plot of expected diffraction efficiency as a function of the grating depth and duty cycle. Notice that a ridge of constant greater than 99% diffraction efficiency is possible for a combination of grating height and duty cycle values. Examination of the e-field distribution generated on these gratings by the incident beam shows standing waves caused by interference from the incident and high-efficiency diffracted wave. These areas of high e-field penetrate into the back side of the grating ridges. It is here that damage to the grating due to multiphoton ionization and avalanche breakdown occurs (Stuart et al., 1996). Analysis shows that the maximum e-field in the solid material is lowest for tall, thin grating lines, so these should have the highest damage threshold (Britten et al., 2003). This has been experimentally confirmed (Neauport et al., 2007). Typical laser damage thresholds for 1780 lines/mm MLD gratings at 1053 nm, 10 ps, about 75°

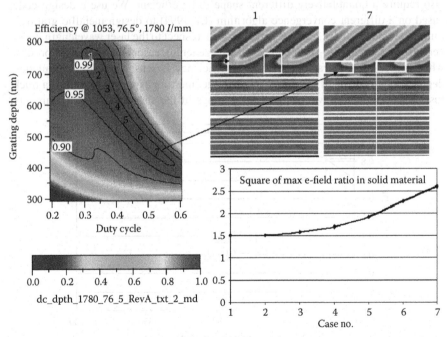

FIGURE 7.3   Left: Contour map of −1 order diffraction efficiency as function of grating depth and duty cycle for a 1780 lines/mm multilayer dielectric (MLD) grating. Right: Outline of grating structure showing layers and grating shape for conditions 1 and 7, both exhibiting more than 99% diffraction efficiency, superimposed with the electric field distribution for each case.

incidence angle, exhibiting greater than 97% diffraction efficiency, are about 5 J/cm^2 (measured for the normal incidence beam). This is more than five times the damage threshold of a similar holographic gold-overcoated photoresist grating that has slightly lower efficiency (Britten et al., 2003; Neauport et al., 2007).

The MLD gratings are not the answer for all short-pulse applications, however. Ultrashort-pulse, high-intensity Ti:sapphire lasers operating below 50 fs typically require a high-efficiency bandpass of up to 100 nm at a central wavelength of about 800 nm. Designers of these lasers demand low line-density gratings to better manage dispersion and provide more alignment tolerance. Figure 7.4 compares diffraction efficiency versus wavelength for optimized MLD and gold gratings at 1480 lines/mm, centered at 800 nm. Not only does the MLD design cover the desired bandpass, but also exhibit very narrowband resonance regions in which the interaction of the impedance in the grating and underlying stack couples light into one or more of the layers, acting as a waveguide. Figure 7.5 shows the electric field distribution at one of these resonance conditions. These types of resonances seem to be ubiquitous regardless of design details of the MLD stack. Although this resonance feature may be useful in some waveguiding applications, it is disastrous for a wide-bandwidth, damage-resistant grating. Therefore, gold gratings remain viable for these applications.

Gold-overcoated gratings operate best in transverse magnetic polarization, in contrast to MLD gratings, which require TE (transverse electric) polarization. They also require a quantitatively different shape to be efficient. We use a design code based on a different convergence algorithm (Li, 1994) to design metallic gratings. Sinusoidal modulations, relatively shallow with respect to the laser wavelength, work very nicely. Certain low-contrast photoresists generate reasonably sinusoidal profiles when processed holographically, but the efficiency is sensitive to the modulation depth, which is difficult to control over a large-aperture exposure format in a thick resist layer. Therefore, we design gold gratings of the shape shown in Figure 7.6,

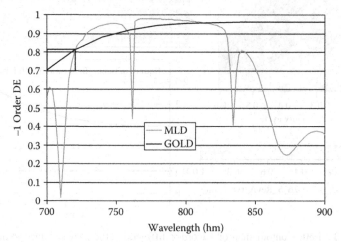

FIGURE 7.4 Comparison of diffraction efficiencies of gold and multilayer dielectric (MLD) gratings at 1480 lines/mm and 36° incidence angle. Notice the drop in efficiency of the MLD grating at 710, 762, and 834 nm. Here, the efficiency is coupled into the stack at the −1 or −2 transmitted order.

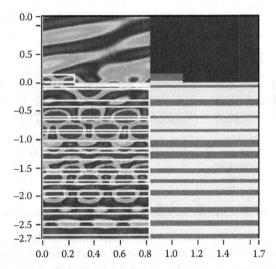

FIGURE 7.5 Electric field distribution in the multilayer stack of a multilayer dielectric (MLD) grating for a resonance condition of Figure 7.4, on left. On right is a pictorial representation of the MLD stack. Dimensions are in micrometers.

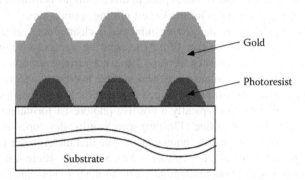

FIGURE 7.6 Optimized profiles for gold-overcoated reflection gratings.

with the modulation depth fixed by the thickness of the photoresist layer (Boyd et al., 1995). Exposure nonuniformities in this design translate into line width variations, to which the diffraction efficiency is relatively insensitive. These gold gratings can exhibit high diffraction efficiency for a very large bandwidth in the near infrared (Britten et al., 1996b).

MANUFACTURE OF LARGE-APERTURE HOLOGRAPHIC PLANE GRATINGS

A process flowchart for manufacture of MLD gratings is shown in Figure 7.7. Dielectric oxide materials are deposited according to the design by a variety of methods. Electron beam deposition is a common method that gives very high laser damage threshold coatings for nanosecond-scale pulses. These coatings are typically

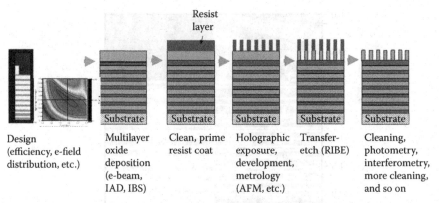

Resist
layer

| Design (efficiency, e-field distribution, etc.) | Multilayer oxide deposition (e-beam, IAD, IBS) | Clean, prime resist coat | Holographic exposure, development, metrology (AFM, etc.) | Transfer-etch (RIBE) | Cleaning, photometry, interferometry, more cleaning, and so on |

FIGURE 7.7   Process flowchart for multilayer dielectric (MLD) grating manufacture.

somewhat porous and so can undergo stress transformations during processing and operation in vacuum as they lose adsorbed water, even to the point of crazing (cracking) under sufficient tensile stress. Ion-assisted e-beam deposition uses an ion beam during deposition to densify the coating. This can drastically reduce the stress evolution of the film during processing and use. Ion beam sputtering can produce very dense films with very low scatter, but at present this technique is limited to apertures on the order of 60 cm, too small for the largest grating applications.

After coating of the stack, the grating substrate is cleaned and primed to facilitate adhesion of the photosensitive film. The priming agents, developed for the semiconductor-processing industry, replace surface hydroxyl groups on the SiO_2 surface with organic groups; otherwise, these surface hydroxyls will react with the base-sensitive photosensitive film and detach it during the development step.

The photosensitive film is typically a positive photoresist formulated for sensitivity to I-line (365-nm) or G-line (436-nm) wavelengths, developed again for the semiconductor industry. A positive photoresist is one that undergoes a photochemical reaction that renders it soluble in a base solution in areas where it has been illuminated with near-UV light but remains as a solid film where not exposed. Meniscus coating (Britten, 1993) has become the method of choice for depositing precise submicrometer-thick photoresist films from liquid solution onto the surface of large, heavy optics.

Exposure of the resist film is done on a large interferometer as shown in Figure 7.8. The largest commercially available class 4 high-coherence-length Ar-ion (351-nm) or Kr-ion (413-nm) lasers are used. We use fused-silica aspheric lenses of 1.1 m in diameter, as shown in Figure 7.8, to generate the two interfering plane waves for our largest format exposures. Alternatively, two large off-axis parabolas can be used in place of the folding mirror/lens combination of this figure. However, the mirror/lens combination results in better diffracted wave fronts and is certainly easier to align and collimate.

A writing process has been developed that uses an interferometrically controlled XY stage to move a substrate under two small interfering beams and pattern a grating by raster scanning (Schattenburg and Everett, 2005). Gratings at 1740 lines/mm approaching 1 m in dimension have been demonstrated. This technology has the

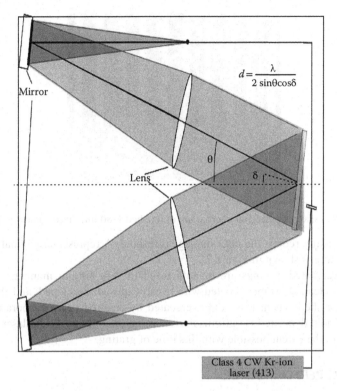

FIGURE 7.8 Large-format grating exposure setup. CW (continuous wave), clockwise.

potential to revolutionize the patterning of gratings to arbitrary scales as well as to write variable line densities, or chirp, across a grating substrate (Pati et al., 2002).

After the development of the exposed pattern using a base solution, the grating undergoes a series of characterization steps. There are more than 600 km of grating lines on an 800 × 400 mm, 1740 lines/mm grating, and all must be within a narrow width and height range for the finished part to be acceptable. We do full-aperture photometry of the photoresist grating at a use angle and wavelength chosen to give high sensitivity to the −1 order diffraction efficiency. The efficiency map generated is used to provide coordinates for submicrometer-scale examination of the grating lines by an atomic force microscope. Thus, areas out of the expected bounds of diffraction efficiency can be examined in detail. The resist coating can be stripped off and the part reprocessed if necessary. If deemed acceptable, it then undergoes ion beam etching to transfer the pattern in resist to the underlying oxide layers. Our ion mill uses a rail transport system to scan a vertically mounted grating back and forth in front of a radio-frequency-generated ion beam of 1 m long by 3 cm wide using a reactive gas $CHF_3/Ar/O_2$ blend to selectively etch SiO_2 with respect to the resist mask. It can process parts up to 2 × 1 m in dimension.

After wet chemical stripping of the remaining photoresist mask, the grating is characterized for diffraction efficiency and diffracted wave front. Then, it undergoes several cleaning steps to remove monolayers of processing residues, which

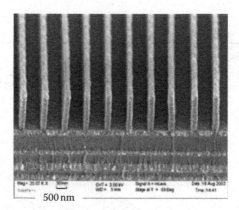

500 nm

FIGURE 7.9 Scanning electron micrograph of finished 1780 lines/mm witness grating.

can dramatically reduce the laser damage threshold. A representative final product microstructure is shown in Figure 7.9.

Gold-overcoated gratings are manufactured in very similar manner. The bare substrate is cleaned, primed, coated with resist, exposed, developed, and character-ized as described. Then, it is gold overcoated by electron beam evaporation. The amount of gold deposited is optimized to increase the duty cycle and laser damage resistance to the extent possible with this type of grating.

GRATING PERFORMANCE

To date, we have fabricated more than 80 MLD gratings at 1740 lines/mm, the larg-est being 807×417 mm in aperture. These represent over 12 m^2 of grating area, with average diffraction efficiency over 96%. Most have been made for pulse compression at 1053 nm at various incidence angles for high-energy petawatt laser systems being fabricated worldwide (reviewed by Hecht, 2006), although several were made for the petawatt field synthesizer pump laser (Fülöp et al., 2007), which operates at 1030 nm, 59°. Figure 7.10 shows efficiency statistics for these gratings.

Of special interest are gratings we have made that can be used in an all-reflective grating interferometer proposed for measuring gravity waves (Sun and Byer, 1998). These require maximal diffraction efficiency in the Littrow order at 1064 nm and must be capable of high average-power operation with no wave front deformation. We have produced gratings of 200×100 mm in aperture that exhibit greater than 99% average diffraction efficiency at 1064 nm at the Littrow angle (67.8°) and that have withstood about 2 kW/cm^2 with no observable change in beam quality (Britten et al., 2007). We have fabricated a similar grating of 470×430 mm in aperture that has an average diffraction efficiency of 99.1% (Figure 7.11). Others have also recently demonstrated MLD gratings with greater than 99% diffraction efficiency (Destouches et al., 2005), but the optic of Figure 7.12 is by an order of magnitude the largest such grating produced.

The diffracted wave front of gratings can be measured interferometrically at the Littrow angle with the grating oriented so that its surface normal is rotated clockwise and then counterclockwise with respect to the incident beam (continuous wave [CW]

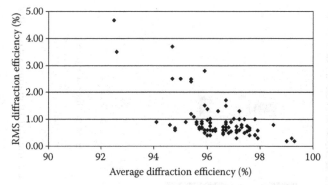

FIGURE 7.10 Average and root mean square (RMS) diffraction efficiency of approximately eighty 1740 lines/mm multilayer dielectric (MLD) gratings representing more than 12 m² of grating area.

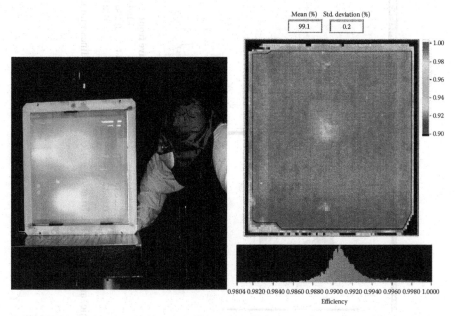

FIGURE 7.11 Photograph (left) and diffraction efficiency map (right) of a 470 × 430 mm, 1740 lines/mm multilayer dielectric (MLD) grating exhibiting greater than 99% average diffraction efficiency at the Littrow angle at 1064 nm, transverse electric (TE) polarization.

and continuous carrier wave (CCW), respectively). The CW and CCW data sets can be subtracted and scaled by half to cancel the surface wave front and produce the holographic wave front. This provides a measure of the errors associated with grating line curvature, chirp, and so on. Similarly, the CW and CCW data sets can be added and scaled to cancel the holographic component and return the surface wave front, which is, of course, related to the flatness of the part. A collection of all wave fronts for a representative grating of 470 × 430 mm in aperture is shown in Figure 7.12. To the best of our knowledge, these gratings have the flattest diffracted wave front available at this time.

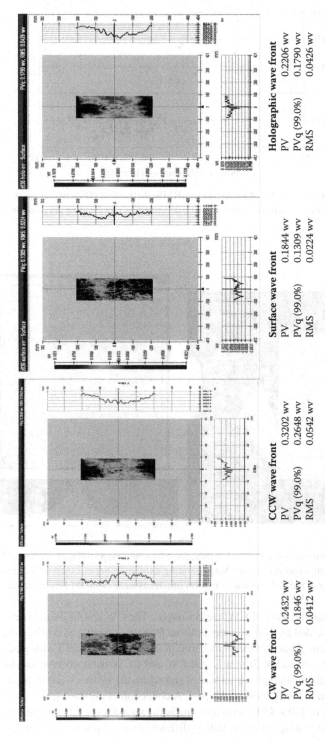

CW wave front

PV	0.2432 wv
PVq (99.0%)	0.1846 wv
RMS	0.0412 wv

CCW wave front

PV	0.3202 wv
PVq (99.0%)	0.2648 wv
RMS	0.0542 wv

Surface wave front

PV	0.1844 wv
PVq (99.0%)	0.1309 wv
RMS	0.0224 wv

Holographic wave front

PV	0.2206 wv
PVq (99.0%)	0.1790 wv
RMS	0.0426 wv

FIGURE 7.12 CW, CCW, holographic, and surface wave fronts of a 470 × 430 mm, 1740 lines/mm multilayer dielectric (MLD) grating measured at 1053 nm at the Littrow angle 66.7°.

CONCLUSION

Intense, short-pulse lasers using large-aperture MLD gratings are just coming online at this writing. These instruments will generate new data on the property of matter at extreme energy densities and further research into nuclear fusion for energy as well as defense applications. New ultrashort-pulse lasers using very large gold-overcoated gratings are being built now as well to probe the behavior of matter at even shorter time scales. At LLNL, we are beginning to build short-pulse capability on the National Ignition Facility (NIF) laser (see LLNL's NIF Web site at https://publicaffairs.llnl .gov/news/news_releases/2007/NR-07-11-05.html). This laser, known as NIF-ARC (advanced radiographic capability), will use four beamlines with 32 gratings to deliver picosecond pulses for x-ray diagnostic generation and research into fast-ignition fusion. This is a burgeoning technological field that promises to yield very exciting science in the coming years.

REFERENCES

Arns, J. A., W. S. Colburn, and S. C. Barden. (1999). Volume phase gratings for spectroscopy, ultrafast laser compressors, and wavelength division multiplexing. *Proc. SPIE*, 3779: 313–323.

Boyd, R. D., J. A. Britten, D. E. Decker, B. W. Shore, B. C. Stuart, M. D. Perry, and L. Li. (1995). High efficiency metallic diffraction gratings for laser applications. *Appl. Opt.*, 34: 1697–1706.

Britten, J. A. (1993). A simple theory for the entrained film thickness during meniscus coating. *Chem. Eng. Commun.*, 120: 59–71.

Britten, J. A., W. Molander, A. M. Komashko, and C. P. Barty. (2003). Multilayer dielectric gratings for petawatt-class laser systems. *Proc. SPIE*, 5273: 1–7.

Britten, J. A., H. T. Nguyen, S. F. Falabella, B. W. Shore, M. D. Perry, and D. H. Raguin. (1996a). Etch-stop characteristics of Sc_2O_3 and HfO_2 films for multilayer dielectric grating applications. *J. Vac. Sci. Technol. A*, 14: 2973–2975.

Britten, J. A., M. D. Perry, B. W. Shore, and R. D. Boyd. (1996b). A universal grating design for pulse stretching and compression in the 800 to 1100 nm range. *Opt. Lett.*, 21: 540–542.

Britten, J. A., H. T. Nguyen, J. D. Nissen, C. C. Larson, M. D. Aasen, T. C. Carlson, C. R. Hoaglan et al. (2007). Large aperture dielectric gratings for high power LIGO interferometry, LIGO Scientific Collaboration Meeting, Baton Rouge, LA, March 19–22 (Lawrence Livermore National Laboratory Report UCRL-PRES-229023).

Britten, J. A., and L. J. Summers. (1998). Multiscale, multifunction diffractive structures wet-etched into fused silica for high laser damage threshold applications. *Appl. Opt.*, 37: 7049–7054.

Chung, T.-Y., A. Rapaport, V. Smirnov, L. B. Glebov, M. C. Richardson, and M. Bass. (2006). Solid-state laser spectral narrowing using a volumetric photothermal refractive Bragg grating cavity mirror. *Opt. Lett.*, 31: 229–231.

Danson, C. N., P. A. Brummitt, R. J. Clarke, J. L. Collier, B. Fell, A. J. Frackiewicz, S. Hawkes et al. (2005). Vulcan petawatt: Design, operation and interactions at 5×10^{20} W/cm^2. *Laser Part. Beams*, 23: 87–93.

Destouches, N., A. V. Tischenko, J. C. Pommier, S. Reynaud, O. Parriaux, S. Tonchev, and M. Abdou Ahmed. (2005). 99% efficiency measured in the −1st order of a resonant grating. *Opt. Express*, 13: 3230–3235.

Efimov, O. M., L. B. Glebov, L. N. Glebova, K. C. Richardson, and V. I. Smirnov. (1999). High-efficiency Bragg gratings in photothermorefractive glass. *Appl. Opt.*, 38: 619–627.

Fülöp, J. A., Zs. Major, A. Henig1, S. Kruber, R. Weingartner, T. Clausnitzer, E.-B. Kley et al. (2007). Short-pulse optical parametric chirped-pulse amplification for the generation of high-power few-cycle pulses. *N. J. Phys.*, 9: 438.

Hecht, J. (2006). Photonic frontiers: Petawatt lasers—A proliferation of petawatt lasers. *Laser Focus World*, 42(8).

Hutley, M. C. (1982). *Diffraction Gratings*. London: Academic Press.

Kashyap. (1999). *Fiber Bragg Gratings*. San Diego, CA: Academic Press.

Kitagawa, Y., H. Fujita, R. Kodama, H. Yoshida, S. Matsuo, T. Jitsuno, T. Kawasaki et al. (2004). Prepulse-free petawatt laser for a fast ignitor. *IEEE J. Quantum Electron.*, 40: 281–293.

Li, L. (1993). A multilayer modal method for diffraction gratings of arbitrary profile, depth and permittivity. *J. Opt. Soc. Am. A*, 10: 2581–2591.

Li, L. (1994). Multiple layer coated diffraction gratings, differential method of Chandezon et al. revisited. *J. Opt. Soc. Am. A*, 11: 2816–2828.

Liao, E. G., M.-Y. Cheng, E. Flecher, V. I. Smirnov, L. B. Glebov, and A. Galvanauskas. (2007). Large-aperture chirped volume Bragg grating based fiber CPA system. *Opt. Express*, 15: 4876–4882.

Loewen, E. G., and E. Popov. (1997). *Diffraction Gratings and Applications*. Boca Raton, FL: CRC Press.

Néauport, J., E. Journot, G. Gaborit, and P. Bouchut. (2005). Design, optical characterization, and operation of large transmission gratings for the laser integration line and laser megajoule facilities. *Appl. Opt.*, 44: 3143–3152.

Neauport, J., E. Lavastre, G. Razé, G. Dupuy, N. Bonod, M. Balas, G. de Villele, J. Flamand, S. Kaladgew, and F. Desserouer. (2007). Effect of electric field on laser induced damage threshold of multilayer dielectric gratings. *Opt. Express*, 15: 12508–12522.

Nguyen, H. T., B. W. Shore, S. J. Bryan, J. A. Britten, and M. D. Perry. (1997). High-efficiency fused-silica transmission gratings. *Opt. Lett.*, 22: 142–144.

Pati, G. S., R. K. Heilmann, P. T. Konkola, C. Joo, C. G. Chen, E. Murphy, and M. L. Schattenburg. (2002). A generalized scanning beam interference lithography system for patterning gratings with variable period progressions. *J. Vac. Sci. Technol. B*, 20: 2617–2621.

Perry, M. D., R. D. Boyd, J. A. Britten, D. Decker, B. W. Shore, C. Shannon, E. Shults, and L. Li. (1995). High efficiency multilayer dielectric diffraction gratings. *Opt. Lett.*, 20: 940–942.

Perry, M. D., D. Pennington, B. C. Stuart, G. Tietbohl, J. A. Britten, C. Brown, S. Herman et al. (1999). Petawatt laser pulses. *Opt. Lett.*, 24: 160–162.

Schattenburg, M. L., and P. N. Everett. (2005). Method for interference lithography using phase-locked scanning beams. U.S. Patent 6882477, April 19.

Shore, B. W., M. D. Perry, J. A. Britten, R. D. Boyd, M. D. Feit, H. T. Nguyen, R. Chow, G. Loomis, and L. Li. (1997). Design of high-efficiency dielectric reflection gratings. *J. Opt. Soc. Am. A*, 14: 1124–1136.

Strickland, D., and G. Mourou. (1985). Compression of amplified chirped optical pulses. *Opt. Commun.*, 56: 219–221.

Stuart, B. C., M. D. Feit, S. Herman, A. M. Rubenchik, B. W. Shore, and M. D. Perry. (1996). Optical ablation by high-power short-pulse lasers. *J. Opt. Soc. Am. B*, 13: 459–468.

Sun, K-X., and R. L. Byer. (1998). All-reflective Michelson, Sagnac, and Fabry-Perot interferometers based on grating beam splitters. *Opt. Lett.*, 8: 567–569.

Wu, Z., C. J. Stolz, S. C. Weakley, J. D. Hughes, and Q. Zhao. (2001). Damage threshold prediction of hafnia-silica multilayer coatings by nondestructive evaluation of fluence-limiting defects. *Appl. Opt.*, 40: 1897–1906.

Yamamoto, R. M., K. L. Allen, R. W. Allmon, K. F. Alviso, B. S. Bhachu, C. D. Boley, R. L. Combs et al. (2006). A solid state laser for the battlefield. 25th Army Science Conference, Orlando, FL, November 27–30 (Lawrence Livermore National Laboratory Report UCRL-CONF-225230).

Yu, J., J. A. Britten, L. J. Summers, S. N. Dixit, C. R. Hoaglan, M. D. Aasen, R. P. Hackel, and R. R. Prasad. (2003). Fabrication of beam sampling gratings for the National Ignition Facility (NIF). OSA Conference on Lasers and Electro-optics, Baltimore, MD, May 6, Paper CFL6.

8 How to Select Optical Mounts

Getting the Most Out of Optical Mounts (Understanding Optical Mounts)

Damon Kopala

CONTENTS

INTRODUCTION

When choosing optics, a common mistake among many engineers is overlooking the integration of optical mounts into their system. There are a variety of mounts available for holding lenses, prisms, mirrors, filters, and other common optical components. Some examples include bar-type mounts, gimbal mounts, adjustable kinematic mounts, fixed mounts, and jaw clamp mounts. When cost is a deciding factor, simple fixed mounts will be more than adequate. However, for applications that require fine positioning, adjustable, stable, kinematic mounts are essential to the integrity of a precision optical system.

OPTICAL MOUNTS SPECIFICS

A three-dimensional rigid body has exactly six degrees of freedom (DOF): X, Y, and Z are translational DOF, and R_x, R_y, and R_z are rotational DOF. A mount is considered kinematic if all six DOF are fully constrained. Most laboratory kinematic optical mounts use the classic cone, groove, and flat constraint system, as illustrated in Figure 8.1, and use two or three adjustment screws. Two adjustment screws, at the groove and the flat, can be used to adjust the rotational DOF. Because the axis of rotation is behind the optic, there will be a slight translation of the optic when an

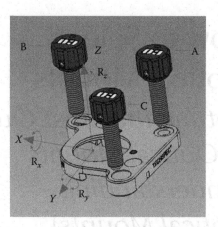

FIGURE 8.1 Top plate and adjustment screws of 1-in. kinematic mount.

TABLE 8.1
Contributions of Individual Errors and Root Sum Square (RMS) Total Error

Movement	Beam 1 (μm)	Beam 2 (μm)
1	1	1
2	2.4	1.6
3	0	0
4	0	3.6
5	1	1
RSS	2.14	4.19

RSS, root sum square.

adjustment is made. A third screw can be placed over the cone to compensate for unwanted translation. The three-screw configuration enables the optic to be rotated as needed using the first two adjustments screws and then returned to its original position along the Z-axis with the third.

To illustrate the importance of choosing the right optical mounts, consider an application that consists of a 25-mm cube beam splitter and a 12-mm focal length lens to combine and couple two 3-mm beams into a 0.12 numbered aperture fiber. To ensure the highest coupling efficiency, assume that both focused spots need to be centered on the fiber to within ±2 μm.

Each component will contribute to the overall positioning error in the system, which can be calculated by the root sum square (RSS) (see Table 8.1). Because each element potentially has six DOF, there are many combinations of movement that can occur. For simplicity, only the following movements are considered:

1. Translation along the optical axis of the lens by 1 μm
2. Rotation of 400 μrad of the focusing lens about a point 4 mm from its nodal point

3. Translation along the optical axis of the beam splitter by 2 μm
4. Rotation of 150 μrad of the beam splitter about the optical axis
5. Translation along the optical axis of the fiber by 1 μm

Clearly, the system in Figure 8.2 undergoes too much movement for beam 2 to maintain the ±2 μm required for high coupling efficiency. Factors that may contribute to misalignment are mount resolution and instability from thermal effects, vibration, and gravity (often referred to as *pointing stability*).

Resolution of kinematic mounts is typically classified into two categories: linear resolution and angular resolution. The thread pitch of the adjustment screws determines the linear resolution, whereas the placement and thread pitch of the adjustment screws provide the angular resolution; 80–100 TPI (turns per inch) adjustment screws are the industry standard. Although higher resolution can be obtained by simply using adjustment screws with a larger TPI count, this is not always optimal because finer threads are easily damaged.

Sensitivity is another parameter, often provided by manufacturers, that is related to resolution. Because most kinematic mounts are manually driven, it is helpful to know the minimum obtainable movement of the optic. In general, fingertips are sensitive enough to resolve a 1° turn of the screw. The movement of the optic that corresponds to a 1° turn is what is used to define sensitivity.

Thermal stability indicates how well the mount will perform when subjected to changes in temperature. For minimum deflection from thermal effects, the mount's coefficient of thermal expansion should be matched to that of the optic. Certain types of stainless steel match up very well to glass and are the preferred choice when thermal stability is of utmost importance. Although aluminum has a higher thermal expansion coefficient, kinematic mounts made from it still perform well in typical laboratory environments.

In addition to movement from thermal conditions, the effects of gravity over time will contribute to the overall misalignment error. Pointing stability is the measure of this error, specified as an angular movement, and is defined at a certain temperature, time lapse, and applied load.

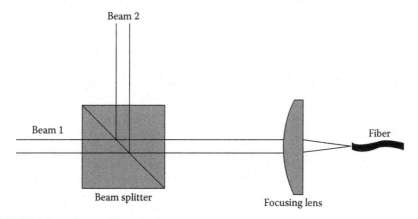

FIGURE 8.2   Fiber coupling system.

Vibration can also degrade optical system performance. Misalignment from low-level vibration will lead to blur in the image plane. Attempting to image with a high-power microscope objective without using a vibration isolation platform will result in this phenomenon. Materials with higher stiffness values have greater fundamental or natural frequencies and faster settling times, resulting in less vibration disturbance.

A big drawback of many mounts is having to drop, rather than place, the optic into position. With the steep cost of precision optics, it is highly undesirable to drop them into place even if it is only from a few millimeters. Choosing a mount with finger cuts incorporated into the design allows for guided placement of the optic, which will reduce chances of chipping and fracture.

CONCLUSION

The key to successfully choosing an optical mount is to prioritize the surrounding requirements. When critical alignment is of primary importance, the user should choose kinematic mounts that offer high resolution and excellent position stability. Thermal stability is increased when all connecting components are constructed from the same material. In the majority of setups, this would consist of stainless steel hardware as most optical benches have stainless steel tops. However, a setup constructed entirely from stainless steel components can prove costly. When cost is more of a concern and alignment is less critical, aluminum is a perfectly suitable alternative.

9 Training Design and Considerations in the Use of Instructor-Led, Web-Based Training of Adults

Matt Vaughn

CONTENTS

INTRODUCTION

Although there are many models of training design, one common model follows the acronym of ADDIE. Each successive initial represents the steps of analysis, design, development, implementation, and evaluation. Rigorous adherence to the principles underlying these five steps is an effective common model path for developing instruction in any context. A full description of how to apply these principles can be found in the book, *The Systematic Design of Instruction* (Dick et al., 2005).

To distill the discussion of the training development process to its basic elements, we look at the who, what, when, where, why, and how of training development. Subsequently, the discussion addresses some of the issues inherent in the use of three common methodologies currently used in safety training: instructor-led training

(ILT), Web-based training (WBT), and on-the-job training (OJT). To introduce how effective safety training is developed and implemented, I first question who will be trained.

WHO

Who will be taking this training? In training terms, this group of people is termed the target audience. Each member of a target audience comes to training with pre-existing knowledge, behaviors, abilities, and attitudes. This information is rarely known or demonstrated. Despite the heterogeneity of any population, prior knowledge of your audience is extremely useful when developing or delivering training.

Your target audience's preexisting behaviors and abilities may align with meeting your instructional goal. These may be as rudimentary as being able to converse in the instructional language, write, or read or as complex as technical proficiency in the safe operation of lasers. However, other behaviors may be counter to your instructional goal. Knowing what behaviors and abilities are present can help when framing course content and instructional methodology.

Prior knowledge of the instructional topic may be found in your target population, ranging from highly detailed procedural knowledge to vague misconceptions of how to safely operate lasers and every possible permutation thereof. Attempting to instruct without knowing the prior knowledge of your audience limits the possibility of linking new information with existing information, correcting misconception, and constructing new understanding.

The attitudes of your audience toward the content you will be teaching, the way in which you deliver the instruction, and even the training organization itself are crucial to the success of your training. Some audiences find PowerPoint lectures boring, while others do not appreciate small group simulations or a presenter with a poor reputation. The tools you use, how you present the content, and who presents the content are likely to be most effective if they meet the expectations of your audience.

As you might imagine, knowing what your audience brings to the table can dramatically alter the instructional methodology.

WHAT

Training in laser safety, as with all training, begins with a clear instructional goal (the *what*). Behaviors that prevent the loss of life, health, and property define safety practices. Therefore, good safety training instructional goals describe observable safe behaviors. Although it may be important for a learner to *know* or *understand* certain information regarding laser operation, ultimately it is the correct application of that knowledge that results in safety. For instance, a laser operator may know how to correctly place beam blocks, but without translating this knowledge into action, an unsafe condition arises. If the operator is to be trained in the correct placement of beam blocks, a *head knowledge* of beam block placement is not adequate. In this example, a clear instructional goal, aligned with the desired outcome, states precisely that on completing the course of instruction the operator will demonstrate the correct placement of beam blocks.

Stating what the learner will be able to demonstrate is a first step in defining the instructional goal. The next step is to define the conditions under which the safe behavior will be performed. The variety of lasers and contexts in which lasers are used creates an array of conditions that may vary the process for correctly placing beam blocks. Unless the planned instruction will cover all contexts for the placement of beam blocks, the instructional goal needs to define the specific performance context (location, laser type, tools available, personal protective equipment needed, etc.).

After defining the desired end behavior and the context in which it will be performed, a final area requires clarification. Suppose on reaching the end of the instruction, the learner declares himself or herself able to correctly place beam blocks in the defined context. From the instructor's perspective, there is little basis for denying the operator's claim because there is no communicated criteria for *correct* placement. To have a basis for evaluating training effectiveness, learner retention, and instructor efficacy, there must be a clear behavioral standard against which to compare the learner's new behaviors.

Before developing any instruction, establish instructional goals that define desired behaviors, the context in which those behaviors will be demonstrated, and the criteria by which the behaviors will be judged. Failing to delineate these three components will result in a breakdown of the instructional process.

WHEN

When should training take place? In asking this question, it may be important to consider factors such as the time of day when training occurs, whether it is delivered inside or outside the work day, how many times a year, and so on. Some of these factors may be outside a training manager's control. However, a careful analysis of when the safe behaviors need to be performed and a subsequent alignment of the training to support the temporal parameters of the job will make for more effective training. For instance, if the performance context requires the behavior be demonstrated 20 times in the course of a morning but never in the afternoon, then the training context is most effective by attempting to mimic this behavior pattern during the morning.

Though the external timing of when training takes place is pertinent to learning, the more complex question addresses when trainees are most receptive to learning. In the field of education, there is a term used to describe the moment at which the learner is receptive due to an opportune set of circumstances. This is termed a teachable moment. Another concept defines the moment when an unexpected outcome occurs. This is called expectation failure (Schank, 2002). In addressing the question regarding when employees learn most effectively, I would state that the most effective learning moment occurs when expectation failure occurs and the learner wants to understand why. Effective training strives to create these moments. Many of the factors influencing when training occurs operate outside the control of trainers and training developers. However, they can influence what happens within the defined time frame of training. The best training is the best training because it generates failures of expectation, drawing the learner into the learning process.

WHERE

There are two wheres of instruction important to training. The first is the context in which the training is used. This is called the performance context. When developing the instructional goal, the performance context is considered to define which environmental factors are critical to the desired behavior. However, beyond these factors, it is useful to know other performance context factors that may influence the learner's ability to put training to use. These factors might include elements such as managerial support, the physical condition of the facility or equipment, and the social dimensions of the work environment. Although not an all-inclusive list, some understanding of whether the training will be supported by the performance context helps making decisions about instructional design.

The second where of training is the training environment itself, also referred to as the learning context. The greater the discrepancy between the learning context and the performance context, the less likely it is that training will transfer to the job. Practical constraints to delivering training in the training performance environment may exist, though it is often the ideal training situation. If the goal is to have beam blocks set in such and such a manner with a certain laser in a certain facility, then the best training will simulate beam block setting in such and such a manner with the desired laser in the desired facility. Although the ability to generalize behaviors to a variety of contexts may be more valuable than knowing one set of behaviors specific to one context, in general the more inclusive the instructional goal is, the more complex the resulting instruction will need to be to successfully meet the goal. Conversely, the more narrowly performance behavior can be defined and then paralleled in training, typically transfer from training to performance and is more direct.

WHY

Why training? Another way to ask this question is, What is determining the need for training? In the safety training industry, regulations and standards often require documented communication of safety processes and policies. In some cases, training might be appropriate in satisfying these mandates. Similarly, there may be organizational goals to reduce incidents, lost workdays, and the like that are based in good business practice. Again, sometimes the pursuit of these goals is served through specific training initiatives. However, it is wise to consider when there is an institutional push for new training, if in fact training will deliver the desired change in behavior. A careful analysis of the factors underlying current unsafe practices may reveal that there are other latent organizational weaknesses that permit an unsafe environment to exist, and training is unlikely to have the desired effect. In creating effective training, it is important to determine early on that training can be effective in addressing a given problem.

HOW

Having laid a foundation of who the target audience is, what specific goal will be addressed by training, and when and where to deliver training, we now have a basis for addressing how training might be implemented. Somewhat anticlimactically, however, there is no single definitive methodology for great training. However, aiming for alignment between performance and training contexts is a useful guiding principle as decisions are made about how to deliver information and practice new behaviors. Three common contexts in which training occurs are ILT, WBT, and OJT. Each method has its strengths and weaknesses and is covered here. However, regardless of the method, there are some common instructional components summarized next that are typically a part of good instruction (Dick et al., 2005).

SUMMARY OF LEARNING COMPONENTS IN A TYPICAL INSTRUCTIONAL STRATEGY

1. Preinstructional activities
   a. Gain attention and motivate learners
   b. Describe objectives
   c. Describe and promote recall of prerequisite skills
2. Content presentation
   a. Content
   b. Examples
3. Learner participation
   a. Practice
   b. Feedback
4. Evaluation
   a. Entry behavior test
   b. Pretest
   c. Posttest
5. Follow-through activities
   a. Memory aids for retention
   b. Transfer considerations

In brief, preinstructional activities set up the delivery of the core content. Gaining the attention of the learner and motivating the learner to engage in the new content is colloquially referred to as *hooking the audience*. This is a prime opportunity in which to create the previously noted *expectation failure*. It is also helpful for the learner to have a clear sense of what he or she will be expected to know and be able to do at the conclusion of the instruction (objectives) and to understand what prior knowledge or skills will be needed to assimilate the new material.

Content presentation contains the new material and examples of how to apply procedures, stories showing how the new behavior is effective, and so on. This leads directly to learner participation in which the new content is manipulated in a manner that aligns with the instructional goal. Feedback is provided in this stage, and the learner is coached toward performing to a certain standard.

Evaluation does not always occur in the three ways noted in the summary, although a posttest is most often present. However, effective training will often evaluate entry behaviors to ascertain whether learners have the basic information needed to assimilate new content. Also, pretesting provides a baseline measure against which final performance can be assessed. This is critical for measuring training effectiveness.

Typically, if only a posttest is given, the end of a course assessment simply measures knowledge transfer under the assumption that learners did not know any of the information coming into the course. This may be reasonable in many cases. However, a good assessment program, aligned with the instructional goal, looks beyond knowledge transfer to behavioral change. As a side note, organizations often look at training effectiveness as it translates to improved organizational performance. For more on this and further discussion of evaluation, see Donald Kirkpatrick and James Kirkpatrick's *Evaluating Training Programs: The Four Levels* (2006).

CONSIDERATIONS IN INSTRUCTOR-LED TRAINING

Instructors do make a difference in training effectiveness. An exceptional instructor is often the essential factor in making a course great. Conversely, a poor instructor can dramatically degrade a learning experience. An instructor's content knowledge and teaching experience are tied closely to his or her effectiveness as an instructor, but of themselves do not guarantee quality instruction. What gives ILT its value is a good instructor's ability to connect content to learners' needs, knowledge, and backgrounds. In addition, as the instructor draws a group of learners into the process of creating meaning from the presented information, there is a synergistic learning dynamic that results, capitalizing on the social aspects of ILT.

Depending on how ILT is designed, some of its strengths lie in using familiar learning strategies, a classroom environment dedicated to instruction, the potential for customized instruction, and access to a master of the material. ILT has an additional benefit of being able to be revised relatively easily. As the content evolves, an instructor can add or subtract material to accommodate the changes. In general, an instructor-led course can be developed in a shorter time frame than instruction delivered through other methods. This is in part due to the instructor's often serving as the expert in the subject matter as well as the de facto instructional designer and developer. Finally, every set of learners in a classroom has a unique mix of characteristics that create sometimes dramatic differences in the target audience. A strength of ILT, with a good instructor, is the ability to flexibly use those differences to enhance the instructional process rather than allowing them to serve as limitations.

The weaknesses of ILT largely tie to not only costs for materials and the instructor's time, but also indirect costs such as room rental, learners' time away from producing a product, and travel costs. In addition, instructors can be in only one place at a time, limiting the number of times a course can be offered in a given time frame. If there are a large number of students to be trained in a short period of time, this can be a substantial drawback. Also once the course is over, it may be difficult to personally access an instructor's expertise, so there is a limited window of time in which learning must be accomplished. Because not all people learn at the same rate, this can also be problematic if new standards of behavior must be established in a short time frame. Finally, the effectiveness of ILT decreases as the number of learners per instructional session increases. Thoroughly understanding our target audience to effectively meet the instructional goal becomes increasingly difficult as the learning population grows in number and inherent diversity. Therefore, the scalability of ILT can become an issue.

CONSIDERATIONS IN WEB-BASED TRAINING

The label *Web-based training* encompasses a variety of strategies for using the communication capabilities of networked computers. For the sake of this discussion, I focus on general issues related to WBT rather than dissecting each variation.

The strength of WBT rests in on-demand access to training, scalability, and long-term cost reduction as compared to ILT. WBT development often requires technically skilled specialists with high pay rates. However, beyond the up-front development costs and those associated with putting a technological infrastructure in place, there are minimal costs associated with the maintenance and ongoing delivery of instruction. Also, there is theoretically no limit to the number of people who can access a Web-based course, assuming the infrastructure can handle the data flow. Perhaps its greatest strength, however, resides in the learner's ability to access information needed when the need is immediate. This is sometimes referred to as *just-in-time* training and aligns with the teachable moment concept introduced earlier.

Major weaknesses of WBT are its reliance on software and hardware that changes regularly, its reduced ability to be revised cheaply and quickly (compared with ILT), and its incapacity to provide specific feedback attuned to the unique needs of the individual learner. The reliance on changing software and hardware creates ongoing compatibility issues, necessitating upgrades to ensure that the training is readily available to a wide variety of learners accessing the information from a variety of workstations. Likewise, because many Web-based courses have customized multimedia assets or learning interactions (video or audio files, animations, branching scenarios, etc.), a content change can involve hours of technical revisions. Web-based instruction tends to be inherently *cookie cutter* in its delivery, unable to morph to the unique characteristics of an individual learner, much less a diverse population of learners. However, this same attribute can be considered a positive characteristic when the content is standardized and must be consistently delivered, as in compliance-driven industries.

Unless the performance environment involves the use of a computer, there can be a considerable misalignment between the performance context and a WBT context. Good instruction strives to align the learning context with the performance context. As the availability of realistic Web-based scenarios and simulations increases, context discrepancy in WBT may decrease. Until that point, however, although there are potential cost savings by using WBT, there may be minimal behavioral transfer from one context to the other. This reiterates the principle that the methodology used to deliver training must align with the instructional goal and that the instructional goal must clearly delineate the context in which the desired behavior is to be performed.

Often, WBT takes place in the context of the learners' typical working environment, and this is indeed a potential strength of the method in that it does not necessitate travel costs that might be associated with ILT. But, the typical work environment also has a host of distractions (phones ringing, colleague interruptions, intermittent computer access, etc.) that can interfere with the effectiveness of the training. There are clearly some trade-offs to consider in choosing WBT and some challenges to address if it is the chosen strategy.

ILT offers a defined time and space in which there is a social expectation to participate in the learning process. WBT, on the other hand, often leaves the learner isolated from a social context, and participation is subject to the individual's intrinsic motivation. Motivating a learner to engage in the learning process in a WBT environment is a major challenge, often addressed through mandates with an associated reward or punishment system. Even so, participation rates are traditionally lower in socially isolated WBT than in more socially contextualized learning. WBT that capitalizes on the use of peer-to-peer technologies has a demonstrably greater effectiveness than individual WBT.

WBT technologies are undergoing a tremendous evolution with a major thrust toward reducing the time needed to generate a training product through the use of *rapid e-learning* software. It will be interesting to see if the ability of WBT to address the diverse needs of large training audiences will evolve to a greater degree also.

ON-THE-JOB TRAINING

Chapter 10 presents the discussion of OJT.

REFERENCES

Dick, W., Carey, L., and Carey, J. O. (2005). *The Systematic Design of Instruction*. 6th ed. Boston, MA: Allyn & Bacon.
Kirkpatrick, D. L. and Kirkpatrick, J. D. (2006). *Evaluating Training Programs: The Four Levels*. 3rd ed. San Francisco, CA: Berrett-Koehler Publishers.
Schank, R. (2002). *Designing World-Class E-Learning*. New York: McGraw-Hill.

10 On-the-Job Training

James Foye

CONTENTS

INTRODUCTION

On-the-job training (OJT) is designed to prepare trainees for job performance through one-on-one training and performance testing conducted by a qualified OJT instructor in the work environment. It provides hands-on experience and has the advantage of training only for tasks that are of immediate need to the trainee.

Shifting from a traditional instructor-led training program to a structured OJT program is often less stressful and easier to manage than you can imagine. Although OJT is not a cure-all for replacing instructor-led or Web-based learning, it is an effective and efficient tool for training new employees or employees switching to a new set of tasks.

Consider a situation in which an instructor-led course is designed for a group of individuals to be trained away from their jobs in a controlled classroom setting and then to return to work after training. Can the managers and supervisors of these individuals be assured that training has occurred and that the principles they learned in the classroom can be applied directly to their jobs? In most cases, managers and supervisors have little knowledge about what occurred in the training their employees received. By contrast, when training is focused on identifiable skills, knowledge, and abilities (SKAs), these same managers and supervisors can be assured that the critical skills needed to perform the job have been achieved. That is what OJT does for you.

MODEL FOR IMPLEMENTING AN ON-THE-JOB-TRAINING PROGRAM

So, what is OJT? How can an effective OJT program be designed and implemented for a laser-related training program? You can conduct OJT by the *follow Joe* method and hope that laser task–related skills and knowledge are transferred as you would expect. Follow Joe is an expression summarizing the training approach many have used for years: "Here's your tool box. Go follow Joe." Traditionally, this approach will work if Joe knows what he (or she) is doing and has the time and ability to do the training properly. In reality, the result is that the worker following Joe becomes a clone of Joe, complete with all of Joe's skills, attitudes, imperfections, and misconceptions. Furthermore, if Joe is practicing unsafe laser work practices, the new worker is just as likely as Joe to be involved in a potentially serious accident or incident.

On the other hand, a formal, structured OJT process is one that is also conducted at the job site, usually on a one-on-one basis, while either performing or simulating the job or task to be learned. Unlike the follow Joe approach, the OJT process provides specific requirements and guidance to both the OJT instructor and the trainee/worker for meeting identified job-related needs. OJT explicitly defines the knowledge and skills required of a new employee, with predictable results.

FORMAL ON-THE-JOB-TRAINING HAS THE FOLLOWING FEATURES

- Training occurs in the workplace.
- It makes use of specifically defined learning objectives.
- Training requires the active involvement of the trainer and the trainee.
- It uses printed materials and job guides.
- It uses a formal, structured approach.

FACTORS THAT DETERMINE A SUCCESSFUL ON-THE-JOB-TRAINING PROGRAM

The ultimate success of any training program requires a strong commitment to training by both line management and training management. The agreement of these groups regarding goals and content of an OJT program is essential for an effective training program. Also, aligning the training goals and developing an effective OJT strategy should be linked to the company's business strategy. These training goals should be clear and specific, stating the desired result of the training. You should also choose the key people who will be conducting the OJT as well as determine who the target audience (trainees) will be. Finally, you should determine ahead of time where the training will take place; for example, where the equipment to be used in the training is located and whether this equipment will be made available for training when it is needed.

The success of any OJT rests on these factors: the ability to correctly identify the key job factors, the identification of the expectations of training performance, the ability of the OJT instructor or subject matter expert (SME) doing the training, the organization of the training materials, the consistency of the training regardless of who is conducting the training, and the relevance of the information. The steps discussed next outline these factors.

STEP 1: COLLECT TASK INFORMATION

An effective method proven to be an accurate and quick way of getting started is to conduct a job and task assessment as a *tabletop analysis* with an assembled group of key people. Simply stated, the tabletop analysis consists of assembling the group supervisor and several workers who are currently performing the job as content experts, also called SMEs. The tabletop analysis strives for multiple positive results:

- The collective knowledge of the group is obtained.
- The group buys into the resulting task list.
- A completed list is obtained of all tasks directly related to knowledge and skills that workers must achieve to perform the work.
- Supplemental processes, materials, or procedures are identified for each task.

A useful and effective method for obtaining the job and task analysis is to provide the group with large, colored sticky notes. Once the group understands its mission and purpose, all members should be asked to write one task on each sticky note and then to write as many job-related tasks as they can possibly think of to perform a specific job. When all in the group have exhausted all of their ideas, they post their notes on a whiteboard (or wall) and sort the tasks according to the sequence in which the job is to be performed. Duplicated tasks are combined or rewritten to make a clearer task statement.

Once this exercise is accomplished, the group should identify any procedure, manual, or related supporting material that will support each task. In particular, the group should also identify potential laser safety concerns and how to effectively mitigate them. The group has now achieved the described positive results of the tabletop analysis. They have the collective knowledge of the group, a complete list of job-related tasks and skills, a list of safety-related issues and mitigations, a list of supplemental procedures or manuals.

STEP 2: OBTAIN FINAL GROUP CONSENSUS

The information the tabletop analysis group generated from the first step should be collected and entered into one of several computer graphic modeling tools, such as Microsoft Visio, where it can be viewed later by the group for accuracy. This serves two purposes: First, it puts the data into a neatly designed graphical image, and second, the data can be reviewed by the group to make any changes that were not apparent or that did not appear in the first step before beginning the next major step of developing the expected performance objectives for each task.

STEP 3: WRITE OBJECTIVES

This third step is critical because it requires fitting each task with a stated performance objective. Objectives are those SKAs that the trainee will be able to demonstrate at the conclusion of training. Constructing objectives is not too difficult because you already have the task statements as your primary parts of the objective. All effective learning objectives have certain characteristics that must be taken into consideration when they are constructed. First, the objective must be *attainable*. Is the objective possible to achieve by the average learner? Second, the objective must be *specific*. Is the wording concise? Has unnecessary and confusing verbiage been removed? Third, the objective must be *clear*. Will everyone interpret the objective in the same way? Fourth, the objective must be *measurable*. Can this objective be measured? How will it be measured? Effective learning objectives can be stated in a variety of formats. The most common format combines the following parts:

- A statement of behavior (action) the trainee must exhibit
- The conditions under which the action will take place
- The standards of satisfactory performance

An effective method to begin defining objectives is to select one of the tasks obtained in Step 1. You can do this by using the primary parts of the task statement. Let us assume that in our processing for developing tasks, one of the tasks identified is to remove an optic in preparation for cleaning. The following statement illustrates this format: "Correctly remove an optic for cleaning using the manufacturer's step-by-step procedures." A primary rule in writing objectives is to be sure that each objective links directly to the job task list. If you come up with unrelated objectives, you most likely did not do an effective job in Step 1 tabletop analysis and task development.

STEP 4: DEVELOP A TRAINING PACKAGE

For repeated use, you can develop a training package that can be used as a template in which all of the data collected in the previous three steps are organized so that the template can be used by any company-authorized SME OJT instructor at any time. The template has the added effect of ensuring that both the trainer and the trainee are using the same guide, ensuring that they are both on the same page and have the same understanding of the expectations of each during the training process. It has an added benefit of assuring management that the tasks are being effectively taught.

The OJT package includes the following pages and headings:

Cover page: It includes the OJT title, date of issue, authorizing names, and names of supervisors or management approving the OJT.
Introduction: It introduces the purpose and scope of the training.
Job description: It provides a complete description of the job, duties, roles, responsibilities, and so on. This can be most useful because it states up

front what the learner should be able to do once qualified. Some of this information can be obtained from the hiring manager's job description, whereas the remaining description is a summary paragraph of the tasks identified in the tabletop analysis.

Instructional time: It indicates an approximate time in hours, weeks, or months it will take to complete OJT under ideal situations.

Prerequisite training: It gives a list of prerequisite courses or reading required of the trainee before beginning this OJT. Prerequisite training ensures that the trainee is prepared to begin the OJT.

List of skills as training objectives: It includes a numbered list of all the skill objectives in the OJT.

List of knowledge objectives: It includes a numbered list of the associated knowledge objectives to support the skill objectives. Listing the objectives up front in the training package ensures that only those tasks previously identified and agreed to by management are those required of an individual to be a qualified worker in a particular job.

Off-normal safety conditions: It identifies what the OJT instructor and the trainee are to do or actions to be taken if a non-normal or off-normal situation should occur while training is in progress.

Hazards and safety awareness: It identifies the hazards associated with the performance of any of the tasks of which the trainee should be aware to work safely.

Knowledge table: It outlines each knowledge skill as taken from the list of knowledge objectives. This includes a sign-off and date section.

Skills table: It outlines each skill taken from the list of skills. This includes a sign-off and date section for a time and date entry that the OJT instructor will use when it is determined that the skill has been completed satisfactorily.

You could use a three-column format for this table. The first column would identify the first and subsequent tasks taken from the list of skills training objectives. The second column would be reserved for *key points* and questions the trainer and trainee should focus on to support that particular objective. The third column would be for the OJT instructor's initials and sign-off date when the trainee satisfactorily has completed the task to the standards given.

OJT sign-off record page: It provides a separate page at the end for the OJT instructor to date and sign after all the tasks have been satisfactorily completed, indicating that the trainee is now fully qualified to begin work under normal supervision. For some critical positions, you may also elect to have a second person or management review signature.

Once the training package is prepared and ready for training, you may want to consider requiring your present workers to demonstrate that they can perform each task required of the job. This will assure management and others in the organization that the job can be performed equally well by anyone doing the job, not just the new people coming into the position.

STEP 5: SELECT AND TRAIN THE ON-THE-JOB-TRAINING SUBJECT MATTER EXPERT TRAINERS

Selected OJT trainers would best be formally appointed and documented by the immediate supervisor or manager. The credibility of the training program depends on the quality of the OJT instructors. If you think of training an instructor from your company training group, consider that the instructor may be an expert on training but will typically not be as knowledgeable or proficient about the specifics of the job as an incumbent SME on the job. In these situations, it is much better to train your SME to be an effective instructor than to train an instructor to be a job expert.

Therefore, consideration should be given to the OJT instructor's current work ethic, skills and knowledge, and competence to perform his or her current job. The instructor must be responsible, use good judgment and safety awareness, and have personal standards of performance and commitment to maintaining high-quality standards.

The OJT instructors must be certified or qualified by your organization or by a certified outside agency. Instructors must be able to perform the tasks themselves, and they and their supervisors must have clear expectations of the roles and responsibilities of the OJT instruction. Although you may want to use OJT trainers certified by an outside training agency, it is better to develop an in-house OJT program designed with the specific requirements agreed to by your management.

STEP 6: CONDUCT THE ON-THE-JOB-TRAINING

Conduct the OJT when all the preliminary work has been accomplished: The tabletop analysis has generated the job tasks list, the knowledge and performance objectives have been written, the procedures and supporting documentation have been identified, and training materials in the form of an OJT guide or package have been developed and prepared.

TRAINER RESPONSIBILITIES

One of the key tasks the OJT instructor must perform when he or she meets with a new trainee is to review the entire training guide with the trainee. The instructor should emphasize the expectations of the training, including the skills the trainee will need to acquire to be fully qualified. The instructor will need to explain how the training will be conducted, how the trainee will be able to successfully achieve each task, and the grading system that will be used as the qualification standard.

This includes describing OJT as a two-part process of training and evaluation. In summary, trainees entering the OJT program need to learn how the training process operates and what will be expected of them throughout the training process.

TRAINEE RESPONSIBILITIES

The trainee's responsibilities are to be a willing participant in the process and to complete each task as outlined in the OJT guide. A trainee should ask questions for clarity when in doubt and keep a copy of the guide to keep track of personal progress; this may include having additional supporting self-study materials.

BEGINNING THE TRAINING PROCESS

The first training session should be an orientation in the actual training setting. The trainee takes part as an observer as the OJT trainer goes over the procedures or supporting documentation while actually performing the job. The trainer is to explain what he or she is doing, why he or she is doing the tasks, and the safety precautions that must be considered to do the job safely. Be careful not to overwhelm the trainee with details on the first day.

HOW THE ON-THE-JOB-TRAINING INSTRUCTOR CONDUCTS TRAINING

1. *Demonstrate how to perform a task.* Using the OJT guide, the instructor should select a simple task and demonstrate how to perform the task correctly, explaining why the task is to be performed, its importance, and the impact to the operation of the part or system if not done correctly.
2. *Allow the trainee to perform part of the task.* In this phase of the instruction, the trainer and trainee are interacting together, with the trainer coaching as necessary.
3. *Allow the trainee to perform the entire task.* In this phase, the trainee is allowed to perform the entire task with coaching as necessary from the OJT trainer. Depending on the task complexity, this may occur several times before the task is mastered.
4. *Evaluate the trainee's performance of a task.* Observe the worker performing the entire task without supervision. For each task, this is the *final test*. When the worker can perform the task without supervision, he or she is considered trained. Trainee guides should be constructed so that each series of tasks has the correct number of places to sign off on the individual tasks or subtasks.
5. *Sign off the training package.* To confirm their competency, the worker should be allowed to perform the task without active supervision. At this point, the trainee's training for that particular task will have been completed, and the instructor can sign the final signature page.
6. *Keep records.* The final signature page should be filed in a secure place. This may be in the form of physical storage or a digital storage system. The training record should be maintained for as long as the organization deems necessary. For some organizations, the minimum time for maintaining training records is 5 years.

TIPS ON FEEDBACK DURING TRAINING

One of the most difficult jobs the OJT instructor faces is providing trainees with feedback about their progress. Giving the impression that the trainee is progressing well when he or she needs more coaching is unhelpful. On the other hand, being too direct or frank about a trainee's progress can easily be mistaken for criticism, especially if there are no established criteria used to evaluate the trainee's performance.

Feedback provides trainees with an idea of how well they are performing the tasks. To develop adequate feedback:

- Provide an immediate and complete answer to the task item after the trainee's completion; all parts of the answer or answers should be provided. If alternative methods for completing the tasks are acceptable, each should be included.
- Promptly give the trainee practical means to better understand a task in which he or she underperformed. For example, go over the task and the procedures to which the task relates and then have the trainee practice under your supervision.
- Provide guidance for remediation. The purpose of feedback is to help trainees learn the material. Therefore, OJT should be designed so that the trainee is led to restudy the information he or she failed to recall, recognize, or perform.

The following criteria will help to measure a trainee's progress. The objective is for the trainee to reach a level 4.0 (see Level Guideline section) before the trainer signs off the task. Ideally, the trainee should demonstrate performance (practice) to a 4.0 performance level three times without trainer intervention. In some critical laser safety or hazardous situations, a higher minimum number of unassisted performances may be necessary.

LEVEL GUIDELINE

1.0	The trainee demonstrated a lack of knowledge about the task or made major deviations or omissions that made accomplishment of the task impossible. The instructor was required to demonstrate proper accomplishment of the task.
1.5	The trainee demonstrated limited knowledge of the task. Although the trainee can begin the task, performance deteriorates quickly and extensive instructor interaction is required to maintain safe accomplishment.
2.0	The trainee has a basic understanding of the task, but errors or deviations are significant and would jeopardize safety or mission accomplishment. Even under ideal conditions, extensive instructor intervention is required for safety or mission accomplishment.
2.5	The trainee made errors or deviations. Limited assistance along with frequent coaching by the instructor was essential for safe accomplishment of the task. The trainee has sufficient systems knowledge to make a correct response when provided coaching by the instructor.
3.0	The trainee accomplished the task successfully, but there were slight errors or deviations that the trainee could not correct. The instructor was required to provide coaching for smooth performance but not for safe mission accomplishment. The trainee can perform under ideal conditions but would have difficulty under adverse conditions.
3.5	The trainee was able to accomplish the task safely and successfully with minor errors or deviations. The trainee was able to correct these minor errors, and no assistance was required from the instructor.
4.0	The trainee performed the task without errors or deviations. No instructor intervention was required. The trainee has progressed beyond mere proficiency and could probably perform well under adverse conditions.

FINAL THOUGHT: OJT SUCCESS

The success of any OJT rests on the ability to correctly identify the key job factors, the identification of the expectations of training performance, the ability of the OJT instructor doing the training, the organization of the training materials, the consistency of the training regardless of who is conducting the training, and the relevance of the information to the trainee and the organization.

Finally, the OJT program quality should be continually monitored. The longer a training program is in place, the easier it is for it to slip from its intended purpose. Any OJT program should be thoroughly reassessed and reevaluated not less than every 3 years. Therefore, establishing and maintaining a system for collecting new job data and trainee end-of-course evaluations are paramount to continuously improving an OJT program.

11 The New Maximum Permissible Exposure

A Biophysical Basis

Jack Lund

CONTENTS

INTRODUCTION

At the core of all laser-safety guidelines is the concept of maximum permissible exposure (MPE), defined as the level of laser radiation to which an unprotected person may be exposed without adverse biological changes in the eye or skin (ANSI, 2007; ICNIRP, 2000; IEC, 2007). Although the concept of MPE is quite simple, the reality is that it is multidimensional, dependent on the wavelength and duration of the exposure and modulated by the presence of repetitive pulses and diffusers or optics, which might increase the size of the beam at the target tissue. The guidelines include several pages of rules, tables, charts, conditionals, and caveats for the computation of the MPE for all reasonably foreseeable laser applications.

Implicit to the definition of the MPE is the assumption that one can know, with some degree of certainty, the threshold level of laser radiation that will cause just perceptible changes to the eye or skin, and further that one can know how the threshold level will vary with exposure parameters. That knowledge has been gained through experimental studies over the past 50 years. Research continues to modify our understanding of relevant dependencies, and as our understanding evolves, the rules evolve. It is anticipated that new guidelines will be published in 2013 by American National Standards Institute (ANSI), International Commission on Non-Ionizing Radiation Protection (ICNIRP), and International Electrotechnical Commission (IEC). The new guidelines will revise the rules for determining the effect of wavelength, exposure duration, retinal irradiance diameter, and repetitive pulses on the MPE. This chapter uses *current* when referring to the MPE or guidelines of the 2007 ANSI standard. It uses *new* when referring to the proposed 2013 revision.

This chapter introduces the reader to the bioeffects database and shows how these bioeffects data relate to the MPE. The retina of the eye is the most susceptible part of the body to laser-induced injury. The chapter is limited to the eye, with primary emphasis on the retinal hazard wavelength range.

NATURE OF BIOEFFECTS DATA RELEVANT TO THE HAZARD OF LASER RADIATION

Researchers have developed a substantial body of data relevant to the threshold for laser-induced retinal injury through a process of introducing carefully controlled and measured doses of laser energy into the eye of an anesthetized animal and subsequently evaluating the exposed area for alteration resulting from the introduced energy. Most commonly, researchers place a number of exposures over a range of introduced energies, correlate the presence of a response to the introduced energy for each exposure, and compute the probability of producing the criterion response as a function of introduced energy using the statistical technique of probit analysis (Finney, 1971; Lund, 2006). The output of the probit analysis includes the ED_{50}, that dose having a 50% probability of producing the criterion response. The ED_{50} is not a threshold; it may not be possible to determine a true threshold for these effects. The ED_{50} should, however, be related to the threshold in a manner that is persistent across exposure parameters and therefore serves as a fair and viable basis for the safety guidelines.

The ED_{50} for laser-induced retinal injury is dependent on a number of factors. Inherent to the laser are wavelength, pulse duration, and pulse repetition rate. The experimental configuration determines the retinal irradiance area and profile, the exposure duration, and the number of pulses. The investigator chooses the criterion for determination of retinal alteration. The visibility of laser-induced retinal alteration varies with the interval between exposure and observation. Early researchers used 5- to 15-minute observation times before standardizing on a 1-hour endpoint. More recently, investigators have augmented the 1-hour endpoint with a second observation at 24 hours after exposure. The later observation typically results in a lower value for the ED_{50}. Early researchers used the rabbit as the animal model but soon adapted the rhesus monkey as a closer match to the human eye. Two distinct areas of the retina of the eye have been variously used for determination of the ED_{50} and these areas produce different results. Within the rhesus eye, the central area of the retina, the macula, is generally more sensitive to laser-induced alteration than the more peripheral extramacular retinal areas (Figure 11.1). The macula tends to be more heavily and uniformly pigmented, and the optics of the eye are better corrected for the macula than for the extramacular regions. As a result, the ED_{50} for macular exposures is generally lower and more reproducible than the ED_{50} for extramacular exposures. But the macula is small, and the amount of macular

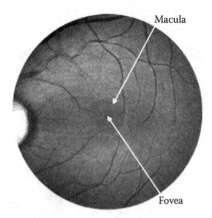

FIGURE 11.1   Rhesus monkey retinal photograph showing fovea, macula, and paramacula.

data that can be obtained from one expensive primate is limited: thus more data exist for extramacular exposures.

Pigmented tissue *in vitro* is also used, not to determine an ocular ED_{50}, but as a vehicle to examine interaction mechanisms and to delineate the relative dependence of injury level on exposure parameters. Retinal explants, in which the preretinal ocular media of an excised eye is removed to allow direct exposure and examination of the retinal tissue, serve to compare the retinal response to the total eye response.

This chapter discusses the wavelength dependence of the ED_{50}, the exposure duration dependence of the ED_{50}, the effects of repetitively pulsed retinal exposures, and the dependence of the ED_{50} on the retinal irradiance diameter. The reader will realize that the parameter space so defined must encompass a very large number of distinct exposure configurations, each of which will result in a unique value for the ED_{50}; further it is highly unlikely that all possible values of the ED_{50} will ever be experimentally determined. In fact, when the guidelines were first drafted in the late 1960s, experimentally determined values for the ED_{50} were scarce indeed. The solution to this problem was to view the eye as a physical system having defined and measured optical, mechanical, and thermal properties.

The function of the eye optics is to image the external world onto the retinal photoreceptors at the focal plane at the back of the eye. Geometric optics then dictate that a collimated laser beam, incident at the cornea, will be focused onto a very small area at the retina: the diameter of that area is limited by the quality of the eye optics and the state of accommodation at the time of the exposure. Immediately posterior to the photoreceptors, and still at the focus of the eye optics, lays the retinal pigment epithelium (RPE), a monolayer of cells containing melanin granules. Melanin is a very strong absorber of optical radiation to the extent that most of the radiation incident on the retina is absorbed in a 5-µm layer of melanin granules. Thus, the collimated laser beam incident at the cornea is concentrated onto a strongly absorbing layer at the retina and injury at that layer results from very small levels of energy incident at the cornea (Figure 11.2).

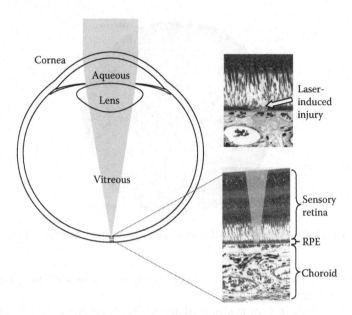

FIGURE 11.2 An incident laser beam is concentrated by the optics of the eye onto the retina. Light is incident from the top and passes through the cornea, aqueous, lens, vitreous, and sensory retina before impinging on the retinal pigment epithelium (RPE) and the choroid. The RPE, a monolayer of cells containing strongly absorbing melanin granules, is the primary locus of laser-induced retinal damage.

Histology shows that threshold laser-induced thermal retinal damage is localized to the RPE. From this observation, one can proceed in two directions. Assume that, for a given exposure duration, retinal injury occurs when the energy absorbed in the RPE reaches some threshold level that is independent of the wavelength of the absorbed energy. The wavelength dependence of the required energy incident at the cornea to produce that level of absorbed energy in the RPE can be estimated based on the spectral transmission and absorption of ocular tissues. Or assume that for a given wavelength, the absorbed energy required to produce retinal injury varies with the exposure duration in a manner that is consistent with the rate of energy deposition, the thermal conductivity in the retinal tissue, and the rate of thermal denaturation of retinal proteins. Then, for exposure durations longer than a few microseconds the level of absorbed energy required to produce retinal injury can be estimated by use of thermal models of laser/tissue interaction. For exposures shorter than a few microseconds, thermal conductivity is too slow to impact the thermal outcome of energy deposition and the absorbed energy required to produce injury is relatively independent of the exposure duration.

Considerations of eye properties were essential to the first iterations of the guidelines and continue to be essential to a system of checks and balances to insure the continued validity of the guidelines. Data have now accumulated, and the interaction models have become quite sophisticated. Still, data must be questioned if they cannot be explained by interaction models, and models must be questioned if they are not consistent with the data.

WAVELENGTH DEPENDENCE: C_A, C_C, AND C_B

An incident laser beam is attenuated by the preretinal ocular media, the cornea, the aqueous, the lens, and the vitreous, before it reaches the retina (Figure 11.2). It is not easy to measure the transmission of these media in the intact, living eye: access to the retinal plane to position a source or detector requires surgical intervention (Dillon et al., 2000). An alternative has been to deconstruct the excised eye, measure the transmission of each of the component parts separately, and compute the transmission of the whole as the product of the transmissions of the parts. The tissue is necessarily *dead* with attendant changes in transmission and scatter of the transmitted light. Scattered light becomes important for the determination of retinal irradiance in that it removes light from the focused beam, thus reducing the retinal irradiance. The primary source of scatter is the cornea, and scatter in the cornea is sensitive to the integrity of its structure. Typically, the transmission of the tissue is measured in a spectrophotometer configured to collect all the transmitted radiation, and thus the measured quantity is the total transmission through the tissue (Figure 11.3). Transmission of the eye based on these measurements will overstate the irradiance at the retina. It is more relevant to measure the direct transmission of the ocular tissues by collecting only that portion that is not scattered outside the focal area after traversing the tissue. The direct and total transmission of the rhesus eye and the human eye are tabulated in the report CIE 203-2012 (Commission Internationale de l'Éclairage [CIE], 2012). The rhesus eye transmission data from that report are shown in Figure 11.4a and are used in this analysis. The choice of rhesus monkey is deliberate as most laser-induced retinal damage thresholds have been determined in that species and the intent is to compare a theoretical action spectrum to the data.

Energy reaching the retina is absorbed in the retina and choroid, where the absorbing tissue of interest is the RPE. Accurate measurement of the percentage of light absorbed in the RPE requires that the RPE be isolated from the other retinal tissues and that all nonabsorbed incident light, including scattered light and

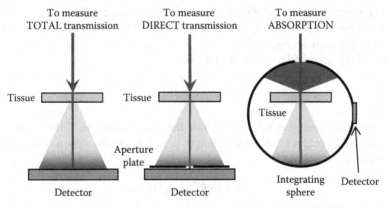

FIGURE 11.3 Configurations to measure TOTAL transmission, DIRECT transmission, and ABSORPTION of tissue.

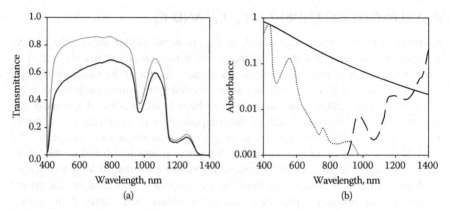

FIGURE 11.4 (a) T_λ, Total transmittance (gray line) and direct transmittance (black line) of the preretinal ocular media of the rhesus eye. (b) A_λ, Absorption of light in the retina of the rhesus eye. Absorbers are retinal pigment epithelium (solid line), blood in the retinal capillaries (dotted line), and water in the sensory retina (dashed line).

transmitted light, be collected and measured. Based on careful measurements of the transmission of the isolated RPE by Gabel et al. (1976, 1978) and supported by other determinations of melanin absorption (Wolbarsht et al., 1981), the RPE absorption is reasonably approximated by the function

$$A_\lambda = 1 - e^{-\alpha_\lambda s} \tag{11.1}$$

where $\alpha_\lambda = \alpha_0 (\lambda_0 / \lambda)^{3.5}$.

The absorption length, s, is 5 µm. A fit to the RPE absorption data is obtained when α_0 is set to 4100 cm^{-1} at the wavelength λ_0 of 380 nm.

Blood is found in the capillaries of the retinal circulatory system resident in the sensory retina. Light must traverse layers of 5-µm-diameter capillaries before reaching the RPE (Lund et al., 2001). The absorption of incident radiation in 5 µm of blood approaches that in the RPE for the wavelength range approximately 440 nm. At wavelengths longer than 1300 nm, light absorption by water in the sensory retina becomes significant (Figure 11.4b).

The energy absorbed by the RPE, Qr_λ, is

$$Qr_\lambda = Qp_\lambda T_\lambda Tb_\lambda A_\lambda \tag{11.2}$$

where Qp_λ is the energy at the cornea within the area of the pupil, T_λ the transmission of the preretinal ocular media at wavelength λ, and A_λ the absorption of the retina at wavelength λ.

$$A_\lambda = A_\lambda(\text{RPE}) + A_\lambda(\text{H}_2\text{O})$$

Tb_λ is the transmission of blood assuming a 5-µm absorption path.

Rearranging,

$$Qp_\lambda = \frac{Qr_\lambda}{(T_\lambda Tb_\lambda A_\lambda)} \tag{11.3}$$

or

$$\frac{Qp_\lambda}{Qr_\lambda} = \frac{1}{(T_\lambda Tb_\lambda A_\lambda)} \tag{11.4}$$

Equation 11.4 is an expression for the relative retinal hazard as a function of wavelength of collimated laser energy incident at the cornea. Compare this function to the wavelength dependence of the retinal MPE, which is governed by the value C_A from 400 to 1150 nm and by the value $C_A \times C_C$ from 1150 to 1400 nm (ANSI, 2007). It can be seen that the wavelength dependence of the MPE fairly represents the wavelength dependence of the expected hazard over much of the retinal hazard region (Figure 11.5). The value of C_C in the current standards results in an MPE that is up to three orders less than the measured ED_{50} for wavelengths between 1150 and 1400 nm. Zuclich et al. (2007) proposed a revision of C_C to bring the MPE more in line with the wavelength dependence of Equation 11.4 and that of the ED_{50}. The proposed revision has been adapted in altered form in the new guidelines. The old and new values of $C_A \times C_C$ are shown in Figure 11.5.

Does Equation 11.3 fairly represent the wavelength dependence of the experimentally determined values for the ED_{50}? Well, it almost does, but not quite. If the eye were a perfect optical system, then the retinal area over which the incident energy was distributed would always be the same, and Equation 11.3 would be sufficient. But, the eye is not a perfect optical system, and aberrations strongly affect the distribution of energy on the retina. Among those aberrations is chromatic aberration. The index of refraction of ocular optics varies with wavelength, and not all wavelengths are in simultaneous focus on the retina. If the eye is in focus for yellow light, then blue light comes to a focus in front of the retina and red light comes to a focus behind

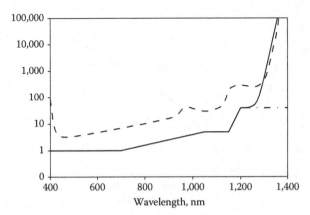

FIGURE 11.5 The wavelength dependence of the relative energy incident at the cornea required to deliver equal absorbed energy in the RPE, $Qp_\lambda/Qr_\lambda = 1/(T_\lambda A_\lambda)$ (dashed line) compared to the wavelength dependence of the maximum permissible exposure as determined by C_A and the current value of $C_A C_C$ (dot–dash line) and the new value of $C_A C_C$ (solid line).

the retina. As a result, given a collimated laser beam incident at the cornea, the irradiated area at the RPE will be larger for blue light and for red light than for yellow light. In effect, the diameter of the irradiated area at the RPE varies with the wavelength of the incident light (Figure 11.6). As shown in section Spot-Size Dependence, the threshold for laser-induced retinal damage becomes larger as the irradiated area of the retina becomes larger. Qr_λ should therefore be expressed by the function

$$Qr_\lambda = Qr_0 \left(\frac{D_\lambda}{D_0} \right)^X \tag{11.5}$$

where Qr_0 is the required energy for a minimum retinal irradiance diameter D_0, and D_λ the chromatic aberration-induced diameter at wavelength λ.

From Equations 11.3 and 11.5,

$$Qp_\lambda = Qr_0 \frac{(D_\lambda / D_0)}{T_\lambda Tb_\lambda A_\lambda} \tag{11.6}$$

or setting the constant $k = Qr_0 / D_0$

$$Qp_\lambda = k \left[\frac{D_\lambda^X}{T_\lambda Tb_\lambda A_\lambda} \right] \tag{11.7}$$

The value of X, which determines the variation of ED_{50} with the retinal irradiance diameter, varies from a value of 2 for exposures of nanoseconds to microseconds to a value of 1 for 0.25-second duration and longer exposures (Lund, Schulmeister et al.,

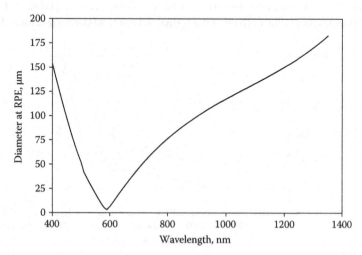

FIGURE 11.6 D_λ – Chromatic aberration-induced variation of laser beam diameter at the retinal pigment epithelium (RPE) of the rhesus eye based on geometrical optics in a simplified eye with the assumption that the index of refraction of ocular tissue varied as the dispersion of water.

2005; Lund et al., 2008) (see section Spot-Size Dependence). Other factors affect the value of D_0. Although the optical quality of the eye will allow the incident laser radiation to be focused to a diameter at the RPE as small as 5–7 µm under optimum conditions, recent research suggests that the threshold for retinal injury does not decrease for image diameters less than 70–90 µm (Lund et al., 2007; see section Spot-Size Dependence).

Figure 11.7 shows a representative exposure system used to measure laser-induced retinal damage thresholds. The laser beam passes through a shutter, which determines the exposure duration. A beam splitter deflects a constant proportion of the pulse energy into a reference detector while the remainder of the energy passes through attenuators and onto a mirror, which directs the laser beam into the eye to be exposed. The mirror can be moved to permit observation and accurately repositioned for exposure. A fundus camera permits observation of the retina and selection of sites for exposure. The fundus camera, mirror, and laser beam are aligned so that the laser energy reflected by the mirror passes through the center of the ocular pupil and strikes the retina at the site corresponding to the crosshairs of the fundus camera viewing optics. Typically, 30 exposures spanning a range of doses are placed in an array in the retina. The exposure sites are examined after exposure and the presence or absence of visible alteration is noted for each site. A number of metrics have been used to determine the presence of alteration. The more sensitive metrics such as measures of visual function and microscopic evaluation of excised tissue are resource intensive and are sparingly used to place a lower bound on the range of introduced energy capable of producing retinal change. The primary metric continues to be the presence of a minimum visible lesion (MVL) detected via ophthalmic examination subsequent to exposure. The response at each site is correlated to the dose at that site, and probit analysis of the dose–response data in four to six eyes yields the ED_{50}.

Generally, the laser is collimated to produce a beam having a beam divergence of less than 1 mrad. The beam diameter at the cornea is limited to 3–4 mm to ensure

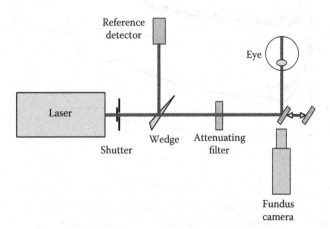

FIGURE 11.7 A typical experimental configuration to irradiate the retina of the eye to a collimated laser beam with carefully controlled and measured dosimetry.

that the entire beam passes through the ocular pupil and impinges the retina. The dose is of necessity measured external to the eye, and the ED_{50} is most commonly expressed as total intraocular energy (TIE), defined as that energy incident at the cornea within the projected area of the ocular pupil. The MPE is provided by the laser-safety guidelines in units of corneal radiant exposure (J/cm^2) averaged over a 7-mm aperture (ANSI, 2007; ICNIRP, 2000; IEC, 2007). When the MPE (J/cm^2) is multiplied by the area of a 7-mm pupil (0.385 cm^2), the TIE (J) introduced into the eye for an exposure at the MPE is obtained.

Tunable lasers have facilitated the experimental study of the wavelength dependence of the ED_{50} for retinal damage. A tunable TI:Saph laser and several fixed-wavelength lasers yielded ED_{50} values for continuous wave (CW) (0.1-second) exposures from 440 to 1320 nm (Lund et al., 1999, 2008; Vincelette et al., 2009). An optical parametric oscillator, capable of providing useful pulse energy throughout a tuning range from 400 to 1200 nm, made it possible to determine the wavelength dependence of retinal damage for Q-switched pulses throughout the visible and near-infrared (NIR) spectrum (Lund et al., 2001, 2003). ED_{50} data resulting from dose–response experiments with these lasers are shown in Figure 11.8. A Qp_λ curve (Equation 11.7) was fitted to the data for each exposure duration by choosing the value of k to match the data at a single wavelength (830 nm) and choosing the value of X appropriate for the exposure duration. The value of D_0 was set to 70 µm in accordance with the available spot-size data. Although not perfect, the match of the data to the theoretical curve is quite good.

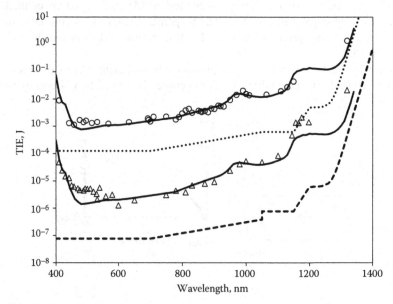

FIGURE 11.8 Wavelength dependence of the ED_{50} for 3.5-nanosecond-duration exposures (diamonds) and for 0.1-second-duration exposures (circles). Qp_λ as defined by Equation 11.7 has been fit to the data. The new maximum permissible exposure for each exposure duration is shown.

Over a broad range of exposure conditions, the experimentally determined ED_{50} for laser-induced retinal damage can be fitted to curves based on Equation 11.7. The data shown in Figure 11.9 include exposure durations from 3.5 nanoseconds to 1000 seconds, wavelengths from 410 to 1320 nm, and retinal irradiance diameters from the minimum the eye will produce to 350 μm (Lund, Edsall et al., 2005). In each case the value of k was chosen to fit the data at 830 nm. The data resulted from a number of experimental studies (Connolly et al., 1975; Ham et al., 1979, 1976; Lund and Beatrice, 1989; Lund et al., 1998, 2006; Onda and Kameda, 1979a, 1979b; Vincelette et al., 2009; Zuclich et al., 1979).

When the wavelength is shorter than 550 nm and the exposure is longer than 10 seconds, laser irradiation can produce photochemical injury to the retina at doses significantly lower than those required to produce thermal injury (Ham et al., 1979, 1982; Lund et al., 2006). It would be convenient if a wavelength could be identified as the transition point between the range wherein photochemical injury dominates and the range wherein thermal injury dominates. Instead, the guidelines require that both the photochemical MPE and the thermal MPE be determined and the lower MPE be used as the exposure limit. Figure 11.10a shows the wavelength dependence of the laser-induced retinal ED_{50} for 1-, 16-, 100-, and 1000-second exposure duration. The data are presented as retinal radiant exposure H_R (J/cm²). The thresholds

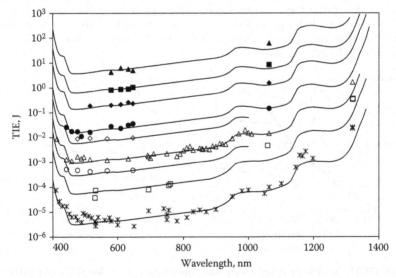

FIGURE 11.9   The wavelength dependence of the ED_{50} for laser-induced retinal alteration for a range of exposure durations with Qp_λ as defined by Equation 11.7 matched to each data set. t is the exposure duration and D is the diameter of the irradiated retinal area. Filled triangle: $t = 1000$ seconds, $D = 350$ μm; filled square: $t = 100$ seconds, $D = 350$ μm; filled diamond: $t = 16$ seconds, $D = 350$ μm; filled circle: $t = 1$ second, $D = 350$ μm; open diamonds: $t = 0.25$ second, $D = 140$ μm; open triangles: $t = 0.1$ second, $D = 30$ μm; open circles: $t = 0.04$ second, $D = 30$ μm; open squares: $t = 600$ microseconds, $D = 30$ μm; cross: $t = 3.5$ nanoseconds, $D = 30$ μm.

for all exposure durations converge onto a common line at the short wavelength end. This line is the threshold of the retina as a function of wavelength for laser-induced photochemical injury and forms the basis for C_B, the blue-light correction factor (Figure 11.10b).

Figure 11.11a superimposes the wavelength dependence of laser-induced thermal retinal injury, converted to H_R and matched to the data for the 1-, 16-, 100-, and 1000-second exposures. In the wavelength range of 400–600 nm both thermal

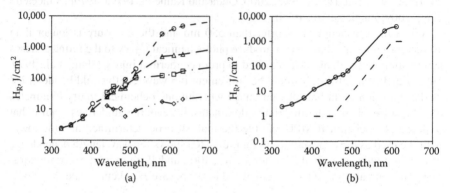

FIGURE 11.10 (a) The wavelength dependence of laser-induced photochemical injury thresholds. (Short dash line, $t = 1000$ seconds; long dash line, $t = 100$ seconds; dash–dot line, $t = 16$ seconds; dash–dot–dot line, $t = 1$ second.) (b) The photochemical retinal injury threshold as a function of laser wavelength (solid line), based on the data of panel a. This curve forms the basis for C_B (dashed line).

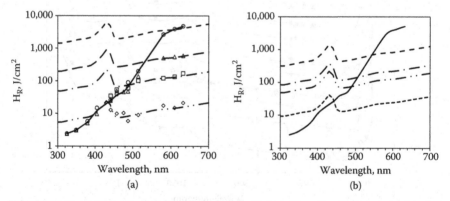

FIGURE 11.11 (a) The threshold for thermal retinal injury (Equation 11.7) compared to the photochemical injury threshold (solid line). The retinal spot size is 325 µm. (Short dash line, $t = 1000$ seconds; long dash line, $t = 100$ seconds; dash–dot line, $t = 16$ seconds; dash–dot–dot line, $t = 1$ second.) The wavelength at which photochemical injury becomes dominate varies with the exposure duration. (b) The threshold for thermal retinal injury (Equation 11.7) compared to the photochemical injury threshold (solid line). The Exposure duration is 16 seconds (dash line, $D = 50$ µm; dash–dot line, $D = 187$ µm; dash–dot–dot line, $D = 325$ µm; dot line, $D = 1700$ µm). The wavelength at which photochemical injury becomes dominate varies with the retinal spot size.

retinal and photochemical retinal injuries are possible. The MPE will be determined by which of these injury thresholds are reached at the lower dose. To the left of the photochemical threshold line the photochemical threshold will be reached first. To the right of that line the thermal threshold will be reached first.

The photochemical threshold occurs at a specific dose (retinal radiant exposure), which varies only with wavelength. As is clear in Figure 11.11a, the time taken to deliver that dose has no effect on the required total dose; the thresholds for each of the exposure durations converge onto the same photochemical threshold curve. The thermal retinal injury threshold does depend on the delivery time, varying as $t^{0.75}$. The effect of this is shown in Figure 11.11a wherein the shorter the exposure duration, the shorter the wavelength at which the photochemical threshold will determine the MPE. At sufficiently short exposures, the thermal threshold will determine the MPE across the entire spectrum.

Nor does the photochemical threshold, expressed as retinal radiant exposure, depend on the diameter of the irradiated area. The thermal threshold does depend on the retinal spot size. For exposure duration longer than 0.25 second, the thermal threshold (H_R) varies as $1/D$. The smaller the spot size, the higher the threshold. The curves of Figure 11.11a are determined for a retinal irradiance diameter of 327 mm $(1/e)$. Figure 11.11b shows the effect of varying the spot size while keeping the exposure duration constant. The photochemical threshold will determine the MPE at a longer wavelength when the spot size is small (50 μm) as compared to the case when the spot size is larger.

These data are accounted for in the safety guidelines through the dual limits for wavelengths shorter than 600 nm (ANSI, 2007; ICNIRP, 2000; IEC, 2007).

The ability to produce retinal lesions has been restricted to the wavelength range wherein the preretinal ocular transmission is equal to or greater than 1%. Lesions have been produced in normal rhesus monkey eyes at 325 and 1330 nm (Zuclich and Connolly, 1976; Zuclich et al., 2007), near the short-wavelength and long-wavelength limits for retinal damage. Increasing the corneal dose to compensate for the preretinal loss does not extend the retinal damage range. Apparently, alteration to the preretinal tissue limits the energy transmitted to the retina. The eye is nonetheless still vulnerable to injury from laser radiation at wavelengths longer than 1400 nm. The injury site is shifted to the cornea and lens, and the doses required to produce injury are higher because the incident radiation is no longer concentrated on the absorbing tissue by the optics of the eye. Although the database for laser-induced corneal injury is sparse compared to the data available for retinal thresholds, enough data are available to match to a curve based on the absorption of corneal tissue (Zuclich et al., 2007). The cornea thresholds can be fit to a curve proportional to the depth at which 95% of the incident laser radiation is absorbed in the corneal tissue (Lund et al., 1981). A continuation of injury level is seen through the transition from retinal injury to corneal injury at 1400 nm with a range at which both retinal and corneal injury are possible. Figure 11.12 compares the MPE to the data for nanosecond-duration exposures from 400 to 4000 nm. The wavelength dependence of the IR MPE is approximated to the corneal 95% absorption curve through a series of step functions (ANSI, 2007; ICNIRP, 2000; IEC, 2007).

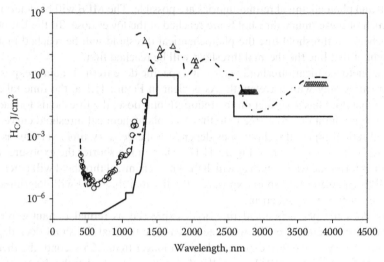

FIGURE 11.12 ED_{50} for laser-induced ocular injury after exposure to nanosecond-duration laser irradiation. The thresholds are presented as corneal radiant exposure. When the injury site is the retina, the incident energy is averaged over a 7-mm-diameter area at the cornea. At 1319 nm both cornea and retina can be injured, but the threshold for retinal injury is lower. The injury site is the retina for wavelengths shorter than 1300 nm and the cornea for wavelengths longer than 1400 nm. Qp_λ as defined by Equation 11.7 is fit to the retinal thresholds. A curve that is proportional to the depth in the eye at which 95% of the incident energy has been absorbed is fit to the corneal thresholds (scale not shown). The new maximum permissible exposure for nanosecond-duration exposures is included.

EXPOSURE DURATION (TIME) DEPENDENCE

Data relating the dependence of the ED_{50} to the exposure duration came earlier and are more abundant than those relating the ED_{50} to wavelength. Almost since laser emission was first achieved, investigators have had access to pulsed and CW laser systems enabling controlled exposure at durations ranging from a few nanoseconds to thousands of seconds; broad wavelength tunability came much later. The wavelength dependence data tend to be variable: laser quality, dosimetry, and detection of effect are not the same in all laboratories and all have improved with time. Data relevant to the ED_{50} for exposure duration shorter than 1 nanosecond are more recent (Cain et al., 1995, 1996, 1999, 2005; Roach et al., 1999).

Laser energy absorbed in retinal tissue heats that tissue. At the same time, thermal conductivity acts to limit the temperature rise in the heated tissue. As with any physical system, the time–temperature history of the tissue is described by the thermal diffusion equation. Solutions pertaining to ocular injury with all the attendant assumptions are well documented and will not be duplicated here (Takata et al., 1974; Thompson et al., 1996).

Consider exposures longer than a few microseconds. Under the influence of constant incident power P, tissue will quickly reach an equilibrium temperature T, and the value of T will be proportional to that of P. If injury could be associated simply with a given temperature T_1, then the dose P_1 required to cause injury would have a constant value independent of exposure duration. The threshold dose expressed as energy ($Q = P_1 \times t$) would vary as $t^{1.0}$, where t is the exposure duration. However, tissue injury results from heat-related denaturation of proteins and depends on both the temperature of the tissue and the time for which that temperature is maintained. A higher temperature will produce injury in a shorter time, whereas a lower temperature must be maintained for a longer period to produce injury. This relationship tends to increase the power at ED_{50} for shorter exposures compared to the power required for longer exposures. As a result, the time relationship predicted by the thermal models closely approximates the dependence shown by the bioeffects data. The values in energy of the ED_{50} for visible laser exposure increase proportional to $t^{0.75}$ for exposure durations longer than about 10 microseconds (Figure 11.13).

For sufficiently short exposures, the energy is deposited in times shorter than the time required to conduct heat out of the exposed area. The temperature of the tissue will then be proportional to the incident energy independent of the duration of the exposure. This condition is referred to as thermal confinement. For exposures shorter than a few microseconds, the temperature required for protein denaturation will exceed the temperature required to produce a phase change of the liquid surrounding the melanin pigment granules, and other damage mechanisms come into play at lower temperatures than required for protein denaturation. In the late 1990s, researchers reported the induction of cell death by microcavitation (bubbles) around melanin granules superheated by incident laser irradiation (Brinkman et al., 2000; Kelly, 1997; Kelly and Lin, 1997; Lin et al., 1999; Roegener and Lin, 2000; Roider et al., 1998). The data relating cell death to microcavitation were derived from laser irradiation of retinal explants or pigmented cells in culture. Microcavitation was detected optically or acoustically and cell death was determined by fluorescent

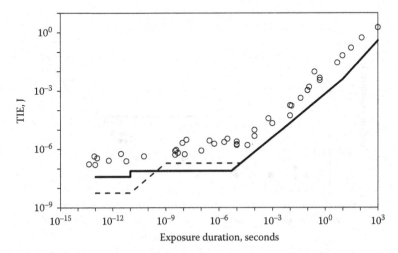

FIGURE 11.13 The time dependence of the ED_{50} for visible laser-induced retinal damage for macular exposures in rhesus monkey. The current maximum permissible exposures (MPEs) (dashed line) and the new MPEs (solid lines) for visible-wavelength laser ocular exposure are included.

assay. Cell death almost always followed the induction of a microcavitation bubble in the cell; the cell membrane is damaged by the pressure wave propagating out from the microcavitation bubble. Gerstman et al. (1996) showed theoretically that bubble formation occurred at lower incident irradiance than required for thermal denaturation for pulses of duration between 1 nanosecond and 1 microsecond. Schüle et al. (2005) experimentally distinguished between thermally induced damage and microcavitation-induced cell death. They decreased the pulse duration from 3 milliseconds and showed that for pulse durations of less than 50 microseconds the damage mechanism at threshold level changes from a thermal one to one based on the formation of microcavitation around the melanosomes in the RPE. Lee et al. (2007) looked more closely at the range from 1 microsecond to 40 microseconds and found a transition at about 10 microseconds. Figure 11.14 includes *in vitro* threshold values from several studies. Each data point in Figure 11.14 represents a threshold for cell death. The open symbols show where either microcavitation was not detected or no measure of microcavitation was attempted. The solid symbols show where microcavitation was detected. For exposures longer than about 10 microseconds, cell death occurs at a lower irradiance than required for microcavitation and thus is considered to result from thermal denaturation. For exposures shorter than 10 microseconds, cell death correlates to microcavitation.

Below about 1 nanosecond, the trend of the *in vivo* ED_{50} is to be increasingly lower than the prediction of constant energy. This trend is explained in part by stress confinement, which results in a shock wave rather than the pressure wave of longer exposures. The greater pressure gradient across the shock wave has a greater potential for cell membrane damage. Ultrashort pulses in the femtosecond duration range

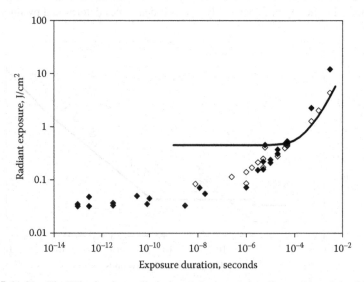

FIGURE 11.14   The ED_{50} for thermally induced damage (open diamonds) and the threshold for microcavitation-induced cell death (solid diamonds) as a function of the exposure duration. Data are obtained for *in vitro* (cell culture or retinal pigment epithelium explant) exposures. The solid line is the predicted cell damage threshold based in thermal models.

produce nonlinear effects that affect the propagation of the laser energy through the transparent preretinal tissue and introduce other interaction mechanisms including plasma formation at the retina (Cain et al., 1996, 2005; Rockwell et al., 2010). That retinal damage is less dependent on melanin granule absorption for the very short exposures is evidenced by the observation that the ratio between the NIR ED_{50} and the visible ED_{50} is less for the shortest exposures.

As more laser-induced retinal injury threshold data have been collected for exposure durations from 1 to 100 nanoseconds, it has become apparent that the current MPE in that range does not afford an adequate safety factor; the safety factor ranges from 2.5 to 4, too low given the uncertainties associated with the ED_{50} for point source, nanosecond-duration pulsed exposures (Cain et al., 1995; Lund et al., 2008, 2011; Zuclich et al., 2008). The new data reflect macular thresholds with a 24-hour MVL endpoint and represent the lowest values for experimentally determined ED_{50}. Advances in experimental techniques and dosimetry contributed to the reduction over earlier values of the ED_{50}. It was apparent that the MPE should be adjusted downward to protect against the new thresholds. Both the current and the new MPE are shown in Figure 11.12. The new MPE for nanosecond-duration pulses has been decreased by a factor of 2.5 to provide an adequate safety margin while the new subnanosecond MPE has been increased to more closely reflect the data.

REPETITIVE PULSES: C_p

Current laser-safety guidelines have adopted an $n^{-1/4}$ relationship for the determination of the MPE for exposure to repetitive pulsed lasers. The exposure for any single pulse in a pulse train shall not exceed the single-pulse MPE multiplied by a multiple-pulse correction factor $C_p = n^{-1/4}$, where n is the number of pulses. The $n^{-1/4}$ relationship, first articulated by Stuck et al. (1978), was not rigorously derived from first principles but is an empirical fit to a body of data. The rule has the advantage that it does not depend on the pulse repetition frequency or the interpulse spacing and is therefore applicable to trains of irregularly or randomly spaced pulses. Commonly this relationship is expressed as

$$MPE(RP) = MPE(SP)n^{-1/4}$$

where MPE(RP) is the maximum permissible exposure for the repetitive pulse train expressed as energy per pulse and MPE(SP) the maximum permissible exposure for a single pulse from the same laser.

Figure 11.15 shows a collection of laser-induced retinal injury threshold data for repetitive-pulse ocular exposures. These ED_{50} data support the idea that the ED_{50}, expressed as energy per pulse, is well represented by a derating factor that varies as the inverse fourth power of the number of pulses in the exposure. On the average

$$ED_{50}(RP) = ED_{50}(SP)n^{-1/4}$$

This result is relatively independent of the wavelength, pulse duration, or pulse repetition frequency of the laser. Models based on a thermal damage mechanism

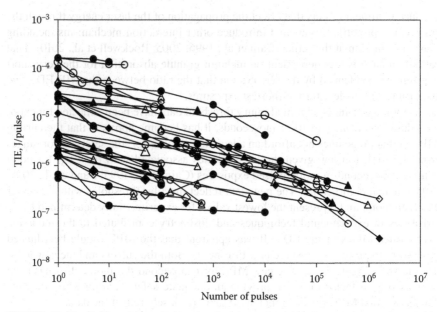

FIGURE 11.15 Collected laser-induced retinal $ED_{50}(RP)$ data for repetitive-pulse ocular exposures. The $ED_{50}(RP)$ expressed as energy/pulse is plotted as a function of n. The solid line has a slope $n^{-1/4}$.

cannot readily explain this result. Additivity of effect requires that each pulse somehow sensitize the exposed retina such that it becomes more susceptible to damage with each cumulative pulse. The fact that the $n^{-1/4}$ dependence is independent of the interpulse spacing essentially rules out the possibility that the mechanism for sensitivity increase is thermal memory. Other sensitizing mechanisms have been elusive.

Menendez et al. (1993) proposed a probability summation model for predicting the threshold for a train of pulses based on the probit statistics for a single pulse. The probability summation model requires that the effect of each pulse in the pulse train be a separate event independent of and unaffected by the effects of other pulses in the train. Until recently, this has been problematic because laser-induced retinal injury was viewed as a thermal effect, and each pulse in the train contributes to the end temperature of target tissue. However, the introduction of microcavitation as a mechanism for cell death opened new possibilities. Lin et al. (1999) noted that "as the laser fluence was increased, more and more cells underwent cavitation," indicating that the threshold for cavitation had a statistical distribution. Roegener et al. (2004) exposed RPE cells *in vitro* to repetitive-pulse trains and observed that "an important finding of this study is that a single cavitation event during a long train of pulses is sufficient to kill the cell." Thus, it would appear that when the mechanism for induction of cell death is bubble formation, the condition of pulse independence is met, and the probability summation model might be appropriate.

Lund (2007) and Lund and Sliney (2013) showed that the probability summation model can lead to an alternate form for the repetitive-pulse correction factor, $C_p = n^{-1/\beta}$. The probit program used to analyze the dose–response data yields not only the ED_{50} but also other parameters of the fit including the slope (SL) of the probability curve (Δprobability/Δdose) at the ED_{50}. The slope is often viewed as a measure of how good the data are: a higher (good) SL indicates a rapid transition from no damage to damage as the dose increases and thus a well-defined ED_{50}; a small value of SL indicates a slow transition over a broader range of doses, thus a more poorly defined ED_{50}. The value of β is related to the experimental values of the slope (SL) of the probit curve fit to the dose–response data as $\beta = SL - 1$. The distribution of values of SL for *in vivo* dose–response experiments is centered at approximately 5; therefore, the value of $\beta = 4$ ($C_p = n^{-1/4}$) used in the safety guidelines is the most probable value of a distribution of values of β based on the collection of bioeffects data.

The idea that the $ED_{50}(RP)$ should decrease simply as $n^{-1/\beta}$ is disconcerting in that it leads to very low values of the $ED_{50}(RP)$ for large numbers of pulses. The solution to the probability summation model leading to $n^{-1/\beta}$ was an approximate solution. Lund et al. (2009) derived exact solutions for the probability summation model, which showed that the $ED_{50}(RP)$ would tend to asymptote to a constant value for large n. The two solutions are compared in Figure 11.16 for $\beta = 4$. These considerations lead to a reanalysis of the data included in Figure 11.15. The data of Figure 11.15 include redundancies (i.e., two or more curves based on the same set of exposures but with different endpoints). It includes data sets that have an average power greater than the CW limit. It includes data sets that contain the ED_{50} for only one value of n in addition to the single-pulse ED_{50}. It includes data that have been determined to be experimentally unsound. Figure 11.17a shows the data that remain when all the superfluous of erroneous data are removed. Also absent in Figure 11.17a are data from a study by Ham et al. (1988). These data will be discussed later in this section. The bulk of the data remaining in Figure 11.17a support the notion that the $ED_{50}(RP)$ does not fall off forever as $n^{-1/\beta}$ but tends to asymptote to a constant value as suggested by the exact solution to the probability summation model. This finding is supported by the data of Figure 11.17b, which shows the dependence of ED_{50} on n for large spot size *in vivo* exposures and for *ex vivo* exposure of retinal explants. A common characteristic

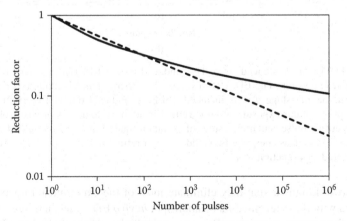

FIGURE 11.16 The multiple pulse reduction factor ($ED_{50}[n$th pulse]/ED_{50}[single pulse]). Dashed line, approximate solution with $\beta = 4$. Solid line, exact solution with $SL = \beta + 1 = 5$.

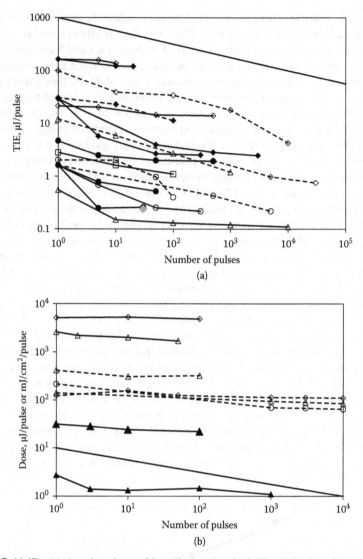

FIGURE 11.17 (a) A reduced set of laser-induced retinal $ED_{50}(RP)$ data for repetitive-pulse ocular exposures. The $ED_{50}(RP)$ expressed as energy/pulse is plotted as a function of n. A line having slope $n^{-1/4}$ is included. (b) Laser-induced retinal $ED_{50}(RP)$ data for repetitive-pulse ocular exposure of large retinal irradiance diameters *in vivo* (solid lines) and for repetitive-pulse ocular exposures of explant retinal tissue *ex vivo* (dotted lines). The $ED_{50}(RP)$, expressed as energy/pulse (solid lines) or retinal radiant exposure/pulse (dotted lines is plotted as a function of n.

of these two designs is that they eliminate most of the experimental uncertainties associated with the determination of small spot *in vivo* ED_{50}, and therefore, the slope of the probit curve is very large. Because of the high value of β, the $ED_{50}(RP)$ falls off very slowly from the single-pulse case. Consideration of Figure 11.17a and b led to the conclusion that the repetitive-pulse derating factor ($C_p = n^{-1/4}$) of the current guidelines imposes an unnecessarily low MPE(RP) for repetitive-pulse exposures. The new guidelines will lower the MPE for single-pulse nanosecond-duration exposures, resulting in an even lower repetitive-pulse MPE(RP). Given these facts, the new guidelines set $C_p = 1$ for most relevant exposure conditions. For some exposure conditions, the value of C_p is reduced, but never lower than the value $C_p = 0.2$. The prospect of very low values of the MPE(RP) for large numbers of exposures has been eliminated without compromising safety.

Data of Ham et al. (1988) are shown in Figure 11.18a. The MPE(RP) in accordance with the new guidelines is a constant value across all values of n. It is evident that the

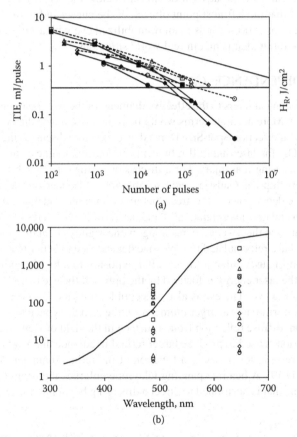

(a)

(b)

FIGURE 11.18 (a) $ED_{50}(RP)$ data for repetitive-pulse retinal exposure (40-microsecond pulses) to 488 nm (solid lines) and 647 nm (dotted lines) laser radiation (Ham et al., 1988). The horizontal line is the new MPE(RP) with $C_p = 1$. The 488-nm thresholds fall below the MPE. Those $ED_{50}(RP)$s below the new MPE(RP) are photochemical in nature. The 647-nm $ED_{50}(RP)$s fall below the MPE(RP) only for exposures of 1000-second duration. (b) The $ED_{50}(RP)$ data for repetitive-pulse retinal exposure (40-microsecond pulses) to 488 and 647 nm laser radiation (Ham et al., 1988) are converted to units of total retinal radiant exposure for the pulse train and compared to the photochemical injury threshold of Figure 11.10. For large n, the 488-nm retinal radiant exposure exceeds the photochemical threshold.

$ED_{50}(RP)$ for these data will fall below the MPE(RP) for large n. Four of the data sets were obtained at a wavelength of 488 nm and, as the authors noted, the thresholds reflect photochemical injury rather than thermal. The average power does not exceed the photochemical MPE for the duration of the exposures, yet the total retinal radiant exposure exceeds the photochemical threshold for large n (Figure 11.18b). The guidelines do not adequately provide for the possibility that the cumulative effect of many short pulses, none of which approach the photochemical limit, might lead to photochemical injury. The other four data sets of Figure 11.18a were obtained at a wavelength of 647 nm where photochemical injury is not a consideration. The average power of the data in the lower of these curves equals or exceeds the CW limit. The other three curves approach but do not fall below the repetitive-pulse limit. It should be noted that the leftmost point of each of the curves in Figure 11.18a represents an exposure of 1000 seconds in an immobilized eye with a fully dilated pupil. This is not a condition that is met in real-world situations.

SPOT-SIZE DEPENDENCE: C_E

Spot size in this context refers to the effective diameter of the laser beam incident at the retina of the eye. More accurate terms would be retinal irradiance diameter or retinal image diameter, and section Spot-Size Dependence therefore discusses the relationship between the ED_{50} for laser-induced retinal injury and the diameter at the retina of the laser beam causing that injury. For a discussion of the effect of the cross-section profile of the spot (top-hat, Gaussian, etc.), see the work of Henderson and Schulmeister (2004). Spot-size dependence is the area wherein the current guidelines are least satisfying. New data and new interpretations of old data now point to a resolution of a situation wherein data and theory conflict resulting in uncertainty about the guidelines. The new guidelines have reformulated the spot-size dependence to better match the data.

The eye is most susceptible to injury after exposure to a highly collimated laser beam because the laser energy is focused by the optics of the eye to a very small area on the retina of the eye. The eye is also susceptible to injury following exposure to a laser beam that irradiates a larger diameter on the retina. Typically, such an exposure results from diffuse reflection from a surface in the field of view. A *Maxwellian view* configuration is used to produce larger retinal irradiance diameter exposures to measure the threshold for injury as a function of the retinal diameter (Westheimer, 1966) (Figure 11.19). A beam-expanding telescope enlarges the beam to overfill an aperture at a lens that is positioned to direct a diverging beam into the eye. The beam

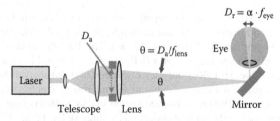

FIGURE 11.19   A *Maxwellian view* exposure geometry for irradiation of large retinal areas.

divergence, θ, is fixed by the aperture diameter and the focal length of the lens while the retinal irradiance diameter is calculated using the relationship $D = f_{me}\theta$, where f_{me} is the focal length of the rhesus monkey eye (13.5 mm). For comparison of the guidelines to the data, the visual angle α is related to D as $\alpha = D/f_{he}$, where f_{he} is the focal length of the human eye (17 mm).

The guidelines specify the MPE as a function of exposure duration and laser wavelength for exposure to a collimated beam, identified in the guidelines as a point source (ANSI, 2007; ICNIRP, 2000; IEC, 2007). A point source has an angular subtense equal to or less than a limiting visual angle, α_{min}, of 1.5 mrad. The MPE for a source subtending a larger visual angle, α, is obtained by multiplying the point-source MPE by a correction factor (C_E in ANSI and ICNIRP, C_6 in IEC), which is a function of α. The current guidelines define C_E to be equal to one for $\alpha < \alpha_{min}$, to be directly proportional to α for $\alpha_{min} < \alpha < \alpha_{max}$ and to be proportional to α^2 for $\alpha > \alpha_{max}$ (Figure 11.20a). α_{max} is defined to be 100 mrad. This formulation for C_E was based on bioeffects data collected between 1960 and 1980 (Lund, Schulmeister et al., 2005). These data were in part puzzling because, while thermal models agreed more or less with the data for longer exposure durations, the concept of thermal confinement seemingly required that the threshold for laser-induced retinal injury for short-duration exposures be proportional to the retinal radiant exposure and thus the ED_{50} expressed as TIE should be directly proportional to the irradiated area (D^2 or α^2).

More recently, Zuclich et al. (2000, 2008) reported ED_{50} data resulting from retinal exposures of 7-nanosecond-duration at 532 nm, which showed that, contrary to the older data, the ED_{50} for laser-induced retinal injury was proportional to D^2 over the range of retinal irradiance diameters from 90 to 1600 μm ($\alpha = 5$–95 mrad). Lund et al. (2007) reported data for the same range of retinal irradiance diameters that show that the ED_{50} for 0.1 second, 514 nm laser-induced retinal injury is proportional to D. On the basis of these new data, it was recognized that the current form of C_E is not an accurate indication of the bioeffects data (Figure 11.20b).

Schulmeister et al. (2005, 2008) presented an examination of the dependence of computer-simulated retinal injury thresholds on retinal irradiance diameter over the range from 30 to 2000 μm and on exposure duration over the range from 1 microsecond to 2 seconds. The computational results were directly compared to and validated by injury thresholds resulting from laser exposure of bovine retinal explants *in vitro* over a range of irradiance diameters from 23 to 2000 μm and exposure durations from 100 microseconds to 1 second. The computed and experimental thresholds, when expressed as total incident energy, vary as D for small spots and vary as D^2 for large spots. The transition between the two zones in terms of D is a function of the exposure duration. The diameter at which the transition occurs is equivalent to α_{max} in that both define the separation between a D dependence of the ED_{50} (α dependence of the C_E) and a D^2 (α^2) dependence. These data, supported by a reanalysis of the collected *in vivo* data, led to a recommendation that the value of α_{max} separating the regime wherein C_E is proportional to α and the regime wherein the C_E is proportional to α^2 should vary with exposure duration. (Lund, Schulmeister et al., 2005; Schulmeister et al., 2011). The new guidelines redefine α_{max}, as shown in Figure 11.21a. The effect on C_E is shown in Figure 11.21b. The value of C_E in the new guidelines accurately follows the experimentally determined spot-size dependence

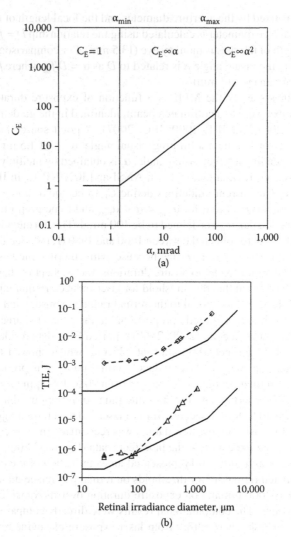

FIGURE 11.20 (a) The dependence of the source-size correction factor C_E on the visual angle α. (b) The dependence of ED_{50} for 24-hour minimum visible lesion endpoint on the retinal irradiance diameter for macular exposure in rhesus monkeys compared to the current maximum permissible exposure. Diamonds; 0.1-second duration, 514 nm. Triangles; 7-nanosecond duration, 532 nm.

for various exposure durations (Figure 11.22). In this figure, the C_E curve has been shifted on the y-axis to coincide with the appropriate data set.

A number of investigations have shown that the incident energy required to produce retinal injury in an intact monkey eye does not decrease for D smaller than about 100 μm, but reaches a minimum at that diameter and remains relatively constant for all smaller diameters (Figure 11.23). These investigations included a number

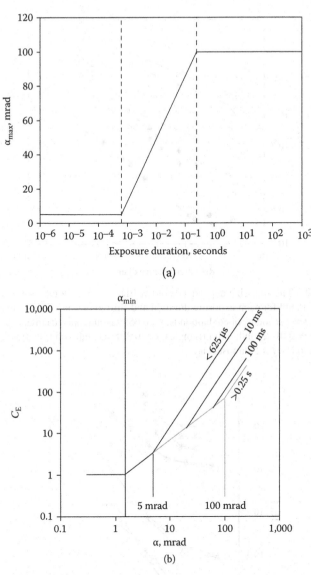

FIGURE 11.21 (a) The dependence of α_{max} on the exposure duration in the new guidelines. $\alpha_{max} = 5$ mrad for $t \leq 625$ microseconds; $\alpha_{max} = 200t^{1/2}$ mrad for 625 microseconds $< t \geq 0.25$ second; and $\alpha_{max} = 100$ mrad for $t > 0.25$ second. (b) The form of C_E for different exposure durations when α_{max} is time dependent.

of wavelengths and a broad range of exposure duration. The underlying cause for this behavior is unclear. Certainly, the optics of the eye can under normal conditions focus a collimated incident beam to a much smaller retinal diameter. Experimental evidence suggests that the higher than expected threshold for collimated beam exposure is a result of the inability of the investigator to visually detect a very small

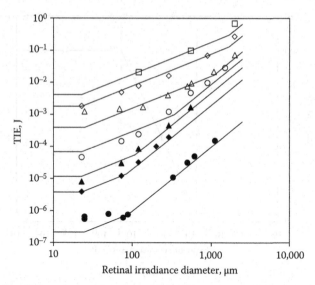

FIGURE 11.22 The spot-size dependence of the ED_{50} for several exposure durations. The new C_E is calculated for each exposure duration and shifted on the y-axis to match the data. Open squares, $t = 2$ seconds; open diamonds, $t = 0.655$ second; open triangles, $t = 0.1$ second; open circles, $t = 0.01$ second; solid triangles, $t = 0.001$ second; solid diamonds, $t = 0.0001$ second; solid circles, $t = 5$ nanoseconds.

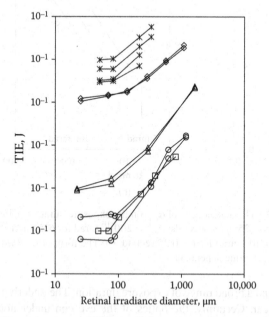

FIGURE 11.23 The dependence of ED_{50} on D for a range of exposure durations and laser wavelengths. Squares: 150 femtoseconds, 1060 nm; open circles: 7 nanoseconds, 532 nm; triangles: 3 microseconds, 590 nm; diamonds: 100 milliseconds, 514 nm; asterisks: 0.125–1 second, 633 nm.

retinal lesion until the laser energy was increased above the true threshold level (Lund et al., 2008; Lund and Sliney, 2013).

Experiments with bovine retinal explants, in which the anterior portions of the eye including the neural retina are removed so that the RPE layer can be directly irradiated, show that the energy required to damage the RPE continues to decrease with the diameter of the irradiated area for irradiance diameters down to 20 μm (Schulmeister et al., 2008). Figure 11.20b shows that the current guidelines provide a margin of safety for 100-millisecond-duration exposures, even given the possibility that the point source ED_{50} might be lower than that obtained through *in vivo* experiments. The current guidelines do not provide a safety margin when compared to the data of Zuclich et al. (2008) for 532-nm Q-switched exposures. The new guidelines will provide a safety margin for that data.

SAFETY MARGIN

The safety margin, defined to be the ratio ED_{50}/MPE and commonly assumed to 10, is best viewed as a measure of confidence in the experimentally determined value of the ED_{50}. That degree of confidence is greater for some combinations of exposure parameters than for others. The least confidence is assigned to the ED_{50} for pulsed exposure to a highly collimated laser beam. The Spot-Size Dependence section discussed uncertainty regarding the true dependence of ED_{50} on retinal irradiance diameter for diameters less than 80–100 μm and left the possibility that the threshold for damage might be lower in the actively blinking, actively accommodating eye of an alert young human. The Repetitive Pulses section raised the possibility that the laser–tissue interactions involving microcavitation are stochastic rather than deterministic. Figure 11.13 shows more scatter in the data for nanosecond- to microsecond-duration exposures as compared to longer duration exposures. On the other hand, the ED_{50} data for exposures at retinal irradiance diameters larger than 100 μm do not suffer from the limitations imposed by aberrations, scatter, and detectability; these data are more reliable and thus a smaller safety margin would suffice. Data for CW exposures are supported by deterministic thermal models and are also more reliable than the pulsed exposures.

The safety margin will also vary with wavelength, because the wavelength dependence of the MPE is a simplified model of the wavelength dependence of the retinal susceptibility to laser-induced injury. The new guidelines have adjusted the wavelength dependence in the 1200- to 1400-nm wavelength range to correct a range of very large safety margins in that range.

CONCLUSION

The body of laser/light-induced bioeffects data is very extensive. The logic and rationale involved in deducing an accurate, consistent, and coherent formulation for the MPE based on these data have been argued and evolved over the last 40 years and continue to evolve as new data challenge the current understanding.

REFERENCES

ANSI. (2007). *American National Standard for the Safe Use of Lasers, Z136.1–2007*. Orlando, FL: Laser Institute of America.

Brinkman, R., Huttmann, G., Roegener, J., Roider, J., Birngruber, R., and Lin, C. P. (2000). Origin of retinal pigment epithelium cell damage by pulsed lase irradiance in the nanosecond to microsecond time regime. *Lasers Surg Med*, 27(5), 451–464.

Cain, C. P., DiCarlo, C. D., Rockwell, B. A., Kennedy, P. K., Noojin, G. D., Stolarski, D. J. et al. (1996). Retinal damage and laser-induced breakdown produced by ultrashort-pulse lasers. *Graefe's Arch Clin Exp Ophthalmol*, 234, s28–s37.

Cain, C. P., Thomas, R. J., Noojin, G. D., Stolarski, D. J., Kennedy, P. K., Buffington, G. D. et al. (2005). Sub-50-fs laser retinal damage thresholds in primate eyes with group velocity dispersion, self focussing and low-density plasmas. *Graefe's Arch Clin Exp Ophthalmol*, 243, 101–112.

Cain, C. P., Toth, C. A., DiCarlo, C. D., Stein, C. D., Noojin, G. D., Stolarski, D. J. et al. (1995). Visible retinal lesions from ultrashort laser pulses in the primate eye. *Invest Ophthalmol Vis Sci*, 36(5), 879–888.

Cain, C. P., Toth, C. A., Noojin, G. D., Carothers, V., Stolarski, D. J., and Rockwell, B. A. (1999). Thresholds for visible lesions in the primate eye produced by ultrashort near-infrared laser pulses. *Invest Ophthalmol Vis Sci*, 49(10), 2343–2349.

CIE. (2012). *CIE 203:2012 A Computerized Approach to Transmission and Absorption Characteristics of the Human Eye*. Vienna: Commission Internationale de l'Éclairage.

Connolly, J. S., Zuclich, J. A., Nawrocki, D. A., Bowie, H. W., Lemberger, R. F., Strickford, J. A. et al. (1975). *Research on the Effects of Laser Radiation*. Brooks AFB, TX: Technology Incorporated, 2nd Annual Report, USAF School of Aerospace Medicine Contract F41609-73-C-0016.

Dillon, J., Zheng, L., Merriam, J. C., and Gaillard, E. R. (2000). Transmission spectra of light to the mammalian retina. *Photochem Photobiol*, 71(2), 225–229.

Finney, D. J. (1971). *Probit Analysis* (3rd ed.). New York: Cambridge University Press.

Gabel, V. P., Birngruber, R., and Hillenkamp, F. (1976). *Die lichtabsorbtion am augenhintergrund*. Munchen FRG: GSF-Bericht A55, Geselschaft fur Strahen- und Umweltforschung.

Gabel, V. P., Birngruber, R., and Hillenkamp, F. (1978). Visible and near infrared light absorption in pigment epithelium and choroid. *International Congress Series No. 450: XXIII Concillium Ophthalmologicum Kyoto*. Amsterdam: Oxford/Excerpta Medica.

Gerstman, B. S., Thompson, C. R., Jacques, S. L., and Rogers, M. E. (1996). Laser induced bubble formation in the retina. *Lasers Surg Med*, 18(1), 10–21.

Ham, W. T., Jr., Mueller, H. A., Ruffolo, J. J., and Clarke, A. M. (1979). Sensitivity of the retina to radiation damage as a function of wavelength. *Photochem Photobiol*, 29, 735–743.

Ham, W. T., Jr., Mueller, H. A., Ruffolo, J. J., Guerry III, D., and Guerry, R. K. (1982). Action spectrum for retinal injury from near-ultraviolet radiation in the aphakic monkey. *Am J Ophthalmol*, 93, 299–306.

Ham, W. T., Jr., Mueller, H. A., and Sliney, D. H. (1976). Retinal sensitivity to damage from short wavelength light. *Nature*, 260, 153–155.

Ham, W. T., Jr., Mueller, H. A., Wolbarsht, M. L., and Sliney, D. H. (1988). Evaluation of retinal exposures from repetitively pulsed and scanning lasers. *Health Phys*, 54(3), 337–344.

Henderson, R. and Schulmeister, K. (2004). *Laser Safety*. Bristol and Philadelphia: Institute of Physics Publishing [now Taylor and Francis].

ICNIRP. (2000). Revision of guidelines on limits of exposure to laser radiation of wavelengths between 400 nm and 1.4 mm. *Health Phys*, 29(4), 431–440.

IEC. (2007). *IEC 60825-1 Safety of Laser Products—Part 1: Equipment Classification, Requirements and Users Guide* (2nd ed.). Geneva: International Electrotechnical Commission.

Kelly, M. W. (1997). Intracellular Cavitation as a Mechanism of Short-Pulse Laser Injury to the Retinal Pigment Epithelium. PhD Thesis, Tufts University, Medford, MA.

Kelly, M. W. and Lin, C. P. (1997). Microcavitation and cell injury in RPE cells following short-pulsed laser irradiation. *SPIE*, 2975, 174–179.

Lee, H., Alt, C., Pitsillides, C. M., and Lin, C. P. (2007). Optical detection of intracellular cavitation during selective laser targeting of the retinal pigment epithelium: Dependence of cell death mechanism on pulse duration. *JBO*, 12(6), 064034.

Lin, C. P., Kelly, M. W., Sibayan, S. A. B., Latina, M. A., and Anderson, R. R. (1999). Selective cell killings by microparticle absorption of pulsed laser radiation. *IEEE J. Select Topics Quantum Electron*, 5(4), 963–968.

Lund, B. J. (2006). *The Probitfit Program to Analyze Data from Laser Damage Threshold Studies, Report No. WTR/06-001, DTIC ADA452974*. Brooks City-Base, TX: WRAIR.

Lund, B. J., Lund, D. J., and Edsall, P. R. (2008). Laser-induced retinal damage threshold measurements with wavefront correction. *J Biomed Opt*, 13(6), 064011.

Lund, B. J., Lund, D. J., and Edsall, P. R. (2009). Damage threshold from large retinal spot size repetitive-pulse laser exposures. *Proc ILSC* (pp. 84–87). Reno, NV: Laser Institute of America.

Lund, B. J., Lund, D. J., and Holmes, M. L. (2011). Retinal damage thresholds in the 1 ns to 100 ns exposure duration range. *Proc ILSC* (pp. 183–186). San Jose, CA: Laser Institute of America.

Lund, D. J. (2007). Repetitive pulses and laser-induced retinal injury thresholds. *SPIE*, 6426, 64625-64621-64628.

Lund, D. J. and Beatrice, E. S. (1989). Near infrared laser ocular bioeffects. *Health Phys*, 56(5), 631–636.

Lund, D. J., Edsall, P. R., Fuller, D. R., and Hoxie, S. W. (1998). Bioeffects of near infrared lasers. *J Laser Appl*, 10(3), 140–143.

Lund, D. J., Edsall, P. R., and Stuck, B. E. (1999). Spectral dependence of retinal thermal injury. *SPIE*, 3902, 22–33.

Lund, D. J., Edsall, P. R., and Stuck, B. E. (2001). Ocular hazards of Q-switched blue wavelength lasers. *SPIE*, 4246, 44–53.

Lund, D. J., Edsall, P. R., and Stuck, B. E. (2003). Ocular hazards of Q-switched near-infrared lasers. *SPIE*, 4943, 85–90.

Lund, D. J., Edsall, P. R., and Stuck, B. E. (2005). Wavelength dependence of laser-induced retinal injury. *SPIE*, 5688, 383–393.

Lund, D. J., Edsall, P. R., and Stuck, B. E. (2008). Spectral dependence of retinal thermal injury. *J Laser Appl*, 20(2), 76–82.

Lund, D. J., Edsall, P. R., Stuck, B. E., and Schulmeister, K. (2007). Variation of laser-induced retinal injury with retinal irradiated area: 0.1 s, 514 nm exposures. *J Biomed Opt*, 12(2), 06180.

Lund, D. J., Schulmeister, K., Seiser, B., and Edthofer, F. (2005). Laser-induced retinal injury thresholds: Variations with retinal irradiated area. *SPIE*, 5688, 469–478.

Lund, D. J. and Sliney, D. S. (2013). A new understanding of multiple pulsed laser-induced retinal injury threshold. *Health Phys*, in press.

Lund, D. J., Stuck, B. E., and Beatrice, E. S. (1981). *Biological Research in Support of Project MILES, Report No. 96*. Presidio of San Francisco, CA: Letterman Army Institute of Research Institute.

Lund, D. J., Stuck, B. E., and Edsall, P. R. (2006). Retinal injury thresholds for blue wavelength lasers. *Health Phys*, 90(5), 477–484.

Menendez, A. R., Cheney, F. E., Zuclich, J. A., and Crump, P. (1993). Probability-summation model of multiple laser-exposure effects. *Health Phys*, 65(5), 523–528.

Onda, Y. and Kameda, T. (1979a). *Studies of Laser Hazards and Safety Standards. (Part 2) Retinal Damage Thresholds for Helium-Neon Lasers*. Fort Detrick, MD: Technological

Research and Development Institute, Japan Defense Agency, US Army Medical Intelligence and Information Agency Translation No. USAMIIA-K9991.

Onda, Y. and Kameda, T. (1979b). *Studies of Laser Hazards and Safety Standards. (Part 3) Retinal Damage Thresholds for Argon Lasers.* Fort Detrick, MD: Technological Research and Development Institute, Japan Defense Agency, US Army Medical Intelligence and Information Agency Translation No. USAMIIA-K9992.

Roach, W. P., Johnson, T. E., and Rockwell, B. A. (1999). Proposed maximum permissible exposure limits for ultrashort laser pulses. *Health Phys*, 76(4), 349–354.

Rockwell, B. A., Thomas, R. J., and Vogel, A. (2010). Ultrashort laser pulse retinal damage mechanisms and their impact on thresholds. *Med Laser Appl*, 25, 84–92.

Roegener, J., Brinkmann, R., and Lin, C. P. (2004). Pump-probe detection of laser induced microbubble formation in retinal pigment epithelium cells. *J Biomed Opt*, 9(2), 367–371.

Roegener, J. and Lin, C. P. (2000). Photomechanical effects—Experimental studies of pigment granule absorption, cavitation and cell damage. *SPIE*, 3902, 35–40.

Roider, J., El Hifnawi, E., and Birngruber, R. (1998). Bubble formation as primary interaction mechanism in retinal exposure with 200 ns laser pulses. *Lasers Surg Med*, 27(5), 451–464.

Schüle, G., Rumohr, M., Hüttmann, G., and Brinkman, R. (2005). RPE damage thresholds and mechanisms for laser exposure in the microsecond-to-millisecond time regimen. *Invest Ophthalmol Vis Sci*, 46(2), 714–719.

Schulmeister, K., Edthofer, F., and Seiser, B. (2005). *Modelling of the laser spot size dependence of retinal thermal damage.* Paper presented at the ILSC 2005 Los Angeles.

Schulmeister, K., Husinski, J., Seiser, B., Edthofer, F., Fekete, B., Farmer, L. et al. (2008). Ex vivo and computer model study on retinal thermal laser-induced damage in the visible wavelength range. *J Biomed Opt*, 13(5), 054038.

Schulmeister, K., Stuck, B. E., Lund, D. J., and Sliney, D. H. (2011). Review of thresholds and exposure limits for laser and broadband optical radiation for thermally induced retinal injury. *Health Phys*, 100(2), 210–220.

Stuck, B. E., Lund, D. J., and Beatrice, E. S. (1978). *Repetitive Pulse Laser Data and Permissible Exposure Limits, Report No. 58.* San Francisco, CA: Letterman Institute of Research.

Takata, A. N., Goldfinch, L., Hinds, J. K., Kuan, L. P., Thomopoulis, N., and WeiGandt, A. (1974). *Thermal Model of Laser-Induced Eye Damage (DDC AD A-017-201).* Chicago, IL: IITRI.

Thompson, C. R., Gerstman, B. S., Jacques, S. L., and Rogers, M. E. (1996). Melanin granule model for laser-induced thermal damage in the retina. *Bull Math Biol*, 58(3), 513–553.

Vincelette, R. L., Rockwell, B. A., Oliver, J. W., Kumru, S. S., Thomas, R. J., Schuster, K. J. et al. (2009). Trends in retinal damage thresholds from 100 millisecond near-infrared laser radiation exposures: A study at 1110, 1130, 1150, and 1319 nm. *Lasers Surg Med*, 41, 382–390.

Westheimer, G. (1966). The Maxwellian view. *Vision Research*, 6, 669–682.

Wolbarsht, M. L., Walsh, A. W., and George, G. (1981). Melanin, a unique biological absorber, *Appl Opt*, 20(13), 2184–2186.

Zuclich, J. A. and Connolly, J. S. (1976). Ocular damage induced by near-ultraviolet laser radiation. *Invest Ophthalmol*, 15(9), 760–764.

Zuclich, J. A., Edsall, P. R., Lund, D. J., Stuck, B. E., Till, S. J., Hollins, R. C. et al. (2008). New data on the variation of laser-induced retinal damage threshold with retinal image size. *J Laser Appl*, 20(2), 83–88.

Zuclich, J. A., Griess, G. A., Harrison, J. M., and Brakefield, J. C. (1979). *Research on the Ocular Effects of Laser Radiation, Report SAM-TR-79-4.* Brooks AFB, TX: USAF School of Aerospace Medicine.

Zuclich, J. A., Lund, D. J., Edsall, P. R., Hollins, R. C., Smith, P. A., Stuck, B. E. et al. (2000). Variation of laser-induced retinal damage threshold with retinal image size. *J Laser Appl*, 12(2), 74–80.

Zuclich, J. A., Lund, D. J., and Stuck, B. E. (2007). Wavelength dependence of ocular damage thresholds in the near-IR to far-IR transition region: Proposed revisions to MPEs. *Health Phys*, 92(1), 15–23.

12 Near-Miss Incidents

Ken Barat

CONTENTS

To many laser users, the concept of tracking near misses is so far from their conscientious it might as well not exist. This author has to say, outside of procedural violation records of laser near misses and any corrective actions from them are far and few between. In recent years, the near-miss tracking has become more real, but still just a twinkle in the eye of laser safety officers (LSOs). As a result, the tracking of near misses does have a concrete value.

WHY TRACK NEAR MISSES?

In simple words, they are a warning that should be taken to heart. They usually represent an unsafe condition or practice, that is, uncorrected can lead to a real injury. Few of these are reported, investigated, or have corrective actions taken. Why? Because it was a near miss no one was hit, someone got away lucky, why cause trouble, why admit a mistake? Why report? How are you going to feel when someone else is injured repeating just what you did, for they did not understand a risk or flaw existed?

Following are near-miss reports, covering procedural, technique, engineering, and human element issues.

Case 1: Human Error and Response

Occurrence Report Number: NA—SS-SNL-5000-2006-0004

The incident occurred at Air Force Facility; November 1, 2006, at 5:45 MST involving Air Force and subcontractor personnel, conducting a laser test.

At a staff meeting on November 2, 2006, the Department Manager was briefed on the previous night's test and told of an equipment anomaly during the trial. Further information was requested from the Air Force operators. On Monday morning, November 6, 2006, the project manager, principle investigator (PI) and the Environment, Safety & Health (ES&H) coordinator reviewed the information and concluded that a laser beam release had occurred when subcontractors were in a nominal hazard zone (NHZ). It was concluded that this constituted a near-miss event and reporting was immediately initiated.

DETAILS

On November 1, 2006, at approximately 5:45 MST, two subcontractors were working in the experimental facility when the beam from the laser (Class 3B) was inadvertently released into the facility. The beam was released by an Air Force safety officer during a safety check, but as a result of not following the correct procedures. The Air Force safety officer was performing a shutter test, without a second beam block in place to prevent the release of the beam. The beam block had been removed during the first safety check, and had not been replaced for the second test. The Air Force safety officer was new (first time working the experiment).

At the time of release, the subcontractors noticed a charge-coupled-device (CCD) camera behaving erratically. The subcontractors asked the Air Force safety officer about a possible release of the laser, to which the response was initially *no*, but, in fact, the laser had been accidentally released.

The subcontractors were later informed that there had been a temporary release of the beam. The subcontractors were not in the beam path; however, they were in the facility, which is included in the NHZ when the laser is present. The beam was shuttered within seconds of the release. They were not wearing the appropriate eyewear because they were not aware of the existence of the presence of the laser. Therefore, they were unaware of the resultant NHZ. The subcontractors were sent for a medical inspection of any potential exposure damage. They received no exposure damage.

MANAGEMENT SUMMARY

During a laser test, a U.S. Air Force officer failed to follow procedures and inadvertently released a Class 3B laser beam within an experimental facility at the Air Force Base while two subcontract employees were working in the building. The two employees were not in the beam path, but were within the laser's NHZ. The employees were not wearing laser eye protection, and were sent to the site medical facility for the evaluation of any potential eye-exposure damage. Although results were negative, it was learned that this event was a near miss and reporting was initiated.

Digging deeper here is a more complete incident summary:

Two subcontractors were working in the experimental facility there when the beam from the laser (Class 3B) was inadvertently released into the facility. The beam was released by an Air Force safety officer during a safety check, as a result of not following the correct procedures. The test director was doing a shutter test, without a second beam block in place to prevent the release of the beam. The beam block had been removed during the first safety check, and had not been replaced for the second test. The Air Force safety test director was new (his first time running the experiment) and was not adequately trained for the job by the Air Force staff.

At the time of release, the subcontractors noticed an instrument (CCD camera) behaving erratically, which they quickly surmised was due to the presence of the laser. They immediately informed the safety director of the release, and instructed him to shutter the beam. The contractors were not in the beam path; however, they were in the NHZ and were not wearing the appropriate eyewear. They do not believe they had any exposure damage; however, they are being sent to medical inspection to confirm that fact. Steve Babcock was a spotter on the range at the time, however, he was not in the NHZ, and was therefore not at risk.

The incident was discussed with the Air Force personnel. A test was performed on Thursday night, after assessing the situation, and taking the initial corrective action of replacing the safety test director with a more experienced individual. We did not report the incident to Operational Reporting Procedure System (ORPS) at that point of time, mistakenly supposing that because it was an Air Force facility, and the Air Force was investigating, that we did not need to do so. After further consideration, they were contacted to seek guidance on whether we needed to report further or not.

CORRECTIVE ACTIONS

1. Air Force safety test director replaced with more experienced Air Force personnel.
2. Individuals with potential exposure risk sent for medical review to assess any health risks (eye damage).
3. Incident reports and suggested improvements solicited from involved parties.
4. An initial suggestion to be implemented, if possible, is the addition of engineering controls in the form of an interlock switch controlled by experimental personnel. This will eliminate inadvertent releases before experimental staff being fully prepared and mitigate potential human communication errors.
5. As a short-term solution, individuals will only open the laser port into the facility when necessary, and will wear goggles always while that port is open.

Case 2: Search Failure and Too Clever Engineering

Occurrence report number SC—TJSO-JSA-TJNAF-2006-0005

INITIAL DESCRIPTION

About 10:30 AM on Friday, December 1, 2006, at a Free Electron Laser Facility, free electron laser (FEL) control room staff noted the presence of a worker in FEL Lab no. 1. Lab no. 1 had received less than 1 minute of FEL light at the time and the lab was immediately saved (laser light delivery stopped). FEL Lab no. 1 had undergone a sweep to remove all workers to initiate *exclusionary state* (no one was allowed in Lab no. 1) laser operations. The experimental configuration was such that the laser light was delivered in an isolated area, and rope barriers were in place to prevent access to the experiment. The worker did not receive any injury.

IMMEDIATE RESPONSE

1. Within a minute of laser beam delivery to Lab no. 1, The FEL control room duty officer saw that the Lab no. 1 lights had been turned on and immediately saved Lab no. 1. This action released the Lab no. 1 magnetic door lock. The worker (a Hampton University technician supporting a laser experiment) then left Lab no. 1. There were no injuries to the worker, and there were no ionizing radiation or radioactive/hazardous materials involved with this *near-miss* event.
2. The duty officer then terminated all FEL operations and contacted the FEL facility manager. The FEL facility manager concurred that all FEL operations should cease. An electronic record was posted stating that no FEL operations were permitted.
3. The FEL facility manager immediately contacted the associate director (AD) for the FEL division, lab senior management, Environment, Safety, Health & Quality Division, and the DOE site office. An event investigation team was named and the team began event to follow up at 11:00 AM.
4. The worker was examined at a clinic by the lab physician during the afternoon. The examination verified that the worker had not received any injury as a result of this event.

Digging deeper here is a more complete incident summary:

The event investigation team's causal analysis determined that the direct cause of the event was an FEL staff member's failure to do a thorough sweep of the Lab no. 1 as required by the Laser Safety Operating Procedure. A properly conducted sweep requires an entire lab area search to verify no one is present. Contributing causes include the following:

1. New equipment had been recently added to Lab no. 1 that reduced the sweeper's field of vision.
2. The technician was using a computer that was not used in normal operations.

3. The sweep is time-limited to 1 minute duration (by the safety system software), which contributes to some urgency in getting the sweep done quickly.

4. Since Lab no. 1 had been in an open state, and this worker was not involved with the lasing aspects of the experiment, no laser-specific training was required. This lab-specific training instructs workers that the crash button on the laser personnel safety system (LPSS) must be pushed as the exit button is inoperable in the exclusionary mode (this is the status immediately after the sweep and before/during laser delivery).

5. The exit buttons are not well engineered from a human engineering standpoint. There is no indication that they do not function when the lab is in exclusionary mode.

Two phases of corrective actions are as follows.

INTERIM ACTIONS

1. Immediate FEL operations stand down.

2. Briefing of all lab operational staff at the lab's two daily operations planning meetings. The outside FEL physics user group also receives a briefing on the event.

3. The FEL lab sweep procedure will be modified until permanent LPSS changes are in place. These interim changes include the use of two people for all presweeps with no time limit. Once the presweep is done, one person performs the sweep procedure called for in the procedure. During the sweep, the sweeper will push the mounted simulated sweep buttons that direct the sweep to all relevant parts of the lab.

4. Personnel will be trained in the new sweep procedure.

5. Review this event with the users involved. The Jefferson Lab director communicated the importance of safety, and the lab's safety expectations to the PIs leading this experiment.

6. A sign will be posted near the exit buttons in all FEL labs indicating which button to use if the exit button does not work.

7. The lab LSO will brief all non-FEL system safety supervisors of this event by December 11. This briefing will also determine if there is any extent of condition applications in other non-FEL activities.

Note: All interim actions were completed by December 4 except for action 7.

PERMANENT ACTIONS

1. Two to four sweep buttons will be placed in each FEL lab to verify that the sweep goes to all portions of the lab. There will be a sweep button in each walk-in hutch as well.

2. There will be a verbal announcement that a sweep is occurring as the sweep begins.

3. There will be a verbal announcement that lasing is to begin after the sweep has taken place and 30 seconds before the beam is provided to the lab.
4. Ensure that FEL users are familiar with this event and applicable FEL protocols prior to initiating experiments.

MORE BACKGROUND DETAILS FOR THE CURIOUS

The FEL is a high-power laser operating at wavelengths between 0.9 and 10 micrometers. It is tunable over each set of mirrors being used. The light is produced in a radiation-shielded vault and transported through an evacuated transport system to one of seven user labs in the FEL facility. Each lab can be swept (cleared of all personnel), locked up, and brought up in a state of *Laser Permit*. If an interlock is tripped or a crash button is pushed, the lab is *crashed* and transitions to an *open* state. If an open state is required, the FEL operator can *safe* the lab using the laser control system. When the lab is in an open state, access is permitted to anyone and the only training required is the training required to be on the accelerator site. When one wants to put the lab into a Laser Permit state, an approved laser user or FEL operator sweeps the lab according to a sweep procedure documented in the Laser Standard Operating Procedure. There are three possible states in a lab when a lab is in laser permit. In *Exclusionary* mode, no one is permitted in the user lab. This is the default state for a lab. In *Hutch* mode, the laser is confined to interlocked hutches inside the lab. In *alignment* mode, the power to the lab is restricted to low levels so that a user with laser-safety eyewear can align the laser beam to their apparatus. Trained personnel (identified by a radio frequency identification [RFID] badge) can enter the lab when it is in alignment mode (if they are wearing appropriate laser-safety eyewear) or in hutch mode. No access is permitted in exclusionary mode. Egress from the lab is accomplished by pushing an exit button on the door or the LPSS. These buttons only function in alignment mode or hutch mode. To gain egress in exclusionary mode, it is necessary to use the crash button on the LPSS chassis. This is taught to all approved laser users.

A sweep consists of presenting one's RFID card, entering the lab, verifying that all personnel have exited the lab, pushing an *initiate sequence* button, exiting the lab, and presenting one RFID card a second time. In this case, the lights were turned off after pushing the initiate sequence button. Thirty seconds after pushing the initiate sequence button, the lab goes into a state of laser permit. During this time a warning beacon is illuminated and a warning alarm is sounded. When the lab is in laser permit, the warning alarm is turned off and the warning beacons stay on.

First, the lab was locked up and saved several times during initial checkout of the experiment's diagnostics.

Subsequently, it was determined that a camera had to be replaced so the lab was saved and three technicians and one of the FEL operators entered the lab to change it out. During the entry, a user associated with the LPSS experiment went into the lab to use a computer along the north wall to check his e-mail. Once the camera was replaced, the three technicians left the room and the FEL operator swept the lab to bring it into laser permit. This was at about 10:28 AM. The start of the sweep was announced by the operator, but the user did not hear him. The FEL operator did not do a thorough sweep and did not see the technician. Following the sweep, he pushed the initiate sequence button, turned off the lights, and exited the lab.

The user realized that the lab was going into laser permit and tried to leave the lab using the exit button on the door to leave.

When the lab is in exclusionary mode, the exit buttons are nonfunctional. Instead, it is necessary to hit the crash button to leave the lab in exclusionary mode, so pushing the exit button did not release the mag locks on the door, and the user was unable to exit. To notify the control room that he was in the room, he turned the overhead lights on and off and then left them on, with the intention of bleaching the camera response to alert the control room staff that there was a problem. He also ensured that he remained at the door facing with his back to any potential stray laser beams. The user had the required Laser Safety Orientation (SAF 1140) and Laser Medical Approval (SAF 114E) to be a user. He did not have, nor was he required to have laser-specific training to be in the lab when in the open state.

The FEL operators sent 25 W of FEL light at 990 nm into the lab. The light was directed through the optical transport onto a power meter for a total of 1 minute (see Figure 12.1). When the FEL operator saw that the lights had been turned on in the lab, he immediately saved the lab. This was at 10:30 AM. When the lab was crashed, the technician was able to immediately exit the lab.

The duty officer then terminated FEL operations and contacted the facility manager. He concurred that all lasing operations should cease and posted an electronic record called an FLOG indicating that FEL operations were not permitted. He then contacted the AD, the lab LSO, and the laser system supervisor (Figure 12.1).

Management Summary

During laser operations in the FEL Lab no. 1, control room staff unexpectedly discovered the presence of a worker in the laboratory after a sweep to remove all workers

FIGURE 12.1 View of Lab 1 from entrance door. Computer workstation is located along the wall on the right side of the lab, behind the magnet setup.

was performed and laser operations were underway. The laser light was in operation for less than 1 minute when the worker was discovered and the laser light was immediately stopped. There were no injuries to the worker. An investigation was initiated.

Lessons Learned, Including ES&H Manual changes (if any):

A persistent problem with administrative controls in safety systems is that their effectiveness depends on the diligence of the person enforcing the control. The importance of this diligence must be continually reinforced over time so that incidents like this do not occur. Engineering controls can force a sweeper to take their time and cover the required territory, but are no substitute for being careful in the sweep. It is also important to continually stress that personnel take whatever time is required to do a careful sweep and not rush the sweep.

Another lesson learned is to consider human engineering when designing the user interface for a safety system. Inexperienced users are reticent to push crash buttons as they think they will break something. They would rather push an exit button. If an exit button is present but non-functional, they will typically be confused and not know what to do. The best solution is for the exit button to be the crash button.

Finally, it is important to consider how things work when things do go wrong. Even if an event is unlikely, it is important that things work as they should during the event.

Case 3: Legacy and Funding Event

During the installation of a new laser interlock system at the ATF, some potential safety risks were discovered in the previously existing system. The interlocks are very complex as there are two Class 4 lasers whose outputs are sent to different places to excite the photocathode that provides electrons for the Linac. They are used in different rooms as part of the various experiments and there are many different configurations on the beam lines. The complexity is reflected in the test procedures where there are 246 steps, many with multiple-step check-offs.

The project had been transferred between groups with varying funding resources. It was determined that both the radiation and laser interlocks were beyond the expertise of the new group and required outside expertise. A memoriam of understanding was developed.

> C-A Department has the responsibility to maintain or modify the ATF access control system hardware and maintenance procedures. This includes, but it is not limited to logic diagrams, wiring diagrams, test procedures, interlocks, interlocked gates, beam crash buttons or cords, reset stations, critical devices and Radiation Safety Committee review of same.

To facilitate this, a complete set of drawings for both systems were requested and received. A management review led to the following memo:

a. Safety Systems
   i. Laser Interlocks: The ATF has a complex set of interlocks that has been weakly supported by the institution interlock group leaving the ATF with a system that is not properly documented for the

system that is in place. *Changes over the years are not documented and the procedures for testing do not properly cover the current configuration. Furthermore, those who designed and implemented the system are refusing to support the ATF even as purchased services. This is a critical situation.* We are actively trying to get the C-A Department to review and take responsibility for the maintenance and testing of the ATF interlocks, but the cost of redoing the system may be prohibitive.

Testing and recertifying the laser interlock system by adding to the test procedures the items known not to have been included, believing the rest of the test procedures to be complete. At the time we did believe that we could not verify the completeness of the test procedures because we were concerned we never had complete drawings. At the time of the turnover from one group to another department, the previous engineer stated that all the changes to the drawings had been completed but changes to the test procedures had not. The changes to the test procedures were completed in July 2004. The system continued to operate believing that all was in order. However, this could not be independently verified. To this end it was decided that the system would have to be replaced, completely re-documented, verified, and certified. Money was allocated for this and with the help of C-AD has now been done as of June 6, 2007.

Interlock Tests

ATF procedures, in compliance with the SBMS subject area, require laser interlocks to be tested every 6 months. Original checklists are kept in the ATF control room with copies sent to the department's Safety and Training Office where the manager of Environment, Health, Safety, and Training program reviews them.

The test procedures were originally developed by the institution and since (working with the C-AD) have been modified to remove testing for devices no longer physically present, add testing for a device added into the interlock chain, and redefine some areas (two separate areas merged into a single area). At that time, the department chair requested assistance from the institution chair who permitted their interlock engineer (person involved in the original design, implementation, and testing) to help facilitate the changes. When the modified procedures were acceptable to both the groups and C-A Departments, they were adopted and published.

Concerns That Arose

In replacing the old laser interlock system with the new one, the plan was to replace the Programmable Logic Controller (PLC) with the same type currently used by the C-AD interlock group for their interlocks to use the same programming language and take advantage of their expertise with this hardware and software. In principle, the new controller boxes had the same inputs and outputs as the old ones, so replacing the new boxes compatible with the new PLC and language should have been easily done and no new wiring was expected. The old bundle of wires simply had to be plugged into the new boxes. After this was done, the system began to be checked and discrepancies were found.

- It was found that the drawings as given from the institution were not complete or accurate. ATF personnel found approximately 30 errors in the existing drawings.
- In addition, it is doubtful that a fully functional test had been done on the status indicator boxes as there was a light-emitting diode (LED) that was wired backward and could have never worked.
- In certain cases, an indicator light, that implied *two* shutters were closed, was lit when only one shutter was actually closed. Other laser status panel *on* LEDs could not be correctly controlled because they were hard-wired to a *shutter open* switch. In order for the LEDs to be correctly controlled, they need to be independently driven by the interlock system.
- The yttrium aluminum garnet (YAG) shutter status of the LEDs on the CO_2 room enclosure did not reflect the position of the YAG shutter because the LEDs were not correctly wired. The shutter *open* LED would light whenever the CO_2 to high-bay pass-through button (means to enter the interlocked area by authorized personnel) was pressed. The shutter *closed* LED would light when the 10-micrometer imminent LED was on. While these wiring problems did not add additional risk to personnel, they are indicative of the lack of a full functional test.
- Finally, it was apparent that the normal *best practices* for color coding different sets of wires were not done as most of the wiring was the same green color.

Safety Implications

The improper lighting of status indicators was viewed by users to determine which laser eyewear to use put them at potential risk for eye damage. This authorized user list was limited in number to those specifically included in the approved Experimental Safety Reviews. These people had the proper training, including the initial eye examinations given to authorized laser users. They are not permitted to do laser alignments or maintenance. Their use of laser light was limited to their experimental equipment.

The incorrect indicators of laser safety shutters state put users and staff at additional risk.

Before the transfer of the ATF to the department, the institution was responsible for the design, documentation and installation, and programming and testing of the laser interlock and radiation systems. Ilan Ben-Zvi, who used to head the ATF, remembers the institution's running functionality tests of the whole system as some points in the past, but it would appear that rewiring must have been done at some point as the backward wiring would have been found at that time as well as the false indication that two shutters were closed when, in fact, the indicator light was lit when only one was closed.

I did not know that the full interlock test procedures did not cover each and every item in the system and that it took five to six days. I was informed this morning that he didn't know this himself until last week. I presumed that when there was a modification, the interlock test procedures tested the whole system.

Analysis to This Point

1. Unlike the interlocks for radiation, I do not believe the areas covered by laser interlocks alone are subject to the Accelerator Safety Order, as Ron Gill (Physics Department's ES&H coordinator) and I discussed this morning.

2. Unlike the area encompassed by the interlocks for radiation (which is an exclusion zone when radiation is present in the Experimental Hall), the laser areas are configured for occupation by properly trained persons (with the proper personnel protective equipment (PPE)) when the interlocks are on and laser light is present in the rooms. This allows properly trained people to make adjustments, align the lasers, maintenance, and approved experimentation.

3. The status indicator lights are not required but are part of the interlock safety system as they get their information from the interlock circuitry, and users will usually use this information to choose the correct eyewear.

4. Laser safety eyewear is intended as an additional control to reduce risk of eye damage and is required by the ANSI (American National Standards Institute) standard. The laser SOP is used to analyze the hazard and specify the optical density. All of that is in order.

5. There is no reason to believe that any user received any eye damage although the potential was there.

6. I believe it is prudent to see if the SDL has a similar situation. Have they had a fully functional test? Do they rely on indicator lights for eyewear? If so, is that system working properly?

7. Looking at the categorizer's procedure, it might come under the following but is at the discretion of line management and not required to be reported: Group 10—Management concerns/issues

    (2) 1–4† An event, condition, or series of events that does not meet any of the other reporting criteria but is determined by the facility manager or line management to be of safety significance or of concern to other facilities or activities in the Department of Energy (DOE) complex. One of the four significance categories should be assigned to the occurrence, based on an evaluation of the potential risks and the corrective actions taken.

    (3) 1–4† A near miss, where no barrier or only one barrier prevented an event from having a reportable consequence. One of the four significance categories should be assigned to the near miss, based on an evaluation of the potential risks and the corrective actions taken.

Present Status

There is a new interlock system in place. This system is fully documented with a new set of drawings, the hardware has been installed by ATF personnel, the controlling software was written by the C-AD interlock group, tested by C-AD interlock technicians and was reviewed and certified by C-AD personnel. The status indicator lights are currently not part of the system but will be added to it, documented, and test procedures will include their testing. Until then, no users will be permitted to be in the laser interlocked areas until the status lights are operational and certified to be working properly.

In the future, any modifications will require a full functional test as well as properly documented updates to drawings.

Case 4: Deliberate Bypassing of a Safety System

During a visit in October 1998, we performed a series of integrated dry runs and discovered that three interlocks on the experiment room door were taped in a bypassed condition. The preliminary inquiry concluded that the bypassing might have taken place the preceding evening, when site personnel were ensuring that two Class 4 lasers were correctly aligned for the integrated dry runs. The test group director has prohibited the operations of x-ray and laser equipment until the incident was fully evaluated. If such equipment is operating while an interlock is overridden, entry into the area will not shut the equipment down, and personal injury could occur.

A follow-up investigation determined that facility workers attempting to identify which interlock was connected to the laser systems had systematically taped the door interlock devices in a bypassed condition, in violation of the approved operating procedure. It was also determined that the facility staff forgot to remove the tape before the dry runs. The review concluded that since the experiment room door had cipher locks and access was administratively controlled, there was no chance of personnel exposure of injury.

Case 5: Wanting to Be Helpful but Just Following Through Not Thinking

September 2006

A plumber asks a researcher (visiting scientists) to let him into an interlocked laser lab. The researcher puts in the access code, allows the plumber to enter, and he returns to his office. Oh yes, the lasers were running in the laser lab. The PI of the lab is in the lab doing a laser alignment. He sees the plumber (without eyewear) and escorts him out.

Case 6: Near Miss at Los Alamos National Lab

September 2000

A service man is working on an LSR flow cytometer (9/20). These units contain from 2–3 Class 3B lasers. He removes the top enclosures and housing over the lasers. The service man leaves the unit in this way and plans to return in a few days to complete the repair. Three days later, a group leader observes two employees standing over the unit, looking inside of it with the laser on.

Case 7: Near-Miss Fiber Optic Example 1

Two workers were moving fiber cables when they inadvertently viewed a green light emitting from one of the fiber cable ends.

Two laser operations were being performed simultaneously:

Workers were in one room attempting to connect the fiber optics to the patch panel.

Laser personnel were in another room aligning their interferometer using the attenuated laser beam.

Following discovery of the laser light, the laser system was de-energized.

The output from evaluation concluded that the workers did not sustain any eye injury.

Case 8: Near-Miss Fiber Optic Example 2

The laser operator had inadvertently left the fiber optic connected to the laser beam.

The only control to keep the laser light out of this pathway was administrative (a checklist), which relied on the laser operator to disconnect or block the laser light from the input fiber. However, the checklist had no step to ensure that the operator disconnected or blocked the laser light.

There was no labeling of the fiber optic cables, patch panel connections, or junction boxes as required.

The laser enclosure lid had no safety interlocks as required.

Neither the workers nor the laser operators had a laser system key control during their work activities. The workers were unaware of laser operations in the laser room (no communication).

Case 9: Near-Miss-Labeled Case 2 Really Class 3B

A labeled Class 2 laser, used in a high-resolution ruby fluorescence high-pressure measurement system, had its beam power level measured and found to be in the Class 3B power output range. The laser was supposed to have a nominal output no greater than 1 mW, but the measured output was 18 mW.

Optiprexx 532 nm laser systems procured before 2007 may be at a higher class power level than advertised.

NEAR-MISS REPORTING

A thorough near-miss investigation has the potential to identify overlooked physical, environmental, or process hazards; the need for new or more engineering controls or safety training; or unsafe work practices. The supervisor of the employee involved in the near miss is responsible for conducting the investigation and, when appropriate, ensuring that corrective actions are taken. The depth and complexity of the investigation will vary with the circumstances and seriousness of the incident. Investigators must maintain objectivity throughout the investigation. The purpose of the investigation is to uncover any factors that may have led to the accident, not to assign blame.

The first is all narrative.

1. Background information: Site location, department, supervisor, date, time, year of work.
2. Witnesses: Self-explanatory.
3. Description of accident: Most near misses result from an accumulation of events. An accurate, factual description of the accident and the events leading to it can be very helpful. This chronological sequence can be studied to determine how each event may have contributed to the near miss. Include photos or drawings of the accident site, if these will be useful to the investigation.
4. Factors: Factors, if any, are the conditions in the workplace or actions that contributed to the near miss of this near miss. Examples might include unguarded machinery, broken tools, and slippery floors, not following established procedures, or insufficient training or maintenance.
5. Corrective actions: List actions or steps that could be taken to control or eliminate the likelihood of a recurrence; include not only those that can be accomplished right away (e.g., providing personal protective equipment, installing a machine guard) but also actions such as changes in policy or providing additional training (Figure 12.2).

DEPARTMENT		PHONE #		Environmental Health & Safety	

1. EMPLOYEE (*last name, first name, mi*)	2. EMPLOYEE ID No.	3. SEX ☐M ☐F	4. AGE	5. DATE/TIME OF INCIDENT

6. TIME IN JOB
☐ Less than 1 mo.   ☐ 6 mos. to 1 year
☐ 1 to 5 mos.   ☐ More than 1 year

7. JOB TITLE AT TIME OF ACCIDENT

8. EMPLOYMENT CATEGORY
☐ Full-time   ☐ Temporary
☐ Part-time   ☐ Student

9. SPECIFIC LOCATION OF NEAR MISS (*bldg., floor, room #, outside*)

10. WITNESS (*list name(s) & phone #*)

11. DESCRIPTION OF NEAR MISS (*Describe sequence of events, including time, date, and location of incident. Attach photos, drawings, or separate page if necessary.*)

12. FACTORS (Why it happened). (*Describe conditions or practices, if any, that may have led to the occurrence of this incident. Attach separate page if necessary.*)

13. CORRECTIVE ACTIONS (Prevention). (*Developed jointly with EH & S.*)

14. REPORTED BY

_____   _____
        Signature   Date

15. DEPARTMENT HEAD/SAFETY MANAGER COMMENTS.

_____   _____
Department Head/Safety Manager   Date
Signature

16. EH & S REVIEW

_____   _____
        Signature   Date

FIGURE 12.2 Near-miss report form sample.

CONCLUSION

The near-miss incident can easily show problems with one's safety culture or that of others. Therefore, the reporting of near-miss incidents should be supported and employees encouraged to do so. In reality, a hard goal to accomplish, but has a positive payoff.

13 Accident Investigation

Bill Wells and Ken Barat

CONTENTS

INTRODUCTION

Although many efforts are directed to prevent laser accidents, we need to have a plan to follow in case an accident happens. One must be prepared to investigate an accident or incident as it is reported (see Table 13.1 for a simple checklist for a laser or any type of accident investigation). Most activities that include lasers are well-thought-out and include measures to prevent, or minimize the damage from a laser accident. For laser accidents in particular, one can take the advantage of this forethought to pinpoint what differed from plan and why as the strategy for conducting an accident investigation.

WHO SHOULD LEAD THE ACCIDENT INVESTIGATION?

The laser safety officer (LSO) should be the lead contact for a laser-related incident. Many LSOs wear several *hats* in addition to that for safety. Therefore, it might be necessary to turn the accident investigation over to another internal person or even an outside party. The ideal would be to have an investigation conducted by someone expert in accident causation; experienced in investigative techniques; and fully knowledgeable of work processes, procedures, persons, and industrial relations environment of a particular situation. Unfortunately, most of us live in the real world. Do not be afraid to call on any resources in your own firm.

The five basic components of an incident review are as follows:

1. Plan for the review.
2. Collect the data.
3. Assemble and review the data.
4. Identify causal factors.
5. Develop corrective actions.

TABLE 13.1

Simple Checklist or Flowchart for Laser or Any Type of Accident Investigation

Incident Investigation Elements	Checklist Items	Considerations
1. Plan the review.	Identify the basic information related to the incident.	For injury incidents, review any initial or medical reports.
	Select the review team.	Team composition will vary based on the category of the injury and the complexity of the investigation. • Supervisor (typically leads the investigation) • Employee • Organization safety leader • Subject matter expert, LSO.
	Preserve the incident site.	Do not disturb the incident site while the review is in progress.
	Schedule the interviews and visit to the incident site to optimize time usage.	Initial report should be completed within 7 days.
2. Collect data.	Observe the incident site and record the conditions.	Do not alter or modify the incident scene. Note if the scene has been altered. Examine the incident scene: equipment, workspace, engineered controls, personal protective equipment (PPE), etc., and take notes. Obtain photographs of the incident scene and when safe, photograph an incident reenactment. Review any written procedures or documents related to the incident. Maintain an open mind as you collect the data.
	Perform the interview process.	Start the interview by explaining the reason for the interview and focus on the positive objective of learning from the incident to prevent recurrence. Ask for a description of what happened from start to finish (sequence of events). Ask questions for clarification. Start at end of sequence and work backward, looking for gaps and asking for other information that has not yet been shared. Close with a thank you and provide information regarding what happens next in the investigation process.

TABLE 13.1   (*Continued*)

Simple Checklist or Flowchart for Laser or Any Type of Accident Investigation

Incident Investigation Elements	Checklist Items	Considerations
	Use these tips for conducting interviews.	Apply *caring objectivity*.
		Avoid asking why and instead ask how and what.
		Do not interrupt.
		Try to stimulate recall.
		Be curious and inquisitive but suspend judgment.
		Use open-ended questions.
		Do not attribute blame.
		Enlist the help of the interviewee to provide information that will prevent recurrence of the incident.
		In a group, only one interviewer at a time should ask questions.
	Interview the person involved (typically the questions are provided).	What was the sequence of events and conditions leading up to and during the incident?
		Was there anything abnormal or any recent changes?
		Were there any work-arounds or task compensations?
		Were there previous related occurrences?
		Are there procedures, and how are they related to the incident?
		What is the training status for the activity?
		How could the accident have been prevented?
3. Assemble and review the data.	Define the incident to be reviewed.	The incident is the reason for the investigation. It is usually the worst consequence. For example, "Technician's hand burned by laser beam."
		For first aid and OSHA recordable cases, the incident typically is the injury.
	Develop an event sequence, or time order of events.	Events: *Who did what? or What did what?* (Use a noun and an active verb.) Each event should include only one action.
		Events should be facts based on evidence gathered. (e.g., "Technician grasped beam-block."

(Continued)

TABLE 13.1

Simple Checklist or Flowchart for Laser or Any Type of Accident Investigation

Incident Investigation Elements	Checklist Items	Considerations
		Make a time-ordered event list, chart, or diagram.
		If a step in the sequence is assumed, identify it as such, or seek additional information to close the gap in information.
		Diagram should include dates and times.
		Use job titles instead of staff names.
	Develop an activity process sequence, highlighting control measures for known hazards.	This is a step-by-step analysis of what is expected in this work activity. It should be at a level of detail that identifies all of the engineered controls (e.g., verify barriers are in place), administrative control measures have been taken (task steps expected for successful completion of the activity), and proper PPE has been used.
		Activity data should be gathered from planning documents gathered during data collection and from interviews of personnel involved in the event.
	Conduct a *Work as Planned vs. Work as Conducted* side-by-side comparison of the activity process sequence and time order of events.	This comparison will pinpoint what control measures (barriers) did not perform as expected. It will identify if the controls were not used or followed, wrong, ineffective, or failed to perform as expected; or if the hazard was never identified. This is an iterative process. When a gap in information is identified, return to the gathering information process to fill the gap.
		This list of differences, or gaps, is the list of causal factors, or what happened to allow the accident to occur.
	Review time order of events, activity process sequence, and causal factors for the incident.	If it tells a reasonable and cohesive story that includes all known facts, this phase of the investigation can be considered complete. If important data gaps exist, collect more information.
4. Analyze causal factors to determine apparent or root causes.	Review the causal factors identified in step 3 and apply appropriate root cause analysis techniques; use these key terms.	For each causal factor, identify the significance of the gap on its consequences. In this step you should identify why the controls identified were not used or followed, wrong, ineffective, or failed to perform as expected; or if the hazard was never identified. The degree of analysis can be proportional to the complexity and severity of the consequences investigated.

TABLE 13.1 *(Continued)*
Simple Checklist or Flowchart for Laser or Any Type of Accident Investigation

Incident Investigation Elements	Checklist Items	Considerations
		Causal factor: Any problem or issue that, if corrected would have prevented the incident from occurring or significantly reduced the incident's consequences.
	For each causal factor, determine apparent or root cause	Apparent cause: The most probable cause(s) that explains why the event happened, that can reasonably be identified, that local or facility management has control to fix, and for which effective recommendations for corrective action(s) to remedy the problem can be generated, if necessary.
		Root cause: The most basic cause or causes that can reasonably be identified that management has control to fix and when fixed will prevent (or significantly reduce the likelihood of) the problem's recurrence.
		If causal factors cannot be identified, more information may need to be collected or assistance from an incident investigator (if the organization has such a person) may need to be taken.
5. Identify corrective actions	Develop corrective actions for the causal factors (for first aid) and root causes (OSHA recordables).	For each of the apparent, or root causes identified in step 4, consider and identify corrective actions.
		Corrective action: A change (procedure, policy, equipment, etc.) that when implemented, corrects the identified causal factor of root cause and thereby prevents (or significantly reduces the likelihood of) the incident from recurring.
	Use objectives of corrective actions.	The corrective actions should be
		• Specific
		• Measureable
		• Accountable
		• Reasonable
		• Timely
		• Effective
		• Reviewed

The interview can be the most informative and challenging part of the entire process. Therefore, it is important to understand the five elements of the incident-review process: planning, data collection, assembly and review of data, performing causal analyses, and developing corrective actions.

Your focal point is to stay on point. Understand the incident so that recurrences are prevented.

- Request a description of what happened (start to finish).
- Ask questions for clarification.
- Work backward through the sequence of events.
- Look for gaps.
- Ask for additional information that has not been shared.
- Close with a thank you and follow-up information.

Typical interview questions are as follows:

- What was the sequence of events and conditions leading up to and during the incident?
- Was there anything abnormal or any recent changes?
- Were there any work-around or task compensations?
- Were there previous related occurrences?
- Are there procedures, and how are they related to the incident?
- What is the training status for the activity?
- How much experience does the organization and individual experience with performing the activity in question?
- Were there any contributing factors?
- How could the incident have been prevented?

SPECIAL NOTE

One issue when interviewing the individuals involved is the common error of thinking you know what happened and asking questions to confirm that, be open-minded.

Interviewing is not a skill the LSO works on or is readily prepared for, hence it is worth repeating guidance on interviewing, so remember the following: Interviewing is art, here are a few do's and don'ts. The purpose of the interview is to establish an understanding with the witness and to obtain his or her own words describing the event.

DO'S...

- Put the witness, who is probably upset, at ease.
- Emphasize the real reason for the investigation, to determine what happened and why, not to find someone to blame or take the fall.
- Let the witness talk, listen.
- Confirm that you have the statement correct.
- Try to sense any underlying feelings of the witness.
- Make short notes or ask someone else on the team to get them during the interview.

- Ask if it is okay to record the interview, if you are doing so.
- Close on a positive note.

DON'TS...

- Intimidate the witness.
- Interrupt.
- Prompt.
- Ask leading questions.
- Show your own emotions.
- Jump to conclusions.

Ask open-ended questions that cannot be resolved by simply *yes* or *no*. The actual questions you ask the witness will naturally vary with each accident, but there are some general questions that should be asked each time:

- Where were you at the time of the accident?
- What were you doing at the time?
- What did you see, hear?
- What were the environmental conditions (weather, light, noise, etc.) at the time?
- What was (were) the injured worker(s) doing at the time?
- In your opinion, what caused the accident?
- How might similar accidents be prevented in the future?

The LSO was not at the scene at the time, asking questions is a straightforward approach to establishing the incident. Obviously, care must be taken to assess the credibility of any statements made in the interviews. Answers to a first few questions will generally show how well the witness could actually observe what happened.

FACTOR LEVELS OF INCIDENTS

Causal factor: Any problem or issue that if corrected would have prevented the incident from occurring or significantly reduced the incident's consequences. A causal factor is identified from the assembled data. It is the stopping point; and the physical and mental condition of those individuals directly involved in the event must be explored. The purpose of investigating the accident is *not* to establish blame against someone, but the inquiry will not be complete unless personal characteristics are studied. Some factors will remain essentially constant while others may vary from day to day: causal analysis of first-aid injuries. Few laser eye injuries fall into this group, but skin burns may.

Apparent cause: The most probable cause(s) that explains why the event happened, that can reasonably be identified, that local or facility management has control to fix, and for which effective recommendations for corrective action(s) to remedy the problem can be generated, if necessary.

Root cause: The most basic cause or causes that can reasonably be identified that management has control to fix and when fixed will prevent (or significantly reduce the likelihood of) the problem's recurrence. For each cause category, one or more basic causes are identified. For each basic cause, one or more root causes are identified (stopping point for causal analysis for Occupational Safety and Health Administration [OSHA] recordable injuries).

Corrective action: A change (in procedure, policy, equipment, etc.) that, when implemented, corrects the identified causal factor or root cause and thereby prevents (or significantly reduces the likelihood of) the incident from recurring. Corrective actions shall be SMARTER:

- Specific
- Measureable
- Accountable
- Reasonable
- Timely
- Effective
- Reviewed

This model of accident investigations provides a guide for uncovering all possible causes and reduces the likelihood of looking at facts in isolation. Some investigators may prefer to put some of the sample questions in different categories; however, the categories are not important, as long as each pertinent question is asked. Obviously, there is considerable overlap between categories; this reflects the situation in real life. Again it should be emphasized that the above sample questions do not constitute a complete checklist, but are examples only.

WHAT ABOUT MANAGEMENT?

Management always has the legal responsibility for the safety of the workplace and therefore the role of supervisors and higher level management, so management systems must always be considered in an accident investigation. Failures of management systems are frequently found to be direct or indirect factors in accidents. Ask questions such as the following:

- Were safety rules communicated to and understood by all employees?
- Were written procedures and orientation available?
- Were they being enforced?
- Was there adequate supervision?
- Were workers trained to do the work?
- Had hazards been previously identified?
- Were unsafe conditions corrected?
- Was regular maintenance of equipment carried out?
- Were regular safety inspections carried out?

USER RESPONSE: DO THEY HAVE AN ACTION PLAN?

One needs to have a plan in place in the event of a laser accident. While our efforts are directed to prevent laser accidents, one needs to have a plan to follow in case an accident happens. More importantly, laser users need to know what that plan is. One solution is to place in every laser work area a poster that provided the names and current phone numbers of emergency personnel and steps to be followed until help arrives. A posted page or section in the laser procedure manual is recommended. The sample poster represented next could also be prepared as a Web page. The advantage of the Web page approach is that it can provide quick updates, such as phone numbers and personnel changes. A disadvantage is that it may not be thought of during the excitement of the moment.

GENERAL GUIDANCE FOR SUSPECTED EYE INJURY

Keep the individual calm.

Preferably keep the individual seated or lying down; avoiding panic or shock is the main goal.

Call for assistance.

If your institution has a central help number for emergencies, call that number. Otherwise, have the number of the local trauma center available; do not count on using or finding the phone book.

If you have a medical clinic on-site, it can be called or notified. If medical facility or security is called first, they should have standing instructions to contact the LSO.

It is important that the medical facility has some understanding of laser eye injuries as well as the laser mechanism.

Transport the person to a medical facility.

Many large organizations have a fire department or security force that would transport the individual; these individuals should be instructed on how to handle a person with an eye injury.

Notifications: Notify the individual's or the area supervisor along with the LSO and others working in the same area or on the same equipment.

MEDICAL FACILITY: DO THEY HAVE AN ACTION PLAN?

Depending on your community, you may have limited choices of where to send an individual with a suspected laser eye injury. It is important that the facility has some understanding of laser eye injuries as well as the laser mechanism. Commonly, once the individual informs the medical staff that he or she works with or around lasers, any injury, particularly retinal, will be assumed to be laser induced. Medical personnel may overlook the fact that many other optical causes or diseases could be the reason for visual problems or defects. Provide the name of a retinal specialist to the individual for further evaluation or follow-up in cases involving visible or near-infrared laser radiation. Not all laser injuries have an immediate effect on vision; consequently, initial and follow-up eye examinations are critical.

14 Laser Accidents Do Happen*

Ken Barat

CONTENTS

---

* For a more complete listing of laser accidents and follow-up activities, see Chapter 9.

INTRODUCTION

Laser technology has been increasing ever since its introduction in the 1960s. This text concerns itself with laser safety, so let us ask the question: Are laser accidents happening, and do we have bodies to count? The simple answer is yes, laser accidents are occurring, and the number is uncertain. One survey of laser accident database demonstrated that more than 1500 cases were registered between 1960 and 2002. The majority of the accidents involves eye injury, approximately more than 70%. For your awareness, a small number of accidents that led to deaths have been reported (approximately 3%). These deaths were due to electrocution or operator room fires, induced by laser radiation. Statistically, this seems like a small number. As in any reporting system, laser-safety professionals believe the number of laser accidents is underreported. Second, trauma to the eye is dramatic, regardless if the effect passes with time.

A number of isolated databases track laser accidents. In the United States, the Food & Drug Administration tracks equipment-related laser accidents. Other federal agencies that track their own accidents are The Department of Energy, Army, Navy, Air Force, and Federal Aviation Administration. In addition, some private industry groups have databases.

Most common factors contributing to laser accidents are the following:

- Unanticipated eye exposure during alignment
- Available laser eye protection not being used
- Poor housekeeping
- Lack of planning
- Equipment malfunction
- Improper high voltage (HV) handling methods
- No protection will be provided for ancillary hazards
- Incorrect eyewear selection
- Inhalation of laser-generated air contaminants
- Viewing of laser-generated plasmas (blue light hazard)
- Improper use of equipment

ALIGNMENT ACTIVITIES

Once the laser beam path is set and established unless, one places themselves a reflective object in the path or approaches the beam with a bend magnetic around their neck, the beam will stay on its appointed path. One can easily see that the overwhelming number of laser accidents or injuries occur when laser optics are being manipulated. Reflections of tools, optical mounts, lens, or mirrors are the chief causes of random or unexpected laser beams. In addition, changes in temperature of the laser area, cooling water, and other factors can induce beam errors. The time and effort spent for performing laser alignment can range from a few minutes to more than 5 hours. Items such as temperature changes in the room, thermal expansion of optics and mounts, and dust play a critical role in how smooth and long an alignment of a laser-based system can be. Whether manipulations are manual or computer controlled, a beam may disappear from the expected path and one has to find it and bring it back into the path. Please remember that laser alignment does not just mean working with visible beams only, but more commonly working with invisible beams. A variety of alignment aids will be needed.

The techniques for laser alignment listed below are to be used to help prevent accidents during alignment of this laser or laser system(s).

PREPARATION FOR ALIGNMENT

- To reduce accidental reflections, watches, rings, dangling badges, necklaces, and reflective jewelry are taken off before any alignment activities begin.
- Use of nonreflective tools should be considered.
- Access to the room or area is limited to authorized personnel only.
- Consider having at least one other person present to help with the alignment.
- All equipment and materials needed are present prior to beginning the alignment.
- All unnecessary equipment, tools, and combustible materials (if the risk of fire exists) have been removed to minimize the possibility of stray reflections and nonbeam accidents.
- Persons conducting the alignment have been authorized by the responsible individual.
- A NOTICE sign is posted at entrances when temporary laser control areas are set up or unusual conditions warrant additional hazard information be available to personnel wishing to enter the area.

ALIGNMENT CONSIDERATIONS

- Whoever moves or places an optical component on an optical table (or in a beam path) is responsible for identifying and terminating each and every stray beam coming from that component (meaning reflections, diffuse or secular).
- All laser users must receive an orientation to the laser use area by an authorized laser user of that area.
- There shall be no intentional intrabeam viewing with the eye.

Coaxial low-power lasers should be used when practical for alignment of the primary beam.

- Reduce beam power with neutral density filters, beam splitters, or dumps, or by reducing power at the power supply. Whenever practical, avoid the use of high-power settings during alignment.
- Laser-protective eyewear shall be worn at all times during alignment.
- Beam blocks must be secured (labeled if possible).
- Have beam paths at a safe height, generally below eye level when standing or sitting. If necessary, place a step platform around optical table.

The laser safety officer (LSO) has authorized eyewear with reduced optical density (OD) to allow the beam spot to be seen. Measures shall be taken and documented to ensure that no stray hazardous specular reflections are present before the lower-OD eyewear is worn. Full protection OD eyewear, as listed in the laser table, is to be worn again once alignment is complete. The reduced-OD eyewear is labeled as an alignment eyewear and is stored in a different location than the standard laser eyewear for this operation.

- Skin protection should be worn on the face, hands, and arms when aligning with ultraviolet (UV) wavelengths.
- The beam is enclosed and the shutter is closed as much as practical during course adjustments.
- Optics and optics mounts are secured to the table as much as practical. Beam stops are secured to the table or optics mounts.
- Areas where the beam leaves the horizontal plane shall be labeled.
- Any stray or unused beams are terminated.
- Invisible beams are viewed with infrared (IR)/UV cards, business cards, card stock, craft paper, viewers, 3 × 5 cards, thermal fax paper, or Polaroid film or by a similar technique. Operators are aware that such materials may produce specular reflections or may smoke or burn.
- Pulsed lasers are aligned by firing single pulses when practical.
- Intrabeam viewing is not allowed unless specifically evaluated and approved by the LSO. Intrabeam viewing is to be avoided by using cameras or fluorescent devices.

ALIGNMENT CONCLUSION

- Normal laser hazard controls shall be restored when the alignment is completed. Controls include replacing all enclosures, covers, beam blocks, and barriers and checking affected interlocks for proper operation.

FAILURE TO WEAR PROTECTIVE EYEWEAR

I would venture to say 99.99% of laser alignment accidents occur with the injured party not wearing laser-protective eyewear. The excuses are common, "I cannot find a pair that fits," "I cannot see the beam," "They are too heavy," "They cost too much," "I know the set up well enough I don't need them," and "I am using visible beams, I will see where the beam goes." Many of these have the similar tone to the ones first heard when seat belts became mandatory.

SOLUTION

The solution lies in two areas, selecting the correct eyewear and peer pressure. If the eyewear selected has a comfortable fit and provides good visual light transmission and protection, many reasons for not wearing eyewear is taken out of the equation.

As to peer pressure, if it is acceptable not to wear eyewear, the practice will continue. Supervisors and the laser-user community in an institution have to police each other.

ELECTRICAL HAZARDS OVERLOOKED

Laser systems and electricity go hand in hand, electric shock is the second most commonly reported incident associated with laser use. Power supplies, flash labs, batteries, capacitors, poor grounding, and accessible wires can be found in majority of laser systems. All need to be addressed and respected. The laser user needs to remember that a current between 6 and 20 mA is sufficient to bring about "Let go effect." Meaning you become part of the circuit and you cannot let go. Equipment circuit breakers are there to protect equipment, not the users.

ADDITIONAL ASSOCIATED HAZARDS AND SOLUTIONS

A long list of additional hazards is associated with laser work, pressure vessels, robotics, and so on. The hazards they present must be evaluated, eyewear does you no good if you are electrocuted or get cancer from agents in a laser plume.

IMPROPER RESTORATION OF LASERS AFTER SERVICING

While good work practice calls for any interlock bypass to be of such a nature the device cannot be restored back to service with them in place, this is not the practice. Items as thin as a paper clip or tape have been used to bypass interlocks, that is, micro switches.

SOLUTION

Involves checklist, for both the service person and user accepting the finished work. Both need to see that all safety systems are operational, (including interlocks and warning lights), that all enclosures and housings are secure.

LACK OF PLANNING

Planning is very important. We should not only have all proper tools available, but also use correct eyewear and set up warning signs and exclusion zone to protect others in the work area.

SOLUTION

Prereading of the procedures and developing checklists are the best cure for a lack of preplanning. Many times a lack of preplanning goes along with rushing, doing that quick task that has led to numerous accidents. Checklists should not only list the activities to be performed but also the required tools to start the task. The drawback of checklists is casual approach by the person or team using them. If an activity is routine, sometimes even when reading the checklist, people are not really reading the list they are anticipating the response. Every pilot knows they should not take off without going through a preflight checklist.

At times, this becomes so routine that expected responses are checked rather than the actual instrument readings. An excellent addition to a checklist is a digital photo of each step.

Checklist for developing a checklist or good questions to ask in making your checklist:

Define the task
    Define checklist's intended user and use
    Know the skill and training level of users
    Study relevant literature and techniques if available
    Discuss with experienced personnel or experts
    Clarify the function and task to be met by the checklist
Checkpoint list
    List descriptors for well-established criteria
    Briefly define each of the checkpoints
    Add descriptors for each checkpoint needed to fill in definition
    Provide definitions for each of the added descriptors
Sort the checkpoints: create a road map to be followed
    Place each descriptor on a separate card
    Sort cards by categories
    Identify the main categories and label
Flesh out categories
    Define each category
    Write rationale for each category
    Define steps of each category
    Present relevant warnings in case of error, which apply in each
        checkpoint
    Present any safety actions to be performed, that is, put eyewear on or
        position beam block
    Review checkpoints
Determine the order of categories
    Decide on order
    Write rationale for each order
    Review order
Initial review of checklist
    Prepare a review version
    Critique with potential users
    Listen to and understand concerns and suggestions, if not ask
        question for clarification
    List issues in need of attention or further development
Revise checklist
    Review concerns
    Rewrite checklist

Checklist format
    Determine with user-preferred format of checklist, its appearance
Evaluate checklist
    Perform a dry run with checklist, or with alignment laser
    Assess whether the checklist accomplishes task while meeting the
        users goal
    Example task may be accomplished but time to completion is excessive
Finalize the checklist
    Consider and address the reviews of the field test (dry run)
    Print off final version
Application of checklist
    Apply checklist for indented use
    Invite users to provide feedback on checklist developer
Review and update checklist
    Have a mechanism to update checklist as conditions change
An excellent addition to a checklist is a digital photo of each step

WEARING THE WRONG EYEWEAR OR IMPROPER FIT

There is not a universal pair of laser-protective eyewear. At least not until the virtual imaging helmet become available. Therefore, the user must pay attention to the specifications of the pair they are using. Is it designed for the wavelengths they may be exposed to? Is it at the appropriate OD to provide protection, full or partial (in the case of alignment eyewear)? Is the eyewear in good physical shape, has its protective features been comprised (scratches, bleaching)? Does it fit their face? Yes, wearing the wrong laser-protective eyewear has injured people (approximately 4% of reported accidents).

SOLUTION

Laser-protective eyewear is labeled or coded with its OD and wavelength range it is designed to protect you from. In some cases, it will indicate if the eyewear is full protection or alignment, continuous wave or pulse.

EYEWEAR FAILURE

I have always found this very curious, if one is dealing with laser beams that can burn through eyewear, and then why are not engineering controls in place to prevent one from being in a situation where direct facial exposure is possible? Other possibility is that the correct eyewear was not chosen; either OD was too low or the eyewears needed to be safety hardened, and was not.

SOLUTION

Inspect eyewear before each use. Look for scratches or signs of physical stress. Read labeling check for wavelength compatibility and correct OD. Evaluate their hazards from the beam; see if remote viewing is the proper safety solution.

FATIGUE AND STRESS

These two factors can make the most diligent laser user a menace to themselves and others.

FATIGUE

My mind clicks on and off. I try letting one eyelid close at a time while I prop the other with my will. But the effect is too much, sleep is winning, my whole body argues dully that nothing, nothing life can attain is quite so desirable as sleep. My mind is losing resolution and control.

Charles Lindbergh about his 1927 transatlantic flight

Although people associate fatigue with the feeling of tiredness, few realize how serious fatigue at the work place can be. In particular, when one is working around photons moving at the speed of light. With very few exceptions, the average adult needs between 7 and 8 hours of sleep within each 24-hour period to be adequately alert for the other 16 to 17 hours awake (that makes up commuting, work, off, and family time). Without the proper amount of sleep, the body cannot and will not function to its potential.

Symptoms of fatigue include a decreased ability to concentrate on multiple tasks, reduced ability to analyze new information, fixation, short-term memory loss, impaired judgment, impaired decision-making ability, reduced ability to think logically, daytime drowsiness and micro-sleeps, reduced motor skills and coordination, reduced ability to think critically, distractibility, reduced visual perception, loss of initiative, personality changes, depression, feeling of indifference to one's performance, and increased reaction time.

The majority of attention to worker fatigue has been directed to two large groups in the transportation industry: truck drivers, pilots and long-trip private driver. Others were second- and third-shift workers.

A study published in the Journal *Sleep* estimated 52.5% of all work-related accidents potentially may be related to sleepiness. After all the investigations were complete, these disasters were attributed to fatigue: the Exxon Valdez oil spill, the Chernobyl power plant explosion, the Challenger explosion, and Three-mile Island and Bhopal disasters. With the rapid increase in technological capability, it is not our machines that fail us, it is the operator. Human errors are the primary cause of 90% of all accidents.

One night of disrupted sleep probably will not result in huge catastrophes, but most people do not have just one night of disrupted sleep, night after night they try to get by on less sleep than their bodies actually need. Sleepiness builds into a sleep debt as the effects of inadequate sleep are cumulative! For example, assume that an adult needs eight hours of sleep each night but only gets seven. By the end of a week, there is a seven hour sleep debt, which is the equivalent of going one full 24-hour period without the proper amount of sleep—in college this is called *an all-nighter*. Now let's figure the sleep debt for an individual who only gets 6 hours of sleep each night (which seems to be more accurate for most Americans), at the end of the week, that sleep debt is 14 hours—or 2 *all-nighters*. Here are just a few effects of sleep deprivation: Bottom line: As fatigue increases your risk on causing an accident increases.

Solution

The solution is simple, but in many a work situation it seems as far away as the end of the day. Take breaks and proper rest. Remember whatever time one thinks is too valuable to break away and take a short rest is extremely long compared with time lost to an accident. Documented cases exist showing operations shut down from 1–7 days to 4 months due to accident investigations and obtaining authorization to operate again.

STRESS

Stress to get the job done, perform for others, and pressures from management can lead to lapses in good judgment, that is, the bypassing of standard control measures in general short cuts. It is taking these short cuts and bypassing standard operating procedures that can cause serious injuries to those that know the right way to do things; it is derivation of the saying *do as I say not as I do.*

What is stress? Finishing a last-minute presentation, getting data for my thesis, or being late for a date?

Webster's defines stress as a physical, chemical, or emotional factor that causes bodily or mental tension and that may be a factor in disease causation. Physical and chemical factors that can cause stress include trauma, infections, toxins, illnesses, and injuries of any sort. Emotional causes of stress and tension are numerous and varied. Although many people associate the term *stress* with psychological stress, scientists and physicians use this term to denote any force that impairs the stability and balance of bodily functions.

A mild degree of stress and tension can sometimes be beneficial. Feeling mildly stressed when carrying out a project or assignment often compels us to do a good job and to work energetically. Likewise, exercising can produce a temporary stress on some body functions, but its health benefits are indisputable. The negative effects of stress appear only when it is overwhelming or poorly managed.

Job stress can be defined as the harmful physical and emotional responses that occur when the requirements of the job do not match the capabilities, resources, or needs of the worker. Job stress can lead to poor health and even injury.

The concept of job stress is often confused with challenge, but these concepts are not the same. Challenge energizes us psychologically and physically, and it motivates us to learn new skills and master our jobs. When a challenge is met, we feel relaxed and satisfied. Thus, challenge is an important ingredient for healthy and productive work.

Some employers assume that stressful working conditions are a necessary evil—that companies must turn up the pressure on workers and set aside health concerns to remain productive and profitable in today's economy. But research findings challenge this belief. Studies show that stressful working conditions are actually associated with increased absenteeism, tardiness, and intentions by workers to quit their jobs—all of which has a negative effect on the bottom line.

Recent studies of so-called healthy organizations suggest that policies benefiting worker health also benefit the bottom line. A healthy organization is defined as one that has low rates of illness, injury, and disability in its workforce and is also competitive in the marketplace.

Solution

Stress solutions can range to re-setting goals to a realistic level, re-evaluating staffing levels and breaking tasks into smaller more manageable sections. Sometimes it is as simple as the realization that just like fatigue if stress is not recognized and dealt with serious accidents can occur, in addition to loss of productivity due to staff illness.

A corporate or workplace culture is—like any other culture—a set of behaviors and codes that people use to govern their interactions with each other. This includes both formal written company policies and informal *rules of the road* that you learn with experience. A workplace culture that allows 12 to 16-hour days and constant pressure to produce will have a history of accidents and employee health problems. In academic and medical settings, the culture is well established that allows and accepts overworking and undervaluing graduate students and residents. Is it any wonder that this population (graduate students) make up such a large portion of those involved in laser accidents and near misses?

Changes in workplace culture must come from the top. Risk management system needs to show top management how much time is lost in sick days and actual productivity with such a culture. The quality of work performed and creativity is the goal, not the number of hours worked.

TO WHOM ARE THESE ACCIDENTS HAPPENING?

Many of the laser accidents do not happen to novice laser users, rather experienced users. Usually, when those with little laser experience are exposed they are similar to the drunk driver victim, the innocent by-stander. Approximately 70% of reported accidents involve the following four groups:

Scientists
Students
Technicians
Patients (Number 1 group for reported deaths)

The remaining 30% of the reported incidents are made up of the following groups:

Industrial workers
Doctor and nurses
Pilots and military personnel
Spectators
Laser show operators
Field service staff
Office staff

THE PERCEIVED INCIDENT

Sometimes the investigation will be due to the perception of an injury. Even if you believe that no exposure occurred, the investigation has to be handled in a professional manner. For the perception, incident might demonstrate a flaw in the safety system, at the very least in training. If the investigation is handled in a manner that makes the reporting party feel as if their concerns are not being treated with respect, future incidents will not be reported. Worst, you may be skipped and they may be reported to a higher level or regulatory agency, all of which may damage the effectiveness of the program.

REAL LASER ACCIDENTS

For a very detailed listing of laser accidents, see *Laser Safety Management*, Chapter 9, Ken Barat, CRC Press. For those who have not yet bought my first book, the next section lists some new accidents:

The rest of this chapter is a random selection of documented laser accidents. I would like to spend a little more time on an incident that contains several of the elements mentioned above as well as an *I do not believe it* factor.

ULTRAFAST LAB ACCIDENT

Basic facts:

1. A student with little ultrafast laser experience is working in a lab as a user.
2. Not a member of the research group.
3. Receives 15 minutes of on-the-job training (OJT).
4. Performs an 800-nm alignment activity without the use of available eyewear (this is how he has observed others working).
5. Key optic is rotating polarizer and reject beam.
6. Has no direct supervisor.
7. Because of items 2 and 6, no one reviews his work plan.
8. When the alignment was completed, he rotates the tube with the beam splitter attached to polarizer to line up one hash mark with zero on polarizer, just because he thinks it looks better that way.

Investigation and additional background material:

- Polarizer-1 with escape window
  Beam tube with escape window
- Rotating mount

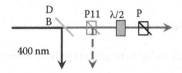

Initially, beam tube was secured to block the γ-ray, which is not used in experiment.

Laser operator wanted P1 polarizer to be orthogonal to incident laser beam and removed eyewear to co-align back-reflected beam from polarizer with incident beam. To do this, operator loosened polarizer and associated beam tube in the rotating mount.

Operator then wanted to get polarizer rotation angle oriented to match 0° marking on rotation mount. Accidental exposure happened when operator rotated polarizer in a way that the escape windows of the polarizer and beam tube aligned, permitting the γ-ray to strike their unprotected eye.

Possible solutions to this incident?

Engineering Controls:

Use a beam tube with no escape window.

Avoid mounting polarizers in rotating mounts, unless absolutely necessary.

Instead, use half waveplates to rotate polarization vector and keep both transmitted and reflected beams in horizontal plane.

Administrative Procedures:

For aligning polarizer normal to beam, three methods can be used with full protection eyewear:

IR sensor card with a hole in it; incident beam goes through hole and view back reflection on sensor card as polarizer pitch and yaw are adjusted.

Cut an IR card so as to have a straight edge of sensitive IR card close to the beam.

Use an iris and IR viewer to see back reflection on the iris.

For performing optics adjustments when there is no need to observe the beam, block the beam upstream.

Not necessary to have polarizer axes oriented at 0° on rotation mount, so do not perform unnecessary optics adjustments where laser beams are present.

Personal Protective Equipment:

Use of full protection eyewear as required by the standard operating procedure.

Accident investigation came up with seven root causes:

1. Inadequate training, in particular OJT
2. Inadequate supervision
3. Inadequate work planning and control
4. Inadequate adherence to requirements
5. Deceptive hazard of a dimly visible beam
6. Inadequate appreciation of out-of-plane beam hazard in using a polarizing beam splitter
7. Inadequate intervention following (prior) safety violations of personal protective equipment

Item 6 is very important that a little more needs to be said as it has been at the center of many a laser injury. *100 mW at 700 nm (nearly visible)* will appear to have similar brightness as *1 mW at 550 nm (green-yellow)*.

Now back to brief summaries of laser incidents.

From the Sky above Us

January 8, 2005

At 10:52 PM EST, an America West Airlines aircraft (AWE 744) was illuminated by a green laser light while at 8000 ft. on approach to Washington Dulles International Airport (Sterling, VA). The cockpit was illuminated for about 5 seconds, after which the pilot experienced annoyance and a very short afterimage.

January 25, 2005

At 7:30 PM PST, a Delta Airlines aircraft (DAL 446) was on approach to Runway 27 at the San Diego (CA) International Airport, when a green laser light flashed the pilots about four times. The pilot reported annoyance, but no visual impairment.

January 28, 2005

At 7:48 PM PST, a United Postal Services aircraft (UPS 905) was illuminated by a green laser light while enroute at 15,000 ft. MSL after departing from John Wayne Airport, San Ana, CA. The exposure lasted 1–2 seconds, and the pilot reported experiencing glare and afterimages lasting 1–2 minutes.

January 29, 2005

At 6:59 PM EST, a Mesaba Airlines aircraft (MES 3253) was illuminated by a red laser light while at an altitude of 2100 ft. on 5-mile final approach to Runway 21L at the Detroit (MI) Metropolitan Wayne County Airport. The pilot reported slight flash blindness and annoyance from the 2- to 3-second exposure.

May 14, 2004

At 9:30 PM PST, a general aviation aircraft was descending for approach from 9000 to 7000 ft. above mean sea level (MSL) near Carlsbad Airport, CA, when the pilot was hit with a green laser light. There was an apparent deliberate intent to track the aircraft for 10 miles to the 2-mile final approach to Runway 25. The pilot reported distraction but no visual impairment.

September 21, 2004

At 8:25 PM CST, an American Airlines aircraft (AAL 576) was on a 4-mile final to Chicago (IL) O'Hare International Airport at an altitude of 2400 ft. when the pilot experienced a laser light illumination of the cockpit. The pilot reported only distraction but no visual impairment.

September 22, 2004

At 9:30 PM MST, a Delta Airlines B-737 aircraft (DAL 1025) was on final approach to Runway 35 at Salt Lake City (UT) International Airport at approximately 1300 above ground level (AGL; 5500 ft. MSL), when a green laser light hit the flight deck. The first officer (FO) was flying the aircraft at the time of the incident. Both captain and FO saw the bright green light coming from the ground at the 1 to 2 o'clock position for a period of about 5 seconds. The captain recognized the light as a laser and turned his eyes away while the FO did not. The FO landed the aircraft safely, but reported some loss of depth perception, causing him to flare too high. After the flight, the FO noticed blurring in his eyes. The next day his vision was more blurred. An ophthalmologist's examination found retinal edema. The FO was unable to fly for about 3 weeks and remained sensitive to bright lights for some time.

December 25, 2004

A Sky West Airlines aircraft (SKW 6166) was illuminated by a laser light approximately 8 miles from Rogue Valley International Airport, Medford, OR. The light initially started as a pulsating white light and changed to a steady green light. The pilot in command reported some eye discomfort but was able to land the plane safely.

December 29, 2004

At 5:35 PM EST, a Cessna Citation, flying at approximately 2500 AGL, was illuminated with a green laser near Rockaway, NJ, causing a temporary loss of vision for both the pilot and co-pilot, but they were able to safely land the plane. Two days later, one of the Cessna pilots joined federal agents in a police helicopter that was investigating the incident. The cockpit of the helicopter was illuminated with a laser light while flying in the vicinity of the previous illumination. The authorities were able to identify the location of the laser light, and an individual was arrested for both incidents. The suspect later pleaded guilty in court to interfering with the pilots of a passenger aircraft.

December 30, 2004

At 5:35 PM EST, a Northwest Airlink/Pinnacle Airlines aircraft was on final to Runway 4 at the Cleveland-Hopkins (OH) International Airport at an altitude of about 4000 ft. when a green laser hit the aircraft and tracked it for about 3–4 seconds. The pilot reported annoyance from the light, but no visual impairment.

December 31, 2004

The pilot of a general aviation aircraft departing from Wiley Post Airport in Oklahoma City, OK, reported that he was hit in the face by a green laser three times. Although the light temporarily blinded him, he was able to continue the flight.

December 31, 2004

An Anne Arundel County, MD, police helicopter was *lazed by a high-intensity green laser* that interfered with cockpit visibility. The pilots determined that the laser came from a group of people standing near the scene of an accident. A 38-year-old man was arrested and charged with reckless endangerment.

December 31, 2004

At 7:26 PM PST, an American Airlines B-757 aircraft (AAL 2083) was illuminated by a green laser light while at an altitude of 4000 ft. on final approach to Runway 25L at Los Angles (CA) International Airport. The laser light tracked the aircraft for about 3 miles, and the FO reported vision problems similar to a camera flash and mild distraction.

Background on Exposure to Pilots

It is worthwhile to note that brief exposures to low-level laser radiation are more likely to result in temporary visual impairment. The severity and duration of the impairment varies significantly, depending on the intensity and wavelength of the light, the individual's current state of light (or dark) adaptation, the use of photosensitizing medications, and even the person's skin pigmentation (eye color). When the human eye is dark adapted (scotopic vision), it is more sensitive to light at 507 nm, while the light-adapted (photopic vision) eye perceives yellow–green light (555 nm) more vividly.

Temporary visual impairment is associated with adverse visual effects that include glare (a temporary disruption in vision caused by the presence of a bright light within an individual's field of vision); flash blindness (the inability to see, caused by bright light entering the eye that persists after the illumination has ceased); and after-image (an image that remains in the visual field after an exposure to a bright light). Although none of these visual effects cause permanent eye injury, the distraction, disorientation, or discomfort that often accompanies them can create a hazardous situation for pilots performing critical flight operations.

In the 1990s, several incidents of illumination of flight-crew personnel were attributed to light from laser demonstrations designed for public amusement or attractions. Subsequently, in 1995, the Federal Aviation Administration (FAA) established flight-safe exposure limits for lasers projected into specific zones of navigable airspace to protect aircraft operations around airports. Development of these exposure limits relied on existing scientific research, along with consultation with industry and governmental laser experts and aviation safety personnel. The resulting guidelines were published in FAA Order 7400.2, *Procedures for Handling Airspace Matters*, Chapter 29, Outdoor Laser Operations (8). The order identifies three zones of protected airspace around airports and assigns specific exposure limits to these zones. Within these zones, laser emissions above that which could cause vision impairment and interfere with normal flight operations are prohibited. These zones include the laser-free zone (50 µW/cm^2), critical flight zone (5 µW/cm^2), and sensitive flight zone (100 µW/cm^2). Incident reports collected by the Civil Aerospace Medical Institute's (CAMI's) Vision Research Team indicate that implementation of these guidelines has resulted in a substantial reduction in accidental exposure of pilots to entertainment and demonstration laser light shows. However, these procedures are only useful when laser proponents comply with applicable guidelines and voluntarily notify the FAA of proposed outdoor laser operations. No order or regulation can prevent thoughtless individuals, criminals, or terrorists from using lasers to interfere with the operation of law enforcement and emergency medical evacuation helicopters or private and commercial aircraft. A database of laser exposure incidents has been maintained at CAMI for nearly a decade. As a result of numerous incidents that occurred in the fall/winter of 2004, there was renewed interest in outdoor laser operations and their impact on aviation activities.

- The so-called *Laser-Free Zone* (LFZ). This is a distance of from 2 to 5 nautical miles around an airport where aircraft should not be exposed to light above 50 nW/cm^2 (the same as 0.05 μW/cm^2). This very low light level, 50 billionths of a watt, "should not cause any visual disruption" according to the FAA.
- The critical flight zone (CFZ), a cylinder encompassing a radius of 10 nautical miles around an airport, up to 10,000 ft. vertically. Light levels should be below 5 μW/cm^2. This will "avoid flashblindness or afterimage effects."
- *Sensitive Flight Zones* (SFZ), with light levels below 100 μW/cm^2. These are optional. Local FAA personnel and the military can designate these zones as needed. An example would be military airspace or a particular air route.

All other areas which are not in the LFZ, CFZ, or SFZ are designated as *normal flight zones*. Within this zone, the laser power must not be an eye hazard. (Technically, power must be below the maximum permissible exposure, the *level of laser radiation to which a person may be exposed without hazardous effect or adverse biological change in the eye or skin*.)

CDRH laser incident reports (CDRH Database Web site)
ENDO laser
Model: Eyelite
Problem: Unintended laser output

The customer reported that during surgery procedure with endo laser in repeat mode, the laser fires twice and stops. When they release the footswitch, the laser fires a third time. Depressing the footswitch again, the same thing happens. They proceeded for 30–40 shots, and then the procedure was aborted. They switched to another unit. Patient outcome was not provided. Additional information has been requested.

TATTOO INCIDENT

MEDLITE C6

Tattoo remover was treating a tattoo and switched the laser wavelength from 1064 to 650 nm by attaching the 650 nm dye conversion handpiece to the end of the articulated arm. She had the patient change the laser safety eyewear to the correct 650 nm glasses, but forgot to change her eyewear to the correct wavelength. When she stepped on the foot switch, she noticed that the 650 nm light was very bright and immediately stopped the laser by releasing the foot switch. She indicates that she saw about 5 flashes before stopping the treatment. After this, she experienced a headache but had no problems with her vision. I advised her to see a physician to examine her eyes for the evidence of damage to the eye. She indicates that the ophthalmologist found no damage and her headache went away the following day.

FEEDBACK FROM MANUFACTURER

Reporter acknowledges that she forgot to change her eyewear to the correct wavelength. She indicates that she was trained on the use of the laser and that the instruction included the necessity to change the eyewear when using the polymer dye conversion handpieces. The manual clearly states in several places to wear the correct eyewear when installing the dye conversion handpieces. The adverse event was not caused by the failure of the system or from inadequate training or instructions for use and hence no corrective action is being taken or is necessary.

Case 1: Medlite C6

According to the allegedly injured party, in 2006, a laser technician practitioner was performing a procedure using a Medlite C6 laser at 6 J/cm^2, 1064 nm, 10 Hz with a 4 mm spot size while compressing the skin with a glass window to force the blood away from the treatment site to minimize the formation of purpura. She reports that during the procedure, she noticed bright spots that caused her to blink. After the procedure, she noticed that there continued to be a *black* spot or shadow in her central vision similar to what happens after you look directly into a light bulb or the sun and then look away. She further reports that 3 days later, she went to see an ophthalmologist who examined her and found bleeding and referred her to a retinal specialist. We have not yet received her formal medical records. Sixteen days later, Ms. Pham sent an e-mail to an independent contractor working with Hoya ConBio, informing her of the adverse event. The independent contractor reviewed the e-mail for the first time and reported the alleged injury to Hoya for the first time after 5 days.

MANUFACTURER RESPONSE

Device not evaluated as there is no defect or problem with the device. The event may be related to a lack of eyewear or the use of incorrect eyewear. The eyewear provided with the C6 laser system is to protect for use with the 1064 and 532 nm wavelengths. This eyewear has a slight amber tint. The reporter requested that a Hoya representative send her a clear set of eyewear without the amber tint. Hoya sent clear eyewear to her, but sent eyewear rated only for 2940 nm. On receipt of the eyewear, she did not check the rating on the eyewear prior to using them. She used them for several months prior to November 2006. She claims to have been wearing this eyewear in 2006 when she allegedly received a back reflection off of the surface of the compression window she was using. The 2940 nm eyewear was removed from the treatment room immediately after the event and replacement eyewear with the correct protection was sent to the institute on December 20, 2006.

Palm burn Catalog Number 840–846

During a surgical case using holmium laser, the single-use 550 duo tome fiber burned through its fiber and insulation causing a burn to the surgeon's palm. The event occurred approximately 20 minutes after the laser/fiber delivery system was in use. There was no harm to patient.

ELECTRICAL SHOCK

Our service technician was working on a Medlite C6 laser at the office of doctor. The laser was reporting error 22, which is *no end of charge* meaning that the HV capacitor is not getting fully charged and the flashlamp is not flashing. He evaluated the system and found the simmer supply was working, the lamp was simmering, but the lamps were not flashing. He decided to replace the HV power supply. He reports that he turned the system off, unplugged the system, and laid down on the floor to remove the power supply. He disconnected the HV cable from the supply that goes to the HV capacitor, the ac input cable to the supply and the last thing he remembered was removing the control cable to the supply and then he received the shock. The office staff heard a loud bang from the room, found him bleeding from the ears, nose, and mouth, and they called the building manager who tried to resuscitate him. The emergency response team resuscitated him and took him to the hospital for treatment.

Manufacturer's Response

They submitted a report on the site evaluation of the system and interviews with the office personnel and with service technician. The HV power supply, silicon-controlled rectifier (SCR) board, and the charge capacitor were returned from the system for further evaluation: the SCR board is intact, the high voltage discharge relay and discharge resistors are intact, and the resistance that was connected to the charge capacitor to discharge the capacitor measured 33 kΩ. This is consistent with the design that has 3100 kΩ resistors in parallel. The HV power supply was opened, and we found that the bridge rectifier on the input ac was burned and all diodes were shorted, which would not allow the power supply to develop an HV output. This is consistent with the report which indicated that when they turned on the system, an error 22 was reported and no voltage was developed on the charge capacitor. In the report, service technician indicated when the system would not flash and was reporting an error 22; he also noticed a burning smell. This is consistent with the burned input bridge rectifier in the power supply. The capacitor was received in good condition. There are some evidences that the threads on one of the posts were damaged, not cross threaded, but the post had some rough spots on the threads. This is consistent with the observation in the report that the nut on the wire going from the charge cap to the SCR board was slightly loose. This is the path to rapidly discharge the capacitor. Note the attached spreadsheet on discharge times for the primary discharge path and the backup discharge path. It was observed that you could move the wire on the post, which was not tight, but it also was not sloppy loose. The diameter of the ring lug on the wire that goes on that post is very close to the diameter of the post so there is little chance that the loose wire could have been really disabled from discharging the capacitor. The backup discharge resistor on the capacitor measures 2.2 MΩ, which is consistent with the design. None of the evidence identifies defective components, parts, or design that would cause the accident to happen. The only possible explanation is that the power supply was able to deposit a charge on the capacitor as it was failing. In addition, the loose nut on the capacitor broke the connection to the fast discharge resistors on the SCR board; as a consequence, the backup bleed resistor was discharging the partially charged capacitor when service technician contacted an HV point when he was removing the low voltage cables from the power supply. The only conclusion is if he had discharged the HV capacitor as he had been recently instructed to do, this accident would not have happened even if one of the system safety discharge circuits was inoperative. Because of the serious nature of the incident, they have taken additional preventative action steps to impress on the service personnel the importance of following a specific safety regimen when working in and around HV. The service bulletin will be incorporated into the service manual, in addition to being sent to the service personnel. Preventive action: a service bulletin has been generated, see attached bulletin 47, which has been e-mailed to all the service engineers and distributors worldwide. This bulletin specifically addresses the safety precautions that must be observed when working in and around HV components. It also states that a minimum wait time should be observed before attempting to discharge the HV capacitor to allow the backup bleed resistor to discharge the capacitor to minimize the risk if the capacitor is not fully discharged. This service bulletin will be sent via UPS with a return receipt tracking requested to verify that all personnel and distributors have received the information. The manufacturing and engineering personnel will be trained in these safety procedures as well.

Case 2: Nd:YAG

A picosecond Nd:YAG pulsed laser operating at 1064 nm was on a laser optics table. The beam was directed from one table to another across an isle. The beam went onto the second table, where it was directed onto a liquid-sample holder. Here, apparently, the beam was bigger than the liquid-sample holder, so the edges of the beam went past the sample bottle and then off that table into the room area where a strip chart recorder was located. A graduate student working on the experiment looked at the strip chart recorder and received about 10% of the beam into the eye. The student reported he "heard a popping sound" which was followed by a white spot in the vision center. The professor took the student to an ophthalmologist for a retinal exam which confirmed the burn exposure. The student did not experience shock. The beam caused a retinal burn. The student now complains that his "eyes get tired" while reading.

Case 3: Electrical Laser Accident

A researcher was working on a home-built 5-W CO_2 laser, replacing and tuning a new tube. The system was obtained from another group. The researcher was unaware that the end caps were energized and, during the tuning process, he made contact with the palm of his left hand on the end cap and his thumb on the (ground) support bar. He received a painful shock from approximately 15,000 V at 20 mA dc. The researcher was treated for injury to his hand and wrist, and the pain (severe) lasted several days. Once again these non-beam hazards can be the deadliest and tend to sneak up on one, for the user is not focusing on them while involved in other activities. This is especially true when working on equipment one is not completely familiar with.

Case 4: Ti:Sapphire Again

A 26-year-old male student aligning optics in a university chemistry research lab using a *chirped pulse* titanium–sapphire laser operating at 815 nm with 1.2 mJ pulse energy at 1 kHz. Each pulse was about 200 picoseconds. The laser beam backscattered off REAR SIDE of mirror (about 1% of total) caused a foveal retinal lesion with hemorrhage and blind spot in central vision. A retinal eye exam was done and confirmed the laser damage. The available laser-protective eyewear was not worn.

Case 5: Technician Injured

A laser lab technician was working without laser-protective eyewear. He was exposed to a single 7 nanosecond pulse at a pulse energy of 10–50 μJ. In the setup, the beam was directed onto a metal *test slide* from the power meter manufacturer. This was used to test whether the beam would harm the power meter. The slide was accidentally tilted so as to reflect the beam into technician's eye (assume about ~4% reflection). At the time of exposure, the person perceived a bright flash that persisted (with eyes closed) as if he had looked at the sun. There was no pain nor did the person go into shock. Eyewear was available but not for the 806 nm wavelength in use.

15 Room Access Interlocks and Access Controls

John Hansknecht

CONTENTS

INTRODUCTION

In a perfect world, all persons working with or around laser systems would be aware of, and respect, every hazard. In this perfect world, every laser system would be enclosed for Class 1 operation, and opening the door to a room containing a laser would not jeopardize the safety of personnel in the hallway. In many cases, the laser safety officer (LSO) can work with the researchers to actually create this perfect world. In other cases, the nature of laser research requires a flexibility that enclosed laser systems cannot provide. In either case, the LSO will need to understand the principles of laser interlocks and access controls.

By definition, the LSO is granted the authority to monitor and enforce the control of laser hazards. This can place the LSO in an uncomfortable position. An LSO has the authority to implement the safety controls, but may have little experience in electrical safety engineering. When he or she mentions the word *interlock*, the laser users will often put up their defenses with thoughts of intrusion, sacrifice, and expense. The intrusion is the feeling that the LSO is placing new demands on the users that they have thus far avoided. The sacrifice is a thought that they must give up flexibility when implementing the safety interlock system and a fear that their experiment will be ruined when the system is tripped. The expense is the monetary cost of the system that takes away from funds that can be used for other research components. These concerns are valid, but not impossible to overcome. It is rare for a newly introduced safety protocol to be greeted with open arms; however, time after time we have observed cases for which the first room interlock installation was fought tooth and nail by the researcher, but after gaining experience with a well-designed system, the other researchers were waiting in line for their own systems to be installed.

BUDGETING FOR ENGINEERED SAFETY

The safety dollar is a rare coin in many facilities. Given this limited resource, the LSO should make every effort to spend it wisely. At the same time, the LSO should realize that the researcher who complains about not having $3000 to $5000 for enhancing safety will easily spend that same amount on a new laser power meter or a single optics translation stage. It really just boils down to priorities.

If a facility wants to use lasers, it must also budget for their safety. When possible, the funding for safety should be placed in the control of the LSO for dedicated safety expenditures. This funding may indeed be taken off the top of funds that were granted to a researcher, but in doing so, the LSO needs to actually spend this money for the benefit of the researcher. This benefits the LSO, the laser researcher, and the facility. The following case is a typical scenario.

Case 1

The LSO approaches Bob (the researcher) and informs him that the facility is adopting a new safety standard. "Bob, to comply with this standard, you're going to have to buy laser eyewear, laser safety signs, curtains, and an access control system. I'll be back in 1 month to check on your progress."

Obviously, this is creating an adversarial relationship between the LSO and the researcher. Even if worded differently, the message is the same. It is also fostering an unhealthy attitude about safety. Every time the researcher hears about some new safety protocol, the first response will be to think, "How much is *this* going to cost me?"

Case 2

The same LSO approaches Bob and informs him of the new safety standard. "And here's what *I'm going to do for you.* To get this going, I'm going to buy you some new eyewear, safety signs, curtains, and an access control system. I'm going to arrange to have this installed and train your people. If you ever need a new sign or safety label, just call me, and I'll have one printed for you. If you ever need new prescription safety eyewear, do the same."

Do you see the difference? This is, of course, an idealized case, but this approach will make the safety pill *much* easier to swallow. It is important to realize that the largest safety expense is an up-front onetime expense that will likely last for decades, so a safety budget should be padded at the front end to account for this.

CREATING THE *PERFECT WORLD,* ONE LASER AT A TIME

Clearly, whenever possible, it is preferable to create a Class 1 enclosure around a laser system. Once enclosed, the laser can operate day in and day out without a need for external safety protocols, protective eyewear, warnings, and so on. The up-front expense of the enclosure may seem high, but the long-term benefit and risk reduction make it a wise investment.

The availability of modern extruded aluminum construction hardware makes the design and construction of an enclosure quite simple. One name brand product known as 80/20®, produced by 80/20 Incorporated, uses the registered slogan: *The Industrial Erector Set.* Indeed, this product can be used by the everyday researcher to build an enclosure without the need for cutting, drilling, or welding. The manufacturer provides hardware cut to the length specified, and an enclosure can be assembled with a simple hex screwdriver. Anodized aluminum panels slide effortlessly into the slots of the extruded structure, forming a light-tight enclosure. The construction of an enclosure does not completely alleviate the responsibility of meeting other safety standards. There are labeling and simple panel interlock requirements that must also be fulfilled. These are generally well within the capabilities of the end user. Of course, there will be scenarios for which the power levels of the laser system exceed the burn-through characteristics of the aluminum. There are laser consultants available to help with these cases.

The large rectangular enclosure that completely surrounds an apparatus is not always the best design. As Figure 15.1 shows, a high-power laser-manufacturing apparatus can be created, which provides Class 1 safety but allows a product to move freely through the process while personnel are free to work in the immediate proximity without danger of exposure.

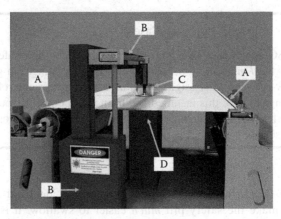

FIGURE 15.1   An industrial Class 4 laser operating in Class 1 conditions. A, Product delivery and take-up spools; B, enclosed laser and beam delivery tubes; C, a shroud made of a safety filter material that completely blocks diffuse and specular reflections above MPE levels; D, a beam dump immediately beneath product. Beam is safely dumped even if product is removed.

INTERLOCKS

DEFINITION

The term *interlock* refers to a hardware device that, when activated, will immediately reduce the laser emission below maximum permissible exposure (MPE) levels. This device may be a mechanically or electrically activated shutter, or it may be an electrical trip of the laser system power source. The interlock is often broadly defined as a room interlock system, by which access controls, door or floor sensors, or motion detectors are interconnected to cause a rapid reduction in laser emission below the MPE when activated.

The interlock is not always obvious. For example, a technician removing an interior cover plate of a common DVD player may release a spring-loaded switch that prevents operation of the laser. This is a hardware-engineered safety interlock designed to protect the technician who has reached the dangerous portion of the machine, but it typically goes unnoticed by the average consumer.

REASONS FOR INTERLOCKS

When dealing with Class 3B and Class 4 lasers, the danger is real, not just a sign on the door. The American National Standards Institute (ANSI) Z136.1 (Section 4.3.9.2, p. 31) definition of the use of the Danger sign reads: "'DANGER' indicates an imminently hazardous situation which, if not avoided, will result in death or serious injury. This signal word is to be limited to the most extreme conditions."

All too often we find the word is overused, leading to a false sense of the meaning and a relaxed attitude. One could even argue that the use of the word is overprescribed within the ANSI standard itself. Clearly, there are differing levels of danger. Something can be dangerous when misused, such as the common laser pointer, or it may be extremely dangerous in its normal operation, such as a high-power laser research experiment. Unfortunately, we have only one danger posting.

Engineered safety interlocks and entryway controls are a method of ensuring that seriously dangerous systems (Class 4) are contained and respected. Proper implementation of the interlocks will limit the laser access to qualified, trained individuals and help prevent harm to those who are not properly trained on the operation and safe use of the laser.

One of the most difficult tasks an LSO will perform is the accurate calculation of the nominal hazard zone (NHZ) for a laser system. To make matters worse, the laser layout in a research or an educational setting can change with every experiment. The release of ANSI Z136.1-2007 has now simplified this task. In lieu of performing the calculations, the LSO can simply declare the entire room as a laser area and provide adequate controls at the perimeter of the room. These controls are most often implemented using interlocks and engineered access controls.

LASER MANUFACTURING REQUIREMENTS

The U.S. Food and Drug Administration's Center for Devices and Radiological Health mandates are released as federal code in Title 21 of the *Code of Federal Regulations* (21CFR). The subsections of this title list the required elements to which manufacturers must comply if they are offering a product for sale in the United States.

The following is a summary of 21CFR chapters related to laser products:

21CFR1000–1005: Broad-scope list of devices regulates the records and reports that must be produced and retained by a manufacturer and the import/export requirements.

21CFR1010: General performance standards, certifications, and variances.

21CFR1040.10: Performance standards for manufacture of laser products. This is where one will find the information pertaining to the laser housing interlocks, remote electrical interlocks, labeling, and power classification. In addition to providing the required information for manufacturers, this standard is a useful reference for the end user when the user is building a Class 1 enclosure around Class 3B or Class 4 laser systems.

21CFR1040.11: Performance standards for specific-use laser products. Medical, survey, leveling, alignment, and demonstration laser products must meet criteria listed in this chapter as well as all criteria listed in 21CFR1040.10.

These chapters are collectively known by the acronym FLPPS, which stands for Federal Laser Product Performance Standards.

All federal regulations are provided free of charge to the public by the federal government. They can be viewed and printed from the Internet. A searchable database can be found at http://www.accessdata.fda.gov/scripts/cdrh/cfdocs/cfcfr/ cfrsearch.cfm.

FLPPS REQUIREMENTS SPECIFIC TO THE REMOTE LASER INTERLOCK

Every commercial Class 3b and Class 4 laser sold in the United States shall include a remote electrical interlock connection. The connection will have two terminals that need to be electrically shorted to each other. If the circuit opens, the laser will not emit a laser radiation hazard above the MPE threshold. The manufacturer is given leeway in choosing the method to be used to accomplish this task. The manufacturer may use an electromechanical shutter or electro-optical modulator behind the laser housing, or may choose to electrically shut down the power supply to the laser. The only real constraints are that the device must be fail-safe and the voltage used for this remote interlock circuit shall be less than 130 V. Although it is unlikely that a modern manufacturer would use this voltage, one needs to be aware of the potential shock hazard that may be present at this connector. If you are unsure of the potential on a particular laser, consult the manufacturer. The fail-safe requirement is an important feature. When a safety circuit has tripped the laser off, the laser must remain off until it has been restarted by an operator even if the safety circuit trip was momentary.

Lasers or Laser Systems Manufactured for *In-House* Use

Every effort should be made to meet the intent of FLPPS when a facility is constructing a laser or laser enclosure for its own use. The FLPPS is intended for the safety of the end user. With the increased availability of high-power laser diodes, any person with a basic understanding of electronics can construct Class 3B or Class 4 lasers without any knowledge of laser physics or laser safety. This places an increased emphasis on the LSO to monitor and enforce laser safety issues.

A facility that fully adopts the ANSI Z136.1 standard is effectively adopting FLPPS. A majority of the FLPPS requirements are repeated (almost verbatim) in Section 4.3 of the ANSI Z136.1 standard. This repeat of the FLPPS requirements within the ANSI standard provides beneficial safety information but can lead to confusion when trying to implement end user–engineered safety controls. ANSI Z136.1-2007 Sections 4.3.1, 4.3.2, 4.3.3, 4.3.4, 4.3.5.1, 4.3.7, 4.3.8, and 4.3.14 are FLPPS specifications that are expected to already exist as a performance feature in a commercially certified product. These items should be verified to exist on receipt of a laser and should be checked periodically to verify their functionality. If the laser or laser system is built in-house, compliance with these ANSI sections is an expectation.

End-User Interlocking Requirements

There is a legal mandate in FLPPS to provide an external interlock on all commercial Class 3B and Class 4 lasers. It is somewhat interesting that there is not a federal legal mandate for the end user actually to use it. In fact, the interlock connector will most likely come from the manufacturer in an electrically shorted state. This is unfortunate because the lasers are often placed into service without consideration of a proper interlock system configuration.

Why is the use of a remote interlock not mandated? There are thousands of possible laser applications. Many applications would not require the connection to an external interlock control system. Instead of mandating the use of an interlock, the ANSI Z136 series of standards provide several methods of instituting an engineered laser safety program.

ENGINEERING CONTROLS

ANSI Z136.1-2007 (Section 4.1, p. 24) states: "Engineering controls (items incorporated into the laser or laser system or designed into the installation by the user) shall be given primary consideration in instituting a control measure program for limiting access to the laser radiation."

Engineered laser safety concentrates on three specific areas:

1. Physical protection from the hazard in the form of curtains, enclosures, or barriers
2. Visual warning of the hazard in the form of signs and electronic warnings
3. Engineered area or entryway safety controls to control access and trip the interlock if needed

PHYSICAL PROTECTION FROM THE HAZARD

Physical protection from a laser hazard is the first line of defense for personnel protection. If a laser system can be operated in a closed environment without the possibility of exposure above the MPE, then the system is Class 1 safe. The most common example of this is the DVD-RW drive in a normal computer. Although it contains a Class 3B laser with powers up to several hundred milliwatts, its operation is safe and requires no outwardly noticeable warnings or interlocks. The designers of this device have ensured that there are no conceivable methods for exposure to the hazard. The same approach can be extended to larger laser systems in the laboratory research, medical, and industrial environments. Unfortunately, as the physical size of a system grows, the risk of human exposure becomes more likely. Service access panels require interlocks, windows require attenuation filters, and curtains require damage threshold analysis. Although it is possible to create large Class 1 laser systems, sometimes the only reasonable approach is to designate a laser room as a laser-controlled area.

In this approach, the perimeter of the room becomes the enclosure. All personnel within the room would wear laser-protective eyewear, and personnel outside the room would not be at risk for exposure. Ultimately, the perimeter of the room must still be evaluated for laser safety with regard to the possibility of generating toxic fumes, burning, or even worse, wall penetration. When designating an entire room as a laser area, one must also ensure that the hazard does not extend outside the door if opened. Several methods can be used to control this hazard, including curtains, barriers, or electronic interlocks. If ordinary drywall is acceptable for the perimeter of the room, then a simple drywall labyrinth at the entryway is an option. The materials chosen to perform these functions must be evaluated for flammability and decomposition product toxicity. Material selection for protection also varies widely with laser wavelength. As an example, many people are surprised to find that common glass can be used as a screen attenuator for a CO_2 laser hazard up to certain power levels. For high-power systems, the process of selecting a barrier may require expertise beyond the capabilities of the LSO and the local researchers. In this case, laser curtain manufacturers should be able to provide recommendations. For moderate- and low-power Class 4 systems, the barrier selection can be as simple as verifying the breakdown resistance of a given material in a controlled manner with the laser in question. Any tight-weave, nonflammable fabric may suffice for a laser curtain if it has been properly evaluated for damage and permeability.

Although an LSO may choose to designate an entire room as an NHZ, every effort should still be made to mitigate the potential for exposure within the room. Beam tubes, beam blocks, temporary barriers, and attenuators can be used to define the direction and power of a beam hazard. As mentioned, the 80/20 extruded aluminum product (or similar products) can be very useful in this regard.

VISUAL WARNINGS

The visual warning is the second line of defense for protecting personnel from a laser hazard. The ANSI standard provides clear instructions describing the text and logos to be used on signage for laser areas (Section 4.7) The design of the sign can follow the ANSI Z535 standard or the International Electrotechnical Commission (IEC) 60825-1 standard. There are companies that sell these signs, but there are also several sites on the Internet that provide free of charge image downloads for printing your own.

Laser areas designated Class 3B should, and areas designated Class 4 shall, have an electronic warning at the entryway to the hazardous area. The ANSI standard suggests that the warning be visible within the area and through laser-protective eyewear. This suggestion comes from the FLPPS requirements for hardware built into the Class 3B and Class 4 lasers, and it requires interpretation.

When an entire room is designated as a laser area, there would be a visible electronic warning outside the entryway to this area. It would not be necessary for the warning light at this entry point to be visible through laser-protective eyewear because a person approaching this door would not have reached the eyewear station. Once inside the room, there should be some method of conveying the laser hazard that *is* visible within the area while wearing the protective eyewear. Most often this illuminated warning is already provided by the laser manufacturer on the laser head or laser power supply in accordance with the FLPPS requirements. There is not a requirement to install another illuminated warning within the area if the laser already provides a warning.

If an area within a room is divided away and designated as a laser area, there will likely be some form of curtain or divider separating the hazardous area from the remaining safe area. All personnel approaching this inner dividing area must be informed of the hazard. It would not be necessary to electronically warn personnel at the outer door to this room because personnel are not in any danger when passing into the safe region of the room.

Ultimately, the judgment is at the discretion of the LSO. There will always be special cases that do not fit within the norm, such as in the case of light-sensitive experiments.

The ANSI Z136.1 descriptions for electronic warning systems are purposely left vague to permit any reasonable method to be used. ANSI Z136.1 gives an example of a red light outside a door, but some facilities may choose a different color because they have designated red for another hazard type, such as high voltage. Many facilities have implemented the illuminated laser warning placard that incorporates the danger warning sign. This is also an option. The warning should be illuminated when a hazard is present, and it should be off or indicate a safe status when the area is safe. Some facilities use simple battery-powered strobes that are manually turned on and off, whereas others use warning strobes built into the interlock system. From an engineered safety standpoint, it is preferable to have the warning connected in such a way that it will automatically illuminate when the area interlock is set for laser operation. This eliminates the potential problem of forgetting to turn it on or off.

There is a common misconception that a laser power supply will turn an external warning on and off when the laser turns on and off. Some laser manufacturers may build in a function that permits this mode of operation, but this is a rare exception. The most common method of controlling the warning is through an interlock control system. In this mode of operation, the interlock is armed, it illuminates the warning, and it gives permission for the laser to operate. When the interlock is disarmed, the warning and the laser are secured. Figure 15.2 shows typical electronic warnings for a Class 3B or Class 4 laser area.

FIGURE 15.2 Typical electronic warnings used to designate a Class 3B or Class 4 laser area.

ENGINEERED AREA OR ENTRYWAY SAFETY CONTROLS

Class 4 laser areas have the potential to be extremely hazardous to untrained and unprotected personnel and require stringent protection. ANSI Z136.1 gives a choice of three methods that can be used to mitigate the hazard:

1. Nondefeatable (nonoverride) area or entryway safety controls
2. Defeatable area or entryway safety controls
3. Procedural area or entryway safety controls

The nondefeatable and defeatable methods use engineered safety devices and are the preferred methods of controlling the laser hazard. The procedural method is not an engineered safety control method and is recommended only for limited specialized applications.

Nondefeatable Area or Entryway Safety Controls

Nondefeatable means *cannot be defeated*. The nondefeatable safety circuit is the safest, simplest, and least-expensive type of engineered safety control. The interlock connection of a Class 4 laser (or shutter at the output of the laser) is connected to a safety circuit that extends to one or more limit switches at the entryway door. In operation, the circuit performs as follows: Open the door, and the laser turns off or the shutter closes.

Figures 15.3 and 15.4 show a greatly simplified version of a nondefeatable interlock configuration. In reality, the connections are more complicated because they must be designed to be fail-safe and fault tolerant. To design for these modes, the system designer must systematically walk through the circuit and ask: What would happen if this single item fails to open or shorts? When the answer does not yield a satisfactory result, the design is reworked. It is easy to see that a short across the main wiring in Figures 15.3 and 15.4 would create an unsafe condition. It is quite common to add redundant circuits as demonstrated in Figure 15.5 for this very reason. In the rare event of a failure of one switch, the second switch will do the job. The safety circuit is also expanded to perform secondary functions, such as control of the electronic laser warning.

As Figure 15.5 shows, a single open or short failure in either circuit will not limit the ability of the system to perform the trip function. It is worthwhile to note that even in this circuit depiction, the final connection to the laser power supply is susceptible to a single-point fault condition if a short circuit appears across the connector. The external interlock system has been designed with fault tolerance, but the interlock provisions from the laser manufacturer still leave a small section of wiring that is not fault tolerant. In most laser applications, this is an acceptable risk, and the

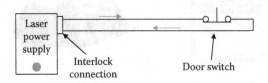

FIGURE 15.3 Simplified electrical schematic of nondefeatable access control interlock with door closed. Laser is permitted to run (or shutter may open).

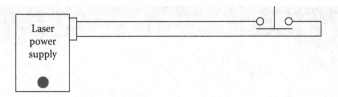

FIGURE 15.4 Simplified electrical schematic of nondefeatable access control interlock with door open. Laser cannot run (or shutter cannot open).

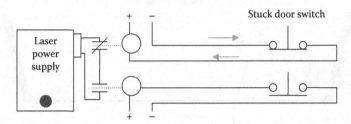

FIGURE 15.5 Redundant monitoring of dual door-limit switches provides enhanced protection.

safety function is routinely checked by a laser safety audit, but when higher levels of fault tolerance are desired, a secondary means of shutdown is used. As an example, the first circuit might be connected to the laser power supply, and the second circuit may control the laser shutter. Thus, if either single circuit fails, the second circuit will still perform the safety function.

This is just the tip of the iceberg when it comes to safety engineering, and many people are unaware of the complexities involved. A complete risk assessment for system design requires the evaluation of many factors, including risk estimation of the severity of potential injury, frequency of exposure, probability of injury, and analysis of personal protection equipment that can complement the safety measures.

How do we *work* with a nondefeatable interlock (Figure 15.6)? The laser operator enters the room and closes the door. The operator then will put on appropriate laser eyewear from the eyewear storage bin. The operator will *arm* the interlock. This will in turn cause the electronic warning outside the laboratory to illuminate and an interlock relay to close, which grants permission for the laser to start. The operator is now free to *safely* perform work with the laser, be it alignment or experiment with exposed Class 4 beams. Personnel outside the door are warned of the hazard by the beacon. If the door is opened, the laser will immediately trip off to prevent exposure. If the door is closed again, the system must be manually rearmed to start the laser.

The number one complaint of any laser researcher is the fact that many lasers are sensitive to thermal transients and have a long warm-up time before reaching a stable operating condition. To satisfy this concern, it is permissible to use a safety shutter connected to the room interlock system in lieu of the laser power supply.

The nondefeatable method sounds restrictive, and indeed it is, but it is also an extremely safe way of doing business. The nondefeatable method can be quite versatile when combined with Class 1 enclosure components. In this scenario,

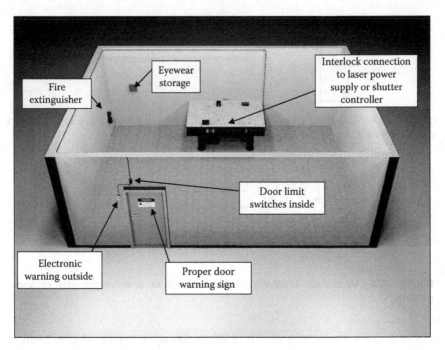

FIGURE 15.6 An overhead view of a typical laser room outfitted with nondefeatable access control.

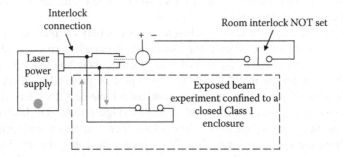

FIGURE 15.7 A Class 1 interlock sensor in parallel with a room interlock.

a simple Class 1 enclosure can be built around the laser table. An interlock switch must be built into the lid or door to this enclosure. The switch must be shorted when the enclosure is securely closed, and it must open if the enclosure integrity is violated. This interlock switch is then wired in *parallel* with the room interlock system, as shown in Figure 15.7.

This system is now significantly more versatile. The operator can close the door, put on safety eyewear, set the room interlock, then open the enclosure and manipulate the experiment. Once the experiment is reconfigured, the enclosure is replaced, and the room interlock can be disarmed. The laser will remain in operation. Because the laser is contained in an interlocked Class 1 enclosure, there is no need for the electronic warning outside the door, and personnel can enter and leave the room as they desire. If someone attempts to open the enclosure without first arming the room interlock, the laser will trip. This mode of operation is ideal for experiments that may change on a daily or weekly basis but are required to run for extended periods without a trip. Indeed, an experiment could run for months at a time if needed. The only event that would trip the laser would be an event that violated the safety protocol.

The nondefeatable entryway control is the ANSI-preferred method of protection. It has the advantage of not requiring complex barriers or expensive laser curtains at the doorway because the door is the interlocked barrier. Commercial nondefeatable laser safety systems start at less than $1000.

Defeatable Area or Entryway Safety Controls

A defeatable safety control is a type of room interlock that allows *authorized trained personnel* to momentarily defeat (bypass) the interlock limit switches at a room entryway to enter and exit the room without interrupting laser operation. To be safe and effective, it is crucial that the level of laser radiation does not exceed the MPE at the entry point. This often requires the installation of barriers or laser curtains at the interior of the entryway. It is important to note that the ANSI Z136.1 standard recommends the defeatable interlock *only if nondefeatable interlocks limit the intended use of the laser or laser system.*

How does it work? As Figure 15.8 shows, the limit switch that monitors the door position can be temporarily bypassed by a relay contactor. A timer circuit holds this contactor closed for a set time period, typically 10–15 seconds. The timer is initiated by an input from one of the two locations, either a push-to-exit button from inside the room or an entry request from outside the room.

Again, this circuit is overly simplified to present the general operating principle. A properly designed interlock system will have checks and balances to ensure that the timing circuit does not leave the interlock in a bypassed mode. If microprocessor controlled, the device should fail in a nondefeating manner if the microprocessor quits. Some *do-it-yourself* systems have used security system keypads with built-in timers.

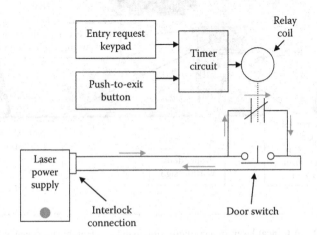

FIGURE 15.8 Simplified electrical schematic of a defeatable access control interlock.

Although failure is unlikely, they are typically not designed with fail-safe circuitry and have the potential to latch in the *bypassed* state that would leave the access control in an unsafe condition.

When comparing the defeatable to the nondefeatable access control, there is very little that has changed in the laser room. The changes are made at the entryway. Most important is the necessity to limit the laser radiation exposure to a level less than the MPE at the entryway. The laser will be running when the door interlock is momentarily bypassed. The curtain shown in Figure 15.9 does not extend to the ceiling and floor or form a complete closure on the left side. This is acceptable if proper hazard analysis has shown that there are no specular or diffuse reflection paths that can divert the beam past the curtain and reach the MPE threshold.

The laser-protective eyewear storage bin has been relocated to the interior of the entryway. It could also be placed outside the main door, but theft and tampering are often a consideration. Following the defeatable entryway protocol, the laser worker would bypass the entry interlock, walk into the protected zone, and put on the laser eyewear before stepping into the main room.

When proper entry protocols are followed, the laser system can run indefinitely. The only conditions that would interrupt operation would be a power failure, a violation of the safety protocol, or a purposeful crash of the system.

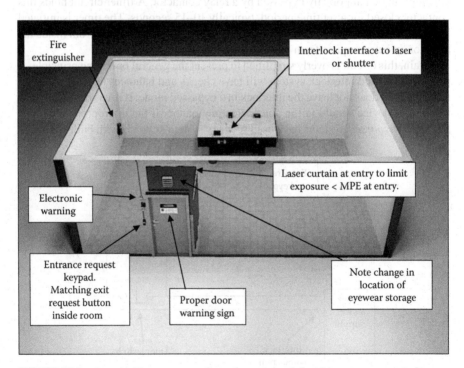

FIGURE 15.9 A typical laser area configuration using a defeatable entryway control.

Hardware choices for defeatable access controls are as follows:

Laser curtain: The method of limiting the exposure at the entryway is not restricted to the use of commercial laser curtains. If it is possible to use barriers or deflectors at the location of the laser table, it would be preferable to placing barriers at the entryway. Trigonometry is our friend in this regard because a small barrier properly placed can create a large protected zone across the room. If analysis of the system will not support these small barriers, then entryway protection will be needed. Remember that drywall or sheet metal labyrinths are often an option. ANSI Z136.1 states (Section 4.6.4, p. 47):

Laser barriers shall be specifically selected to withstand direct and diffusely scattered beams and shall exhibit a damage threshold for beam penetration for an exposure time commensurate with the total hazard evaluation for the facility and specific application. Important in the selection of the barrier are the factors of flammability and decomposition products of the barrier material.

Entryway keypad or button: The ANSI standard does not specify a particular electronic method for granting access to the laser area. Keypads, electronic card readers, cipher locks, and key switch bypass certainly provide a superior level of administrative control. The key or code can be restricted to personnel who have been trained and authorized for access. Some facilities have used a simple pushbutton to activate the time delay for entry. This is less than ideal because it does little to limit access. The *secret* button is only a secret for a couple of days.

To magnetically lock or not to magnetically lock: Most commercial systems provide the option of magnetically locking the access door to the laser area. One could argue that the magnetic lock actually does very little from a laser safety standpoint; after all, the laser will trip off if the door is opened without first inputting the interlock bypass request. The opposing argument in favor of the lock is the fact that it guarantees that the door will not be opened unless the bypass request is given. This is a higher level of security, and it serves the function of limiting *accidental* trips of the interlock system. It is not difficult to imagine a researcher deep in thought reaching for the doorknob and opening the door before realizing the experiment was just tripped off because the exit button was not pushed.

The magnetic lock is chosen over other types of lock because it is a fail-safe device. It needs electricity to lock the door. When power fails or an electronic interlock circuit drops to a safe mode, the door unlocks. The magnetic lock is relatively simple to add to a door, and the electronic control makes it simple to add to an interlock system for access control. That being said, the magnetic lock also presents additional challenges that must be taken into account.

Looking at Figures 15.10 and 15.11, we can see the typical required elements when using a magnetic lock. Every door that has been magnetically locked must have a crash button on both sides of the door that will break power to the lock in the event of an emergency. These crash buttons must actually be in series with the power delivery

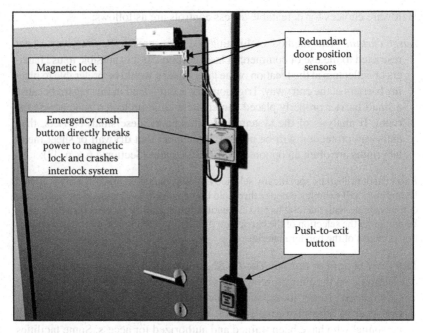

FIGURE 15.10   Detail of a defeatable access control exit.

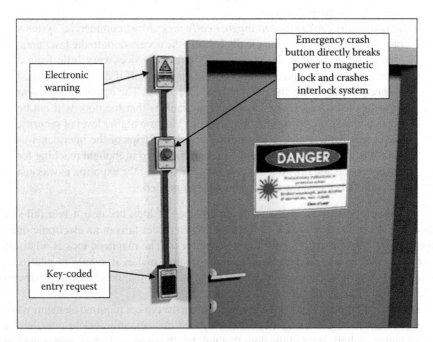

FIGURE 15.11   Detail of a defeatable access control entrance.

to the lock, not simply telling a microprocessor that *it* should release power. There are several reasons for this requirement. One reason is in the unlikely event of an electrical malfunction of the interlock circuitry. Another is the need for emergency personnel to enter the room in case of fire or injury. Depending on your location, there may be a requirement to connect the magnetic lock circuit to the building fire alarm circuitry to automatically unlock the door in the event of fire. Finally, there are some localities that specify a *no prior knowledge* requirement for exiting a door with a magnetic lock. This requirement specifies that a single action will release the door. Figure 15.10 depicts a dual-action scenario. The push-to-exit button is depressed to release the lock, and the doorknob is turned to actually open the door. There are several options available to meet the single-action requirement, including crash bars with a built-in switch or *look-down* motion detectors that will automatically issue an open request when a person approaches the door. Each of these options adds to the complexity and cost of the system. In some cases, the requirement can be dropped by simply explaining the situation to the building authority who is asking for the no prior knowledge operation that the persons working in the laser area when the interlock is set will be highly trained or, at the very least, escorted by a highly trained individual; the training will include laser safety and operation of the laser safety system; and the premise of no prior knowledge is not realistic in this situation because everyone will have exposure to, and knowledge of, the proper operating procedures.

The defeatable access control is significantly more versatile than the nondefeatable version. The added versatility comes at a price. If a laser area has multiple doors and the laser user has a need for a defeatable access, it is recommended that one door is selected for defeatable access, and the remaining doors are configured for nondefeatable operation. This is safer and more economical. Commercial defeatable interlock systems start at about $2500.

Procedural Area or Entryway Safety Controls

Where safety latches or interlocks are not feasible or are inappropriate, for example, during medical procedures and surgery, the following shall apply:

The ANSI Z136.1 (Section 4.3.10.2.2, p. 34) standard is clear. The procedural method is to be used only where the nondefeatable and defeatable interlocks are not feasible or are inappropriate. If someone is blinded by a laser at your facility, it would be a tough sell to tell an Occupational Safety and Health Administration investigative team that your budget was your reason for using the procedural method.

The example of the surgical laser is indeed one of the few appropriate cases for the use of a procedural method. The interlock exception is being granted to lasers used in medical procedures for the following reasons:

1. No one wants to inadvertently secure a laser in the middle of a medical procedure by tripping an interlock.
2. The flow of traffic in and out of a surgical area cannot be impeded when life safety is involved, so door locks or entryway switches are inappropriate.

The procedural entryway safety control method uses the following assumptions:

1. The laser is under the active control of an operator who can quickly secure the laser in the event of an emergency (supervised laser operation). This assumption is *not* stated as a line item in the ANSI standard, but can be inferred from the suggestion of a surgical laser.
2. All personnel are adequately trained and provided with personal protective equipment on entry.
3. The exposure must be less than the MPE at the entryway. Curtains, laser type, and room design are evaluated to meet this condition.
4. A visible or audible signal is provided at the entryway to indicate the laser is energized and operating at Class 4 levels.

Other than the surgical laser situation, there are few valid scenarios that can justify the use of the procedural area or access control protocol for a Class 4 laser area. It is worthwhile to note that even within ANSI Z136.3 (*Safe Use of Lasers in Health Care Facilities*), Sections 4.1 and 4.8 state that when a laser is not being used on a human patient, it shall be interlocked in accordance with the more stringent requirements of ANSI Z136.1.

ENGINEERED SAFETY TRAINING

Engineered safety training is a small subset of the overall training provided to the laser worker. Understanding the complexity of a laboratory can be a daunting challenge to a new student or employee. The laser worker must be taught about the particular nuances of the working environment. Lasers are often used in research laboratories containing a multitude of dangers. These systems often have multiple separate warning systems and emergency controls. For example, a room may contain a laser system, cryogenics, vacuum systems, and high voltage. Each system has its own dangers, so one might find warning strobes for laser hazard, oxygen-deficiency hazard, high-voltage hazard, and more. Beyond the individual hazards, there are often secondary hazards that appear through a combination of systems. For example, when a sufficiently high voltage is applied to a vacuum chamber, it is possible to generate X-rays. This new ionizing radiation hazard requires another safety protocol that must be enforced.

The training must address each of the hazards. It must explain the process to mitigate or eliminate the hazard in an emergency situation. If crash switches and warnings are a part of the various safety systems, it is vital that the switches be labeled or color coded to match the hazard type (Figure 15.12). There have been incidents in the past when users have been reluctant to crash an interlock system because they were unsure of the end result of depressing the poorly labeled crash button. The worker must be taught never to be afraid of pressing a crash button, even if it will result in downtime for the process or experiment. A crashed system is better than a person harmed.

FIGURE 15.12 An example of a crash switch with a purpose that is well defined by text. There can be no mistake that this button is intended to crash or *safe* the laser system.

ENGINEERED SAFETY SYSTEM SELECTION

It has been a common practice for many years to have on-site staff or students design and install laser interlock systems. This practice was viewed as a cost-savings measure and was often performed out of necessity because of the limited number of commercial system manufacturers. Be wary of technicians or students offering to design and install a system just because they tell you they can save you $1000. The time and effort required to reinvent the wheel is a resource expense. Sometimes, the costs involved actually end up exceeding the cost of a commercially available system. As these systems age and the original designers move or retire, the expansion, modification, or repair of the systems is placed in jeopardy if the system is not properly documented. If there is a commercially available product that fits your requirements, it will be worth the investment in the long run. A well-designed commercial system will last for the life of the laboratory and will be supported in long term with spare parts and documentation. The number of commercial manufacturers of laser interlock systems is growing, and products are becoming more versatile, so the argument to build your own system is getting weaker. Whichever route is chosen, it is imperative that the system is well documented.

The ultimate decision for selection of an engineered safety system should be a joint decision between the LSO and researcher. There will undoubtedly be situations for which commercially available systems are not adequate for the safety integrity level required. The LSO is encouraged to seek assistance from safety system manufacturers, peers, or professional safety consultants when there is any doubt over the proper course of action.

The following is a list of commercial laser interlock system and component manufacturers:

Manufacturer	Telephone	Web Address
CAL-AV Labs Inc.: Specialist in laser warning signs and sign controllers	USA 520-624-1300	http://www.cal-av.com/signs.html
Kentek Laser Corp.: Full-service laser safety provider	USA 800-432-2323	http://www.kentek-laser.com/
Lasermet Ltd.: Full-service laser safety provider	UK +01-202-770-740	http://www.lasermet.com
Laser Optical Engineering Ltd.: Specialist in laser interlock systems	UK +44 (0) 1509 223 948	http://www.laseroptical.co.uk/
Laser Safety Systems LLC: Specialist in laser interlock, access control systems, and electronic area warnings	USA 757-229-6109	http://www.lasersafetysystems.com/
Photonic Solutions PLC: Specialist in laser access control systems	UK +44 (0)131 664 8122	http://www.psplc.com
Rockwell Laser Industries Inc.: Full-service laser safety provider	USA 800-945-2737	http://www.rli.com/

16 Laser Eyewear

Ken Barat

CONTENTS

INTERESTING EYEWEAR FACT

In 1962, Dr. Harold Straub of the U.S. Army Harry Diamond Laboratory, developed the first laser eye protector by installing a 2 × 4-inch, blue-green glass (Schott-type BG–18), filter plate in a standard acetylene welding goggle frame.

The most common misconception laser users have about laser-protective eyewear is that it is the first line of defense against laser radiation. In reality, it should be the last line of defense. Beam containment will do more of a laser user than laser-protective eyewear.

If you can eliminate the possibility of eye damage because of enclosing the laser beam path such that NO radiation exposure to the eye is possible than do so. Although critically important, the implementation of laser-protective eyewear is always understood to be the second line of defense.

Laser-protective eyewear has a valuable role to play in laser safety and presents many challenges to the user and Laser Safety Officers (LSO). The remainder of this chapter will deal with the selection and use of laser-protective eyewear.

Laser-protective eyewear comes in two flavors, full attenuation and alignment eyewear. By full attenuation I mean this eyewear will completely block the transmission of a direct exposure laser beam from penetrating the eyewear. Conversely, alignment or partial attenuation allows an individual, while wearing laser eyewear, to have some visibility, which means some of the beams energy will pass through the laser-protective eyewear (Figure 16.1).

Frequently, one encounters cases where an LSO recommends, and researchers are then supplied with full attenuation laser eyewear, which subsequently is *underutilized*

FIGURE 16.1   Laser eyewear.

because of research conditions where partial attention is required for the proper execution of laser-related applications.

To talk about laser-protective eyewear, one really needs to understand two terms: optical density (OD) and maximum permissible exposure (MPE). OD is a filtration factor and MPE is like the speed limit. Your eye can accept irradiance up to and including the MPE without damage. The higher the exposure or irradiance over the MPE the greater the damage until a threshold is reached (one you do not want to reach), where the damage itself moderates the energy and damage (Table 16.1).

OD is a parameter for specifying the attenuation afforded by a transmitting medium. It is in log units; therefore goggles with a transmission of 0.000001% can be described as having an OD of 8.0. Logarithmic expression of OD can be described as follows:

$$OD = \log_{10}\left(\frac{M_i}{M_t}\right) \qquad (16.1)$$

where M_i is the power of the incident beam and M_t the power of the transmitted beam.

Thus, a filter that attenuates a beam by a factor of 1,000 or 10^3 has an OD of 3, and one that attenuates a beam by 1,000,000 or 10^6 has an OD of 6. The OD of two highly absorbing filters stacked together is essentially the sum of two individual ODs. When optical aids are not used, the following relationship may be used when radiant exposure (H) and irradiance (E) are averaged over the limiting aperture for classification:

$$OD_{req} = \log_{10}\frac{(E \text{ or } H)}{MPE} \qquad (16.2)$$

where the radiant exposure (H) or irradiance (E) is divided by the MPE.

When the entire beam could enter a person's eye, with or without optical aids, the following relationship is used:

$$OD_{req} = \log_{10}\left[\frac{0 \text{ or } Q_0}{AEL}\right] \qquad (16.3)$$

where AEL is the accessible emission limit (i.e., the MPE multiplied by the area of the limiting aperture) and 0 and Q_0 are the radiant power or energy, respectively.

TABLE 16.1

Code Definitions

Testing Conditions for Laser Type	Typical Laser Type	Pulse Length (s)	Number of Pulses
D	Continuous wave laser	10	1
I	Pulsed laser	10^{-4} to 10^{-1}	100
R	Q Switch pulsed	10^{-9} to 10^{-7}	100
M	Mode-coupled pulse laser	$>10^{-9}$	100

TYPES OF LASER SAFETY EYEWEAR

GLASS

Glass laser eyewear is heavier and more costly than plastic, but it provides better visible light transmittance. There are two types of glass lenses, those with absorptive glass filters and those with reflective coatings. Reflective coatings can create specular reflections and the coating can scratch, minimizing the protection level of the eyewear.

POLYCARBONATE

Polycarbonate laser eyewear is lighter, less expensive, and offers higher impact resistance than glass, but allows less visible light transmittance.

DIFFUSE VIEWING ONLY

As the name implies, diffuse viewing only (DVO) eyewear is to be used when there is a potential for exposure to diffuse reflections only. DVO eyewear may not provide protection from the direct beam or specular reflections.

ALIGNMENT EYEWEAR

Alignment eyewear may be used when aligning low power visible laser beams. Alignment eyewear transmits enough of the specified wavelength to be seen for alignment purposes, but not enough to cause damage to the eyes. Alignment eyewear cannot be used during operation of high-power or invisible beams and cannot be used with pulsed lasers.

LASER SAFETY EYEWEAR FOR ULTRAFAST (FEMTOSECOND) LASERS

Temporary bleaching may occur from high peak irradiances from ultrafast laser pulses. Contact the manufacturer of the laser safety eyewear for test data to determine if the eyewear will provide adequate protection before using them.

LABELING OF LASER SAFETY EYEWEAR

Laser safety eyewear shall be labeled with the OD and the wavelength(s) the eyewear provides protection for. Additional labeling may be added for quick identification of eyewear in multiple laser laboratories.

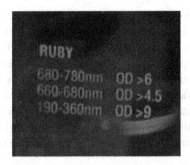

INSPECTION AND CLEANING OF LASER SAFETY EYEWEAR

Laser safety eyewear should be inspected periodically for the following:

1. Pitting, crazing, cracking, and discoloration of the attenuation material
2. Mechanical integrity of the frame
3. Light leaks
4. Coating damage

Follow manufacturers' instructions when cleaning laser safety eyewear. Use care when cleaning eyewear to avoid damage to absorbing filters or reflecting surfaces.

CONSIDERATIONS IN CHOOSING LASER-PROTECTIVE EYEWEAR

The most important considerations for picking eyewear are listed below. There may be other considerations.

1. OD requirement of eyewear filters at laser output wavelength(s)
2. Comfort and fit of eyewear with no peepholes
3. Visible light transmission requirement and assessment of the effect of the eyewear on the ability to perform tasks while wearing the eyewear
4. Need for prescription glasses

The following factors are included to calculate the OD of the filter:

1. Largest laser power and/or pulse energy for which protection is required
2. Wavelength(s) of laser output
3. Exposure time criteria (e.g., 0.25, 10, 100, or 30,000 seconds)

COMFORT AND FIT

Comfort and fit is a personal preference. Consider overall comfort when evaluating in terms of short, moderate, or protracted wearing times. If a pair of protective eyewear fits poorly, it will not work properly. Moreover, the likelihood of its use decreases. This is true for a respirator, facemask, or laser-protective eyewear.

One size does not fit for all. Users do not want uncomfortable eyewear that is too loose, too tight, too heavy, fogs up, or slips. The effort spent in finding proper fitting eyewear is well worth the time.

To help with fitting loose eyewear, you may need to place a strap across the back to keep the frame tight if necessary. Another option is to use flip-down eyewear over a user's own glasses so the eyewear is familiar. Manufacturers offer a range of options in sizes, including new eyewear for slim faces to very large faces. There are options for fitting different nasal profiles, including flat or low nasal profiles, and combinations for small faces with flat nasal profiles. Adjustable temple lengths are also helpful, as well as temples with gripping ends. Bayonet temples (the straighter temple) also help in fitting large faces. Choices of laser-protective eyewear have come a long way. All users should be able to find an usable pair.

OPTICAL DENSITY

Optical Density	% Radiation Transmission
1	10%
2	1%
3	0.1%
4	0.01%
5	0.001%
6	0.0001%
7	0.00001%

FULL ATTENUATION

Without exception, for Class 4 lasers and Class 3B lasers (when the MPE limit is exceeded) it is recommended to provide full attenuation laser-protective eyewear in: all UV (i.e., nominal 190–380 nm), ocular focus near IR nonvisible (i.e., nominal 700–1400 nm) wavelengths, as well as mid to far IR regions. The logic in doing so is quite simple, if one cannot see the beams and they exceed the MPE limits, then there is no reason to do anything other than fully attenuate those same wavelength regions.

Moreover, in the visible regime (i.e., nominal 400–700 nm) when the detection of the termination point of the visible laser wavelength is NOT required for one's application, then full attenuation of these same visible wavelengths is also recommended.

The LSO is tasked with recommending proper eyewear selection for the wavelength or wavelength region in question to meet the required OD for each laser application(s).

Once the small source intrabeam OD for each laser wavelength or wavelength region has been posted, various other ancillary conditions emerge that may both positively (or negatively) impact the intended use of the chosen laser-protective eyewear.

To state the obvious, to be effective, laser eyewear MUST be worn. As readily apparent and obvious as that comment is, the single most prevalent cause—by far—of all laser-related eye injuries is the fact that laser-protective eyewear, although typically available and appropriate to the prevailing laser application, was not worn.

Why? This is where many of the ancillary features such as weight considerations between glass and polycarbonate lenses, *acceptable* versus *unacceptable* visual luminous transmittance (VLT), subjective preferences of comfort and fit, prescription lenses (Rx) capability, propensity of eyewear to fogging, peripheral visual capacity or lack thereof, and so forth, come into play.

Visual Light Transmission

Undoubtedly, VLT and fit are the two most compelling features in the usage or aversion to usage of laser eyewear. Simply stated, VLT is the mean average percentage of the entire visible spectrum, as weighted for blue spectral responsiveness, which is NOT being filtered by these same lenses. Repeatedly, experience has indicated that the higher the VLT, the higher the likelihood of eyewear usage and consequently laser eyewear safety compliance.

In many research and academic circumstances, overhead room lights may be turned off for a variety (e.g., beam collimation, alignment, etc.) of conditions and VLT in these circumstances becomes of pre-eminent concern. Moreover, laser-related electrical hazards, which have caused serious injuries and may include death, must be fully considered in light of diminished visual acuity due to a loss of VLT when wearing laser- protective eyewear. Lest we forget, although laser radiation can blind you, electricity can kill you.

In addition, the distinction between OD and VLT, especially in full attenuation conditions, are sometimes misunderstood or misrepresented. Assumptions abound that a higher OD *necessarily* implies a reduction of VLT. However, the reduction of VLT is directly correlated to a higher OD only when visually limiting ODs are directly attributable to the visible (i.e., nominal 400 nm–700 nm) region only.

In laser eyewear attenuation conditions in UV, near-, mid-, and far-IR regions, or a multiwavelength combination thereof, there are relative instances where one may encounter eyewear that possesses: high OD, low VLT; high OD, high VLT; low OD, low VLT; low OD, high VLT. In my estimation, any eyewear possessing a VLT at less than 15%–20% is dangerously close to creating a loss of visual acuity where other potential (notably electrical) dangers become considerably more likely.

Therefore, in seeking full attenuation laser eyewear with appropriate OD values for one's application(s), increasing VLT may require certain trade-offs. Typically, this is the decision juncture at which one considers the use of plastic versus glass lenses.

Polycarbonate lenses are lighter in weight than glass lenses. As such, polycarbonate lenses have inherent (and perfectly logical) preference for the user; especially in conditions where protracted usage is required. There are certain common and very prevalent wavelength regions (notably Nd:YAG @ 1064 nm) where glass lenses have higher VLT than polycarbonate lenses. In this instance, the trade-off is of course that although one is increasing the VLT, they are simultaneously increasing the weight of the eyewear and thereby potentially diminishing the perceived comfort of the eyewear.

Fortunately, various manufacturers of both glass and polycarbonate eyewear have noted the general preference for polycarbonate eyewear and have made significant strides in increasing their products' VLT in near IR and certain other visible wavelength regions. Much of this improved VLT is because of eyewear that obtains all or some of its OD through the reflection of the laser beam. Yes, sufficient energy can be reflected off the eyewear to injure someone else if conditions are right.

These requirements should be used to aid in the selection of appropriate eyewear/materials. VLT shall be computed and shall be available for the light- (photopic) and dark- (scotopic) adapted eye (see American National Standards Institute [ANSI] Z136.7 standard). Please keep in mind that normally the minimum acceptable photopic luminous transmittance is approximately 20% for most applications. Sunglasses are typically 12%–18% photopic luminous transmittant. Normally, the minimum acceptable scotopic luminous transmittance is approximately 20% for unaided applications.

Adequate OD at the laser wavelength of interest shall be weighed with the need for adequate visible transmission. The minimum adequate acceptable photopic and scotopic VLT is approximately 20% for most applications. At VLT levels less than 20%, other nonbeam hazards may exist by virtue of diminished visual acuity.

COMFORT AND FIT

Comfort and fit considerations are the wholly subjective and depend entirely upon individual preferences that each wearer maintains concerning how a specific set of laser-protective eyewear feels when worn. Comfort and fit primarily center upon personal preferences issues like: overall comfort when evaluated in terms of short, moderate, or protracted wearing times.

Overall, if a pair of protective eyewear does not fit properly, it not only cannot perform its function to the required specifications but also likelihood of it being used decreases. This is true for a respirator, facemask, or laser-protective eyewear.

Users want their eyewear to be as natural an extension of their faces as possible. They do not want to be constantly reminder they are wearing eyewear by it being too loose, too tight, too heavy, fogging up, slipping, or other well-known problems.

Therefore, effort placed in finding proper fitting eyewear is well worth the time. One size does not fit all. One solution maybe to place a strap across the back to keep the frame tight is necessary. Another solution maybe is to flip-down on one's own glasses. Manufacturers offer a range of options in sizes, including new eyewear for slim faces, and for very large faces. There are options for fitting different nasal profiles, including flat or low nasal profiles, and combinations for small faces with flat nasal profiles. Adjustable temple lengths are also helpful, as well as temples with gripping ends. Bayonet temples (the straighter temple) also help in fitting large faces. Choices of laser-protective eyewear have come a long way. All users should be able to find just that right pair.

ULTRAVIOLET 200–266 NM BEAMS

Always wear gloves and long sleeves when aligning UV beams to prevent skin exposure. Skin exposure to lasers could lead to possible skin cancer.

1. Use CaF_2 substrate for transmissive optics to prevent nonlinear absorption of red fluorescence with high-energy, high-power UV beams. Red fluorescence ultimately leads to permanent increase of optical transmission loss (brownish coloring).
2. Use fused silica substrate for reflective optics to reduce coating absorption.
3. Aluminum-coated gratings, even when coated against oxidation, will degrade rapidly when used for UV high-energy beams.
4. Remember: the lower the wavelength, the smaller the spot size for a given focal length lens/optic. When looking at beam profile on camera, ensure ALL harmonics are filtered out.

ULTRAFAST OPTICAL PARAMETRIC AMPLIFIER (166 NM to 20 μM) BEAMS

1. For NIR and IR beams, liquid crystals papers (from Thorlabs or Edmunds Optics) can be very helpful to detect the position of far IR beams, outside range of conventional beam viewers.
2. Do not be fooled by harmonic components.

800 NM BEAMS

No wavelength has been involved in more laser eye injuries in the past 10 years as the Ti:S 750–850 nm beam. The eyes lack perception of this wavelength band—less than 1% of these photons are perceived by the eye. A user may see a faint dot giving the false impression of low power.

1. For alignment of an 800 nm compressed beam (peak power), you can use a white bleached business card (while wearing eyewear) to see the SHG (second harmonic generation [blue color]) beam on the card.
2. When aligning compressed or very intense large diameter beams, use the SHG on a white business card to center the beam on alignment irises. Center the beam on the iris looking at the throughput beam (symmetrically clipped SHG blue beam).
3. When aligning small diameter beams, use an IR viewer to look at the concentric beam around the hole of the iris or use an orange card looking at the throughput beam.
4. Beware of the secondary lasing cavity caused by back reflections when introducing reflective surfaces in a pumped amplifier with flat (not Brewster) Ti:S crystals (valid for other type of gain medium).
5. ALWAYS use a MINIMUM number of mirrors to realign an amplifier.
6. White thin ceramic plates are useful for finding the beam. They are safe with both low- and high-power beams.

Case 1

A 26-year old male student aligned optics in a university chemistry research lab. He used a *chirped pulse* titanium–sapphire laser operating at 815 nm with 1.2 mJ pulse energy at 1 kHz. Each pulse was ≈200 ps. The student was not wearing protective eyewear.

The laser beam backscattered off the REAR SIDE of a mirror (about 1% of total) and caused a foveal retinal lesion with hemorrhage and blind spot in central vision.

A retinal eye exam was done and confirmed the laser damage. Protective eyewear could have prevented the injury.

Case 2

A postdoctoral employee received an eye exposure to spectral radiation from an 800 nm Class 4 laser beam. The extremely short pulse (100 fs) caused a 100-μm-diameter burn in the employee's retina. The accident occurred shortly after a mirror was removed from its mount and replaced with a corner cube during a realignment procedure. Although the beam had been blocked during several previous steps in the alignment, it was not blocked in this case. The employee was exposed to laser radiation from the corner cube mount when he leaned down to check the height of the mount.

Neither the employee who was injured nor another employee who was working on the alignment was wearing the appropriate laser eye protection. The researcher may have underestimated the hazard because the visible portion of the 800 nm beam only represented 1%–2% of the beam.

FLASH LAMP YAG HIGH-ENERGY 532 NM BEAMS

1. Always align beams at LOW POWER (detune the Q-switch [QSW] timing instead of LAMP timing to reduce green).
2. Always verify the YAG beam profile PRIOR to sending it to a Ti:S crystal or other crystals. Hot spots will likely cause severe irreversible damages to the crystal lattice or the crystal coating. Dummy testing on sapphire crystals can be an inexpensive way to ensure integrity of the Ti:S when pumped.
3. White ceramic is the preferred permanent beam block material for YAG energetic beams.
4. Practical *short-term* beam blocks for YAG 10Hz green beam are white packing foams, which diffuse the green powerful beams temporarily during specific and approved alignment procedures.
5. DO NOT USE black anodized metal surface as beam blocks. Photographic *burn paper* or nondeveloped black photo paper can be used to visualize the beam quality. Make sure to put the paper into a clear plastic bag to avoid debris blasts and avoid over-exposure. Beware of plastic bag laser reflections. Using back burns can help maintain information contained in the burn mark.

YAG/YLF HIGH-POWER 532/527 NM BEAMS

1. Wear Lawrence Berkeley National Laboratory-approved alignment goggles that allow you to see a faint green beam. Goggles are very useful for avoiding burns during alignment.
2. Remember that high power high rep rate beams will ablate black anodization of most beam blocks, leaving residues onto optics nearby.

DAMAGE THRESHOLD CONSIDERATIONS

Once one finds the appropriate eyewear with adequate OD to achieve full atten-uation and *suitable* VLT, there is yet another trade-off hurdle to ponder, namely damage threshold considerations. As a general *rule of thumb*, polycarbonate eye-wear can withstand approximately 100 W/cm^2 of direct incident laser radiation for approximately 10 seconds duration prior to *damaging effects* noted on the lenses. Conversely, glass eyewear can withstand approximately 10 times ($\approx$1000 W/cm^2) the value of polycarbonate laser eyewear for the same time duration.

With the assumption that a collimated, focused beam is impinging upon a dis-creet, nonwavering point on the polycarbonate or glass lens, polycarbonate lenses are prone to exhibiting sequentially: a superheated plasma effect at the surface of the lens, degradation of the absorptive dyes (with possible carbonization and darkening effects noted), the emission of smoke, possible noxious odors, the emission of flame and potential ultimate penetration of the lenses. Conversely, glass lenses are prone to *catastrophic* degradation effects where the accumulation of irradiant energy results in loss of integrity with effects noted as: a popping sound when the beam strikes the glass lens with potential *spider vein* crazing and with sufficient accumulation of energy, a complete shattering of the glass lens.

Generally speaking, these physical effects for both polycarbonate and glass lenses have readily apparent visual and auditory correlates that forewarn the wearer of an impending damage threshold danger. However, they do come into consideration when one is deciding upon which trade-offs to implement to optimize the likelihood of eyewear suitability and will also be discussed when ultrafast pulse considerations are presented later in the chapter.

SIDE SHIELDS

The ANSI Z136.1 standard, in "Factors in Selecting Appropriate Eyewear" mandates one to *consider* side shields, overall, the presence of side shields is not an issue that can be considered and then decided against. Rather, even though they may impair peripheral vision, I am of a mind that the presence of side shields is mandatory and be commensurate with the level(s) of optical density that the main viewing lenses provide.

The ANSI Z136.1 standard Safe Use of Lasers does not require laser-protective eyewear to be ANSI Z87 compliant. ANSI Z87 is the standard for safety eyewear; the most common element is impact resistance. Therefore in evaluating ones eye-wear needs, the question of impact resistance needs to be addressed. Simply, is it needed or not. If not, no further action is needed; if the LSO hazard evaluation is, yes it is required, one has three choices:

1. Obtain a pair of laser eyewear that is compliant with Z87 (most polymer eyewear are compliant).
2. Wear safety glasses over the laser eyewear.
3. Have glass laser eyewear hardened to meet Z87.

Choice 2 can affect comfort or the ease of wearing the protective eyewear and general vision, whereas choice 3 will affect the cost of the eyewear.

PRESCRIPTIONS

Now eyewear for prescription wearers has several options. These include eyewear with prescriptions ground into the glass laser lens, eyewear that holds prescription inserts, and eyewear with flips, with polymer prescriptions in the base or the flip. For ground lenses, the frame selections have widened to include titanium frames and frames with adjustable temples.

WEIGHT

Weight of eyewear is a particular concern in the consideration of acquiring multi-wavelength or prescription eyewear. Depending on wavelength combination 7 mm of glass is not unheard of. This thickness of glass, which is two to three times a normal prescription eyewear, may prove to be uncomfortable for a user to wear for extended periods. This can lead to a lack of productivity or times of no eye protection. Some breakthrough in polycarbonate prescription flips and over-glasses may help improve this item.

LABELING

ANSI Z136.1 and International Electrotechnical Commission require laser-protective eyewear to be labeled with the wavelength and OD it is intended for. The laser eyewear manufacturer will imprint on the eyewear the most common range of wavelengths and OD for a particular pair (Figure 16.2). For the vast number of laser users, this is satisfactory. Always remember the guarantee of protection is only made for the wavelengths imprinted on the eyewear frame. Even curves for the eyewear are just a generalization. Unless you know the lot number and have curves for that run, only

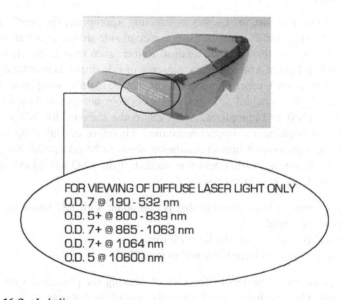

FOR VIEWING OF DIFFUSE LASER LIGHT ONLY
O.D. 7 @ 190 - 532 nm
O.D. 5+ @ 800 - 839 nm
O.D. 7+ @ 865 - 1063 nm
O.D. 7+ @ 1064 nm
O.D. 5 @ 10600 nm

FIGURE 16.2 Labeling.

the imprinted OD is guaranteed. A small segment of users are using the eyewear for wavelengths not listed on them. Curves and other documentation provided by the eyewear manufacturer or distributor will show the OD at the desired wavelength. To be compliant the facility LSO will have to label the eyewear or post the information where the eyewear is stored and have a way to identify which pair is which.

ULTRAFAST LASERS

Testing by the Army branch at Brooks Air Force base has shown a nonuniform bleaching effect on standard laser eyewear against ultrafast pulses. This relates back to the relaxation time of the absorption molecules. Not all eyewear for ultrafast pulses demonstrate this effect, but a significant number do, which makes it a real concern. Therefore, ultrafast laser users who wish for full protection will need to check with the manufacturer of the eyewear for their testing results to verify suitability of the eyewear for their use. Usually, the manufacturer can provide a sample piece of the lens for testing with a power meter in the actual application, to verify the appropriateness of the lens in question.

It is imperative to recognize that if one is using ultrafast lasers (particularly regeneratively amplified sources) there exists the potential that OD values may be compromised should femtosecond (fs) beam exposure to one's laser eyewear occur. Should temporary or permanent loss of OD (and commensurate exposure levels in excess of applicable MPE values) occur as a consequence of these conditions, obvious detrimental eye safety effects become plausible. The core safety issue surrounding laser-protective eyewear and femtosecond-lasers is as follows: in certain ultrafast (fs) operating conditions, saturable absorption effects with calculable losses in purported OD values of the femtosecond-subjected laser eyewear have been observed.

It is the intention of ANSI committees involved in this matter that the underlying mechanisms of the degradation effects so noted be investigated and, to the greatest extent possible, elucidated for everyone's general understanding.

ADDITIONAL CONSIDERATIONS

Other important consideration is antifog capabilities, especially for goggles. Multiwavelength operations have special questions, as the more wavelengths you try to remove with one pair of eyewear, typically the darker the eyewear gets. You can try flip options or more than one pair to alleviate this problem. Laser inscribed markings (printed ones wash off when cleaned) also help the longevity of the eyewear, as well as UV inhibitors to prevent darkening over time in polymer eyewear. Finally, cost is important, but you also must consider what is the cost of an eye?

SATURABLE ABSORPTION

Certain dyes used to absorb laser radiation may undergo saturable absorption (aka, induced transmittance or transient photobleaching) where the ability to absorb radiant energy decreases with increasing radiant exposure or peak irradiance. When this occurs, the OD may decrease providing less protection to the user. This has been reported for both glass and polycarbonate filters for certain pulsed lasers and is associated with high values of peak irradiance. Lasers evaluated were pulsed (Q-switched and ultrashort pulses) titanium sapphire and neodymium:YAG (1064 nm and 532 nm) lasers.

ANGLE OF EXPOSURE

On the basis of the composition of the laser-protective-eyewear filter the angle of exposure can have an effect on the effectiveness of the eyewear filter. Dielectric coatings on laser-protective eyewear are designed to deliver the labeled OD within a set angle of acceptance (similar to the acceptance angle of an optical fiber. Laser radiation incident upon the eyewear outside that angle will yield a different OD. The obliqueness of the angle may or may not limit the laser radiation entering the pupil.

CLEANING AND INSPECTION

Periodic cleaning and inspection shall be performed on the protective eyewear to ensure they are maintained to a satisfactory condition. The frequency of the safety inspection should be once per year, or as determined by the LSO. This includes the following:

1. Periodic cleaning of laser eyewear. Care should be observed when cleaning lenses of protective eyewear to avoid damage to the absorbing and reflecting surfaces. In some uses (e.g., surgery) eyewear may require cleaning (and sterilization) after each use. Consult eyewear manufacturers for instructions for proper cleaning methods.
2. Inspection of the attenuation material for pitting, crazing, cracking, discoloration, delamination or lifting of dielectric coatings, and so forth.
3. Inspection of the frame for mechanical integrity.
4. Inspection for light leaks and coating damage.
5. Inspection of goggles for loss of ventilation port plugs, deformation of the face piece, and stretching of the head strap.

Eyewear in suspicious condition should be tested for acceptability or discarded. If, upon inspection, the LSO is unsure of the severity of these defects, as they relate to efficacy of use, the LSO should contact the manufacturer for guidance and recommendations as to replacement or acceptability of current laser-protective eyewear usage.

USE WITH HIGH-POWER LASERS AND EYEWEAR LIMITATIONS

The following section is from the ANSI Z136.8 standard for laser safety in research and development. The key in this section is that if you have a high power laser or a beam of very high irradiance, counting on laser-protective eyewear as one sole or primary safety control is foolish. You are saying "I hope if there is a stray reflection, it hits my eyewear and I can get out of the way before it burns through." Engineering control measures shall be implemented with high-power, multikilowatt laser beams, unless impractical, control measures may be used. Personal protective equipment, in the form of laser eye protection, may be inadequate to protect the user from serious ocular exposure from such laser beams. In addition, if the multikilowatt laser beam does not strike the laser-protective eyewear, the skin of the face may receive a significant injury (e.g., third-degree burn and laceration from facial motion during exposure).

Most of the radiant energy absorbed by the filter is transformed into heat. If the radiant flux is quite high, as it would be for multikilowatt beams, the heat may fracture a glass lens or melt polycarbonate. If the radiant energy is concentrated in a small diameter spot, then enhanced heat transfer may result in damage to the surrounding matrix material. The latter may occur for radiant power much less than a kilowatt.

It is even likely possible for powerful lasers that the filter material, glass, or plastic, may be damaged in a time period that is shorter than the time base used to determine the MPE. This is particularly true as the radiant exposure increases. Guidance on typical laser-induced damage threshold levels may be found in ANSI Z136.7 (latest revision). For polycarbonate, these values are 10 J/cm^2 (exposure [t] < 10^{-3} seconds) and 300 $t^{0.5}$ J/cm^2 (exposure [t] $\geq$ 10^{-3} seconds). For glass, the values are 1 J/cm^2 (exposure [t] < 10^{-6} second) and 1000 $t^{0.5}$ J/cm^2 (exposure [t] $\geq$ 10^{-6} second).

Users of laser-protective eyewear shall be trained to understand potential early signs of damage. These may include, but are not limited to, smoke, flame, incandescence, and luminescence.

PARTIAL ATTENUATION (AKA ALIGNMENT EYEWEAR)

ALIGNMENT EYEWEAR

Many users confess they take off their eyewear because they cannot see the beam. Alignment eyewear provides a solution to this problem. For alignment of visible beams, conditions may arise that require the user to see the beam through their protective eyewear (cases where remote viewing is not possible). In these situations, the use of alignment eyewear can be approved by the LSO. Alignment eyewear is assigned an OD lower than that which would provide full protection from a direct accidental exposure. For continuous wave lasers, the alignment OD shall reduce irradiance to between Class 2 and Class 3R level. For pulse lasers, the alignment OD shall be no less than the full protection OD – 1.4.

The purpose of alignment eyewear is to allow the user visualization of the beam while lowering the intensity of any beam that is transmitted through the user's eyewear to a Class 2 level. To address this issue there is an existing European Norm that recommends OD for alignment eyewear versus the output of lasers used.

Scale Number	OD	Max Instantaneous Power Continuous Wave Laser (W)	Maximum Energy for Pulsed Lasers (J)
R1	1–2	0.01	2×10^{-6}
R2	2–3	0.1	2×10^{-5}
R3	3–4	1.0	2×10^{-4}
R4	4–5	10	2×10^{-3}
R5	5–6	100	2×10^{-2}

Therefore, for *alignment* laser eyewear to be effectively utilized, preferentially all of the following conditions should be in place: (1) administrative liability acknowledgment and acceptance of same, (2) acknowledgment of potential hazards with the utilization of eyewear that does not protect one against small source intrabeam or specularly reflected exposures and finally, (3) collaborative agreement between the LSO and researcher(s) of alignment eyewear safety protocols and appropriate alignment laser-protective eyewear. Once these preliminary *philosophical* protocols are established, the implementation of alignment eyewear can proceed forward.

The trouble with EN207 is on the pulse side, it does not cover today's laser pulses of nano-, pico-, and femtoseconds. My experience has been a decrease of 1.4 OD is the maximum for alignment eyewear used for pulsed lasers.

LOW-LEVEL ADVERSE VISUAL EFFECTS

At exposure levels below the MPE, several adverse visual effects from visible laser exposure may occur. The degree of each visual effect is strongest at night and may not be disturbing in daylight. These visual effects are as follows:

1. *Afterimage.* A reverse contrast, shadow image left in the visual field after a direct exposure to a bright light, such as a photographic flash. Afterimages may persist for several minutes, depending upon the level of adaptation of the eye (i.e., the ambient lighting).
2. *Flashblindness.* A temporary visual interference effect that persists after the source of illumination has been removed. This is similar to the effect produced by a photographic flash and can occur at exposure levels below those that cause eye injury. In other words, flashblindness is a severe afterimage.
3. *Glare.* A reduction or total loss of visibility in the central field of vision, such as that produced by an intense light from oncoming headlights or from a momentary laser pointer exposure. These visual effects last only as long as the light is actually present. Visible laser light can produce glare and can interfere with vision even at exposure levels well below those that produce eye injury.
4. *Dazzle.* A temporary loss of vision or a temporary reduction in visual acuity.
5. *Startle.* Refers to an interruption of a critical task due to the unexpected appearance of a bright light, such as a laser beam.

SURPRISE FACTORS THAT AFFECT LASER EYEWEAR COATINGS

A common question is "Can I stand in front of any beam?"; the simple answer is you should never stand directly in the path of the beam and the other answer is NO. We are now in the realm of laser filter damage threshold (maximum irradiance) (Figure 16.3). At very high-beam irradiances, filter materials that absorb or reflect the laser radiation can be damaged. It, therefore, becomes necessary to consider a damage threshold for the filter. Typical damage thresholds from Q-switched, pulsed

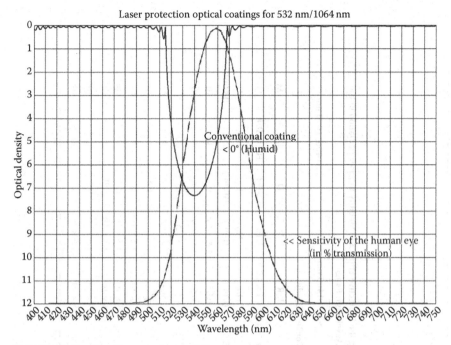

FIGURE 16.3   Humid versus dry conditions.

laser radiation fall between 10 and 100 J/cm^{-2} for absorbing glass, and from 1 to 100 J/cm^{-2} for plastics and dielectric coatings. Irradiances from continuous wave (CW) lasers, which would cause filter damage, are in excess of those that would present a serious fire hazard, and therefore, need not be considered. Personnel should not be permitted in the area of such lasers. A few words about international standards, EN stands for European Norm, in U.S. terms a standard (Table 16.2).

EN 207

Laser eye protection products require direct hit testing and labeling of eye protectors with protection levels, such as D 10,600 L5 (where L5 reflects a power density of 100 MegaWatt/m^2 as the damage threshold of the filter and frame during a 10-second direct hit test at 10,600 nm). Filter and frame must both fulfill the same requirements. It is not acceptable to select glasses according to OD alone. The safety glasses must be able to withstand a direct hit from the laser for which they have been selected for at least 10 seconds (CW) or 100 pulses (pulsed mode).

EN 208

This norm refers to glasses for laser alignment. They will reduce the actual incident power to the power of a Class 2 laser (<1 mW for CW lasers). Lasers denoted as Class 2 are regarded as eye safe if the blink reflex is working normally. Alignment

TABLE 16.2

Scale Numbers

Power and Energy Density (E, H) for Testing the Protective Effort and Stability to Laser Radiation in the Wavelength Range

Scale Number	Maximum Spectral Transmittance for Laser Wavelength $\tau(\lambda)$	180–315 nm			>315–1400 nm			>1400–1000 nm		
		D $>3\times10^{-4}$ L_D (W/m²)	I, R 10^3 to 3×10^4 H_{IR} (J/m²)	H $<10^{-3}$ L_H (W/m²)	D $>5\times10^{-4}$ L_D (W/m²)	I, R 10^9 to 5×10^{-4} H_{IR} (J/m²)	M $<10^{-9}$ H_M (J/m²)	D >0.1 L_D (W/m²)	I, R 10^9 to 0.1 H_{IR} (J/m²)	M $<10^{-9}$ L_M (W/m²)
					For Test Condition					
L1	10^{-1}	0.01	3×10^2	3×10^{11}	10^2	0.05	1.5×10^{-3}	10^4	10^3	10^{12}
L2	10^{-2}	0.1	3×10^3	3×10^{12}	10^3	0.5	1.5×10^{-2}	10^5	10^4	10^{13}
L3	10^{-3}	1	3×10^4	3×10^{13}	10^4	5	0.15	10^6	10^5	10^{14}
L4	10^{-4}	10	3×10^5	3×10^{14}	10^5	50	1.5	10^7	10^6	10^{15}
L5	10^{-5}	100	3×10^6	3×10^{15}	10^6	5×10^2	15	10^8	10^7	10^{16}
L6	10^{-6}	10^3	3×10^7	3×10^{16}	10^7	5×10^3	1.5×10^2	10^9	10^8	10^{17}
L7	10^{-7}	10^4	3×10^8	3×10^{17}	10^8	5×10^4	1.5×10^3	10^{10}	10^9	10^{18}
L8	10^{-8}	10^5	3×10^9	3×10^{18}	10^9	5×10^5	1.5×10^4	10^{11}	10^{10}	10^{19}
L9	10^{-9}	10^6	3×10^{10}	3×10^{19}	10^{10}	5×10^6	1.5×10^5	10^{12}	10^{11}	10^{20}
L10	10^{-10}	10^7	3×10^{11}	3×10^{20}	10^{11}	5×10^7	1.5×10^6	10^{13}	10^{12}	10^{21}

glasses allow the user to see the beam spot while aligning the laser. This is only possible for visible lasers (according to this norm *visible lasers* are defined as being from 400 to 700 nm). Alignment glasses must also withstand a direct hit from the laser for which they have been selected, for at least 10 seconds (CW) or 100 pulses (pulsed mode).

EN 60825

Requires that laser safety eyewear provide sufficient OD to reduce the power of a given laser to equal to or less than the listed MPE levels. It allows specification according to ODs in extreme situations, but recommends the use of EN 207 with a third party laser test. In neither standard is a nominal hazard zone allowed; the only consideration is protection against the worst-case situation such as direct laser radiation.

17 Introduction to Laser Beam Profiling

Larry Green

CONTENTS

INTRODUCTION

Lasers produce light with many characteristics that are unique compared to other light sources. For example, the monochromatic nature of a laser beam means that it emits photons at a single wavelength with very little energy at wavelengths other than the central peak. The temporal nature of a laser beam enables it to vary from a continuous wave (CW) to an extremely short pulse providing very high power densities. The coherence of a laser enables it to travel in a narrow beam with a small and well-defined divergence or spread. This allows a user to exactly define the area irradiated by the laser beam. Because of this coherence, a laser beam can also be focused to a very small and intense spot, resulting in a highly concentrated energy density. This property makes the laser beam useful for many applications in physics and chemistry and in medical and industrial applications. Finally, a laser beam has a unique spatial intensity profile that gives it very significant characteristics. The beam profile is the quantitative pattern of the distribution of the irradiance across the beam.

CASE FOR LASER BEAM PROFILING

There are many laser applications in which the beam profile is of critical importance. Because the beam profile is important, it is usually necessary to measure it to ensure that the proper profile exists. For some lasers and applications this may only be necessary during design or fabrication of the laser. In other cases, it is necessary to monitor the laser profile continuously during the process. For example, scientific laser applications often force the laser to its operational limits, and continuous or periodic measurement of the beam profile is necessary to ensure that the laser is still operating as expected. Some industrial laser applications require periodic beam profile monitoring to eliminate scrap produced when the laser beam quality degrades. In medical laser applications, the practitioner often does not have the capability to tune the laser. The laser manufacturer measures the beam profile in the design phase to ensure that the laser provides reliable performance at all times and is responsible for guaranteeing that the spatial beam profile remains stable. However, there are medical uses of lasers, such as photorefractive keratotomy (PRK), for which periodic checking of the beam profile can considerably enhance the reliability of the operation. PRK is an example of laser beam shaping, a process by which the irradiance of the laser beam is changed along its cross section. For this laser beam shaping to be effective, it is necessary to be able to measure the degree to which the irradiance pattern or beam profile has been modified by the shaping medium.

The importance of the beam profile is that the total energy density at the target, the spatial uniformity, and the collimation of the light are all affected by it. In addition, the propagation of the beam through space is significantly affected by the beam profile. Figure 17.1 shows three laser beam profiles among the variety of spatial profiles that can exist. Because almost every laser has a unique beam profile, it is essential to measure the profile to ensure that the spatial distribution is appropriate for the application.

BRIEF HISTORY OF SPATIAL LASER BEAM PROFILING

Several nonelectronic methods of laser beam profile measurement have been used ever since lasers were invented. The first and by far the simplest and least expensive method is observance of a laser beam reflected from a wall or other object. The problem with this method is that the response of the human eye is logarithmic and can differentiate only among orders of magnitude differences in light irradiance. In a single order of magnitude, a well-trained observer can distinguish only 8–12 shades of gray. Thus, it is nearly impossible for a visual inspection of a laser beam to provide anything even approaching a quantitative measurement of the beam size and shape. The beam width measurement by eye may have as much as 100% error.

Figure 17.2a is a photograph of a HeNe laser beam reflected from a wall (photographic film has even less dynamic range than the human eye). The image appears to be a very intense beam at the center, but a very large amount of structure far out from the center. This structure, however, which one might mistake as part of the laser beam, is less than 1% of the total energy in the beam. Yet, the eye and the film

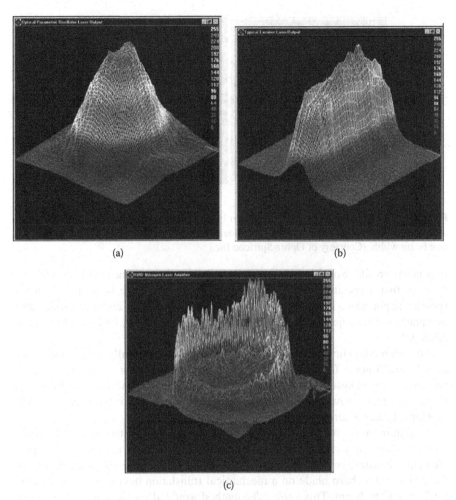

FIGURE 17.1 Various laser beam profiles: (a) HeNe, (b) excimer, and (c) nitrogen ring laser. Each has a unique and different spatial beam profile. (Courtesy of Ophir-Spiricon Inc.)

give erroneous importance to this region of low energy. In addition, the eye cannot really distinguish structure in a laser beam with less than 2-to-1 magnitude variation.

Burn paper, photographic film, and some polymers, such as Kaptan, have also been used for making beam profile measurements. Figure 17.2b illustrates thermal paper irradiated by a laser beam. The burn paper typically has a dynamic range of only three levels: unburned paper, blackened paper, and paper turned to ash. Sometimes, skilled operators can distinguish in between these levels and increase the dynamic range to five levels. The main objection to this manual method is that the spot size is highly dependent on the duration of the irradiance of the beam. With longer exposure times, the center of the burn spot may not change, but the width of the darkened area could change by 50% or more.

Wooden tongue depressors and burn spots on metal plates have also been used because the depth of the burn can often yield additional insight into the laser

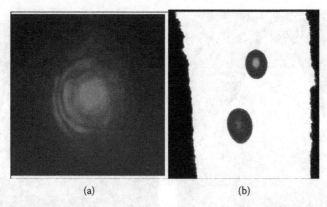

(a)   (b)

FIGURE 17.2 (a) Photograph of a HeNe laser beam as (b) it appears on a scattering surface and burn spots on paper. The human eye is logarithmic and will incorrectly determine the true beam width. (Courtesy of Ophir-Spiricon Inc.)

irradiance profile. Sometimes, operators learn from experience with these burn spots so that a specific pattern is indicative of an acceptable beam profile for a specific application. However, this measurement system is archaic, crude, and nonquantitative; subjective to the experience of the operator; and therefore quite unreliable.

Another method involves the use of fluorescing plates, typically used to view CO_2 laser beams. Energy from the laser quenches the fluorescing property roughly proportional to the intensity, and this pattern can be viewed by the technician. However, fluorescing plates have limited dynamic range, which adds to the dynamic range problem already described when viewing reflected beams.

As requirements for laser beam profiling became more stringent, electromechanical methods were developed. They generally involved the use of a single-element electronic or thermoelectronic detector to measure the total power of the beam and a sharp blade on a mechanical translation device that could accurately block the beam. This *knife-edge* method would allow the scientist to record the points on the translation stage at which the power reaching the detector was reduced by known amounts, typically 10% and 90% of the total. The difference in the location of the knife edge at these two points was then given as the beam width.

Once this method was in general use, several other *calculated* beam widths were defined, such as full width at half maximum; $1/e$ and $1/e^2$; percentage of peak; and percentage of energy. The proliferation of these measurements was that they all yielded different beam widths, even on the same laser beam. This prompted one researcher to comment that "Measuring a beam width is like trying to measure the width of a cotton ball with a caliper."

The International Organization for Standardization (ISO) has standardized the measurement of beam width under their specification 11146, which has made the measurement of laser beam characteristics unambiguous and identical for all beams. The standard measurement $D4\sigma$, commonly called the *second moment*, is clearly defined and can be made on all types of laser beams.

SPATIAL PROFILING REQUIREMENTS

Of course, modern beam-profiling techniques make spatial laser beam profiling quite simple compared to the early days of the laser. However, unless one takes particular care in defining the conditions under which the measurement is to be made, the results will not be representative of the laser beam. In general, the following must be given thorough consideration before a spatial beam measurement is performed:

- The type of laser to be measured.
- What kind of laser is it?
- What is the main wavelength of the laser?
- The operating condition of the laser.
- Is the laser running in pulsed mode or CW mode?
- The total average power of the laser.
- If CW, then what is the total average power (expressed in watts)?
- The energy per pulse of the laser.
- If the laser is pulsed, then what is the energy per pulse (expressed in joules)?
- In addition, what is the pulse rate (expressed in pulses per second or hertz)?
- What is the pulse duration, expressed in time units (milliseconds to femtoseconds)?
- The anticipated beam diameter (expressed in millimeter).
- If the beam is parallel (e.g., unfocused) or focused.

Once these issues have been properly addressed, it is then possible to define the instrumentation that can be used to spatially profile the laser.

TYPES OF BEAM-PROFILING SYSTEMS

MECHANICAL SCANNING DEVICES

One of the earliest methods of measuring laser beams electronically was using a mechanical scanning device. This usually consists of a rotating drum containing a knife edge, slit, or pinhole that moves in front of a single-element detector. This method provides excellent resolution, sometimes to less than 1 μm. The limit of resolution is set by diffraction from the edge of the knife edge or slit, and roughly 1 μm is the lower limit set by this diffraction. These devices can be used directly in the beam of medium-power lasers with little or no attenuation because only a small part of the beam is impinging on the detector element at any one time. Most of the time, the scanning drum is reflecting the beam away from the detector.

However, mechanical scanning methods are suitable only for CW (not pulsed) lasers. They also have a limited number of axes for measurement (usually two) and integrate the beam along those axes. Thus, they do not give detailed information about the structure of the beam perpendicular to the direction of travel of the edge. However, rotating drum systems with knife edges on seven different axes provide multiple axis cuts to the beam. This can assist somewhat in obtaining more detailed information about the beam along the various axes. These beam profile instruments are adaptable for work in the visible, ultraviolet (UV), and infrared (IR) by using different types of single-element detectors for the sensor. In addition, software has been developed that will display two-dimensional (2D) and three-dimensional (3D) beam profiles as well as fairly detailed quantitative measurements from the scanning system. This software now exists in the PC-based Windows operating system for easy use.

A variation of the rotating drum system includes a lens mounted in front of the drum (Sasnett, 1990). The lens is mounted on a moving axis and thus focuses the beam to the detector at the back side of the drum. By moving the lens in the beam, a series of measurements can be made by the single-element detector that enables calculation of M^2. A photograph of a seven-axis measuring instrument and readout is given in Figure 17.3.

For electronic laser beam profile analysis, it is nearly always necessary to attenuate the laser beam, at least to some degree, before measuring the beam with an electronic instrument. The degree of attenuation required depends on two factors. The first is the irradiance of the laser beam being measured. The second is the sensitivity of the beam profile sensor. Figure 17.4 shows a typical setup for when the maximum amount of attenuation is required before the sensor measures the beam.

Typically when measuring the beam profile of higher-power lasers (i.e., in excess of 50–100 W), the irradiance of the beam is sufficient to destroy most sensors that might be placed in the beam path. Therefore, the first element of Figure 17.4, the beam-sampling assembly, is typically used regardless of the beam-profiling sensor. It should be noted, however, that there are some beam-profiling sensors that can be placed directly into the path of a high-power beam of 10 kW for short periods of time under very specific operating conditions.

For mechanical scanning instruments, the beam-sampling assembly is usually sufficient to reduce the signal from high-power lasers down to the level that is acceptable by such instruments. For mechanical scanning instruments, lasers up to 50 W can be measured directly without using the beam-sampling assembly because the rotating drum blocks the energy from the laser during most of the duty cycle of the sensor; therefore, high power is not placed on the sensing element.

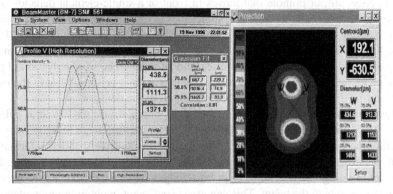

FIGURE 17.3 Typical readout from seven-axis system.

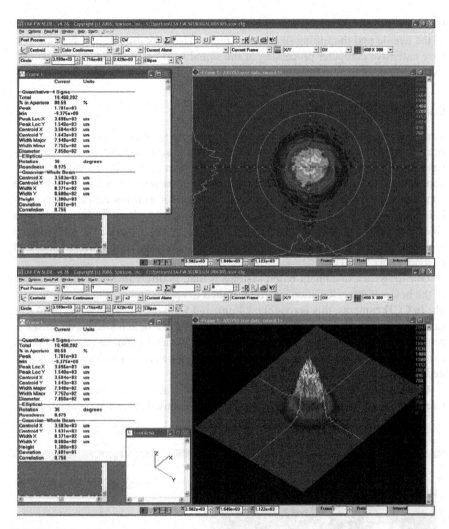

FIGURE 17.4 Schematic diagram for beam profiling.

Another mechanical scanning system consists of a rotating needle that is placed directly in the beam. This needle has a small opening that allows a very small portion of the beam to enter the needle. A 45° mirror at the end of the needle reflects the sampled part of the beam to a single-element detector (see Figure 17.15 later in the chapter). The needle is both rotated and translated across the beam to sample it, much as a fax machine scans a document. The advantage of a rotating needle system is that it can be placed directly in the beam of high-power industrial lasers for near- and mid-IR wavelengths. It introduces a small distortion in the process beam, so many users restrict its use. In addition, the translation of the needle can be made extremely small for focused spots or large for unfocused beams. It has, however, some characteristics of the rotating drum mentioned. Specifically, it is not very useful for pulsed lasers because of the synchronization problem between the position of the needle and the timing of the laser pulse. These rotating needle systems also have extensive computer processing of the signal with beam displays and quantitative calculations in Windows-based systems.

CAMERA-BASED IMAGING SYSTEMS

Camera-based imaging systems are used to provide simultaneous, whole 2D laser beam measurements in real time. Unlike mechanical scanning instruments, camera-based systems work with both pulsed and CW lasers. Most laser beam-profiling applications occur from the UV (193-nm) region to the near-IR at 1.1 μm. For these applications, silicon-based cameras are ideal because of their low cost and high resolution. Figure 17.5 illustrates the detail in the structure of a laser made capable with charge-coupled device (CCD) technology. Other types of cameras will image in the x-ray and deep UV regions, the far-IR above 1.4 μm into the terahertz band, and beyond, above 2000 μm.

One drawback of camera-based systems is that, in direct imaging, the resolution of the beam width calculations is often limited to approximately the size of the pixels. In CCD cameras, this is roughly 4.4 μm, and for most IR cameras the pixel size ranges from 25 to 100 μm. However, a focused spot can be reimaged with lenses to provide a larger image for viewing on the camera, which provides a resolution down to approximately 1 μm. In this case, the resolution for a camera-based system is limited by diffraction in the optics.

Cameras are now available with a variety of interfaces to connect to a computer. Current computer software packages provide very detailed 2D and 3D beam displays,

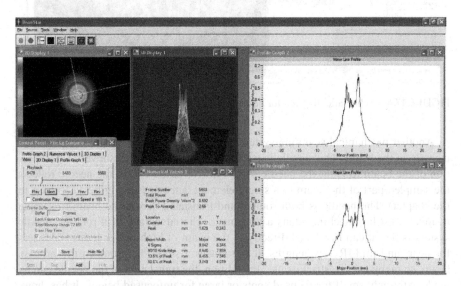

FIGURE 17.5 Highly structured laser beam measured with a charge-coupled device camera and shown in both two and three dimensions. (Courtesy of Ophir-Spiricon Inc.)

as shown in Figure 17.6. The software (Figure 17.7) also provides very sophisticated numerical analysis of the beam profile.

A camera-based beam-profiling system includes a computer; an interface (Frame grabber, FireWire, USB-II, or similar) to digitize the signal; and software to control the camera, display the beam profiles, and make quantitative calculations (Figure 17.4). A silicon CCD camera can be used from 190 to 1300 nm, covering the UV to the near-IR range. A pyroelectric, lead sulfide (PbS or Vidicon), indium gallium arsenide (InGaAs), or other camera is used for wavelengths to about 2.2 µm. CCD cameras have extremely sensitive sensors and can measure nanowatts of power. Other types of cameras can image from about 1 µW. Therefore, laser beam attenuation is almost always needed before it enters the camera. Often, the beam is either too large or too small, and beam sizing optics or other techniques must be used to size the beam for the camera. Figure 17.4 illustrates schematically the steps necessary for proper beam profiling.

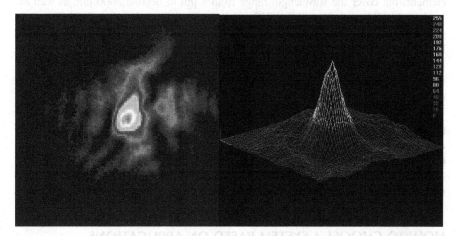

FIGURE 17.6   Examples of scientific-based beam-profiling software. The two-dimensional and three-dimensional images are displayed in real time, giving the user a powerful intuitive tool for visualizing and diagnosing the beam structure. (Courtesy of Ophir-Spiricon Inc.)

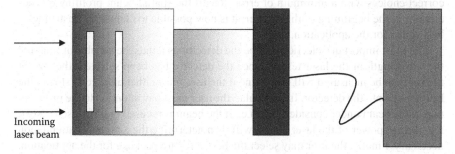

FIGURE 17.7   Example of industrial beam-profiling software. (Courtesy of Ophir-Spiricon Inc.)

WAVELENGTH DEPENDENCIES OF CAMERA-BASED SYSTEMS

Selection of the proper camera technology for each application is critical to proper imaging. Each type of camera has advantages and disadvantages for various applications. The most common type of camera for laser beam diagnostics uses a silicon-based sensor. These cameras consist of two types, charge injection devices and CCDs. Silicon-based cameras cover the wavelength range from 190 nm to 1.3 μm when the protective glass window is removed, which would otherwise attenuate the UV and also cause interference fringes in the image because of the parallel surfaces of the window. These cameras are fairly inexpensive, and because they cover the visible region, they are the most common cameras used. At slightly longer wavelengths, lead sulfide (Vidicon) sensors cover the region from the visible to 2.2 μm. A newer solid-state camera technology based on an InGaAs sensor also covers the range from visible to about 2 μm. The most common cameras for IR lasers of wavelengths longer than 1 μm are pyroelectric solid-state cameras. These are fairly low resolution at 124 × 124 pixels with 100-μm elements but cover the wavelength range from 1 μm to beyond 2000 μm, as well as work in the UV range from 190 to 350 nm. In addition, ferroelectric (microbolometer) sensors have been tried for the range of 4–12 μm, but their baseline characteristics have not been satisfactory for quantitative measurements, and they are seldom used for laser beam profiling. Finally, there are many cooled IR cameras that can be used for the wavelength range from 1 to about 12 μm. In general, their initial cost and the use of liquid nitrogen make them unsuitable for all but the most demanding applications.

In addition, there are many considerations when choosing a camera for beam profiling, especially when using pulsed laser sources in the near-IR region (1064 nm) as many sensors will produce a ghost image when the photons from the laser excite both the charge-coupled sensor and the silicon substrate. This technology is constantly evolving, and the user is advised to contact the major beam-profiling vendors to ensure that the latest camera sensors are used for best results.

HOW TO CHOOSE A SYSTEM BASED ON APPLICATIONS

The real challenge in beam profiling is how to determine the optimum combination of attenuation, sizing, sensors, and analysis for each application. Fortunately, Figure 17.4 enables both the experimentalist and industrial user alike to make the correct choices with a minimum of error. Recall the spatial beam-profiling considerations at the beginning of the chapter; it is now possible to choose the instruments best suited for the application.

The most important selection will be the detector as it must be capable of imaging the wavelength of the laser tested. Once the detector has been selected, the size of the beam to be measured will determine if the user must either enlarge or shrink the beam to fit onto the detector. In particular, the user must pay attention to the propagation of the beam under consideration (i.e., if the beam is focused or unfocused). Next, the average power of the laser output will then determine the amount of attenuation necessary. Finally, the user may select the best software package for the application.

The market for laser beam profiling is quite active and crowded. For this reason, the user should rely on major manufacturers of these instruments for their expertise so that a valid measurement may be made. If adherence to the ISO specification is required, it should be the responsibility of the experimentalist to verify that the vendor understands this requirement.

TEMPORAL BEAM ANALYSIS

Another critical parameter of the laser is its temporal profile, that is, the power or energy delivered during a single pulse of the laser. The temporal profile can be as important in laser processing as the spatial profile. Most laser manufacturers can shape a temporal laser pulse. This process has many applications in industry, such as treating metal components to improve strength and corrosion resistance. Other applications of temporal pulse shaping include drilling via holes in multilayer printed circuit boards. It follows that one needs to measure the temporal profile to verify that the laser is responding properly to the input waveform.

For most lasers, this measurement is fairly simple to perform. All that is required is a fast photodiode with a response that lies in the wavelength range of the laser and a suitable recording device such as an oscilloscope.

WIDTH MEASUREMENTS AND WHY THEY ARE IMPORTANT

The most fundamental laser profile measurement is the beam width. It is a measurement of primary significance because it affects many other beam parameters. The beam width gives the size of the beam at the point where it is measured. However, this can be significant in terms of determining the correct diameters of the elements in the optical train. Measurement of beam width is also a part of measuring the divergence of the laser beam, which is significant in predicting what size the beam will be at some other point in the optical train. The beam width is critical for the performance of most nonintegrating beam-shaping systems. Statistical measurement of the width of the beam is also a significant factor in determining the stability of the laser output. Finally, measurement of the beam width is essential in calculating the M^2 of the laser. In spite of its significance, the beam width is sometimes a very difficult measurement to perform accurately.

Recent ISO standards (Sasnett and Johnston, 1991; Siegman, 1990a–c, 1993; Woodward, 1990) have defined a second moment beam width, abbreviated $D4\sigma$, that for many cases gives the most realistic measure of the actual beam width. The equation for the second moment beam width is given in Equation 17.1. Equation 17.1 is an integral of the irradiance of the beam multiplied by the square of the distance from the centroid of the beam and then divided by the integrated irradiance of the beam. This equation is called the second moment because of the analogy to the second moment of mechanics and is abbreviated $D4\sigma$ because it is the diameter at $\pm 2\sigma$, which is equivalent to $\pm 1/e^2$ for Gaussian beams. This second moment definition of a beam width enables a user to accurately predict what will happen to the beam as it propagates, its real divergence, and the size of the spot when the beam is focused.

$$D4\sigma_x = 4\left(\frac{\iint (x-X)^2 E(x, y, z)\,dx\,dy}{\iint E(x, y, z)\,dx\,dy}\right)^{1/2} \tag{17.1a}$$

$$D4\sigma_y = 4\left(\frac{\iint(y-Y)^2 E(x, y, z)\,dx\,dy}{\iint E(x, y, z)\,dx\,dy}\right)^{1/2} \qquad (17.1b)$$

where $(x - X)$ and $(y - Y)$ are the distances to the centroids in the X and Y axes, respectively.

Sometimes, there are conditions of laser beams for which the second moment measurement is not an appropriate measurement. This is particularly true when there are optical elements in the beam smaller than twice the $1/e^2$ beam width, which cause diffraction of part of the energy in the beam. This diffraction will put energy further out into the wings of the beam, which, when measured by the second moment method, will cause a measurement of the beam width much larger than is significant for the central portion of the beam. In Equation 17.1, the $(x - X)^2$ term overemphasizes small signals far from the centroid. This requires judgment on the part of the users regarding whether measurement of this diffracted energy is significant for their application. If the diffracted energy, which typically diverges more rapidly than the central lobe, is not significant, it is possible to place a physical or software aperture around the main lobe of the beam and make second moment measurements only within this aperture and disregard the energy in the wings. However, if the application is dependent on the total amount of energy and it is important to know that part of this energy is diffracted, then one would want to place this aperture such that it includes all the beam energy in making the calculation.

When using a camera-based profiler, a number of camera characteristics must be carefully considered and accounted for in accurately measuring beam width. Among these considerations is the signal-to-noise ratio, that is, the magnitude of the beam signal relative to the background noise in the camera. The amount of attenuation used for the camera is usually adjusted to enable the peak pixel in the camera to be as near to saturation as possible without overdriving the camera. If the beam width is very small in a very large field of camera pixels, the beam may represent a very small amount of signal compared with the random noise of all the pixels. Proper treatment of this noise must be performed.

The camera baseline offset is another factor that must be accurately controlled. Because the energy of a laser does not abruptly fall to zero but trails off to a width roughly four times the standard deviation, or twice the $1/e^2$ width, there is a lot of low-power energy that must be accounted for in accurately measuring the width of the beam (Roundy, 1993, 1996; Roundy et al., 1993). (The percentage of total energy in a Gaussian beam is 68% in $\pm 1\sigma$, 95% in $\pm 2\sigma$, and 99.7% in $\pm 3\sigma$. Nevertheless, experiments have shown that as an aperture cuts off the beam at less than $\pm 4\sigma$, the measured beam width begins to decrease.) Correct and incorrect baseline controls are shown in Figure 17.8a through c. In Figure 17.8a, the baseline is set too low, and the digitizer cuts off all the energy in the wings of the beam. The beam is seen to rise out of a flat, noiseless baseline. This means that without the wings of the laser beam, a measurement would report a width much too small. In Figure 17.8b, the baseline offset is too high, as seen by observing the beam baseline relative to the small corner-defining mark. In this case, the software will interpret the baseline as part of the laser beam. A calculation of beam width will be much too large. In Figure 17.8c,

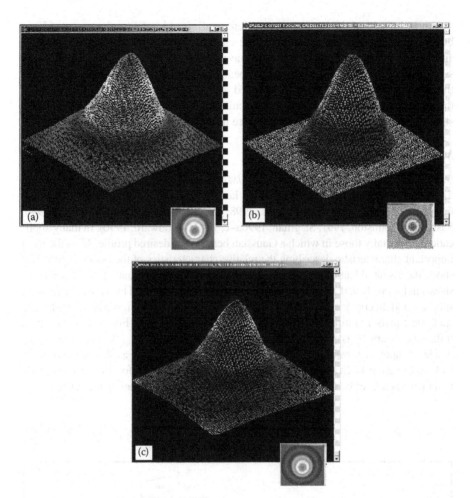

FIGURE 17.8 (a) The baseline is well above the zero mark, and the beam width is calculated 24% too large. (b) The beam rises out of a flat background, which shows clipping of the wings of the beam by the analog-to-digital converter. The width is calculated 26% too small. (c) The baseline is calibrated correctly, as seen by the wings just going to zero. (Courtesy of Ophir-Spiricon Inc.)

the baseline is set precisely at zero. Both positive and negative noise components are retained out beyond the wings of the beam where there is no beam energy. The software will interpret the average of the positive and negative signals as nearly zero.

Because the energy of the beam in the wings can have a significant effect on the width measurement, it becomes necessary to be able to characterize the noise in the wings of the beam. Both the noise components that are above and the noise components that are below the average noise in the baseline must be considered. The noise below the average baseline is called *negative noise*. Any software that does not correctly account for this negative noise will not yield correct beam width calculations.

Since the size of the beam measurement is affected by the total amount of laser beam energy relative to the noise of the camera, it has been found that software apertures placed around the beam can have a very strong effect in improving the signal-to-noise ratio. For a nonrefracted beam, an aperture approximately two times the $1/e^2$ width of the beam can be placed around the beam, and all noise outside the aperture can be set to zero in the calculation. This greatly improves the relative signal-to-noise ratio when small beams are measured in a large camera field.

M^2 AND WHY IT IS USEFUL

Increasingly, M^2, or k in Europe ($k = 1/M^2$), has become important in describing the beam propagation characteristics of a laser beam (Borghi and Santarsiero, 1998; Chapple, 1994; Herman and Wiggins, 1998; Johnston, 1990, 1998; Lawrence, 1994; Sasnett and Johnston, 1991; Siegman, 1990a–c, 1993; Woodward, 1990). In many applications, especially those in which a Gaussian beam is the desired profile, M^2 is the most important characteristic describing the relative characteristics of the beam. Figure 17.9 shows the essential features of the concept of M^2 as defined by Equations 17.2a and b. As shown in Figure 17.9, if a given input beam of width D_{in} is focused by a lens, the focused spot size and divergence can be readily predicted. If the input beam is a pure TEM$_{00}$, the spot size equals a minimum defined by Equation 17.2a and d_{00} in Figure 17.9. However, if the input beam D_{in} is composed of modes other than pure TEM$_{00}$, the beam will focus to a larger spot size, namely, M^2 times larger than the minimum, as shown mathematically in Equation 17.2b and d_0 in Figure 17.9. The ISO definition for the beam propagation factor of a laser beam uses M^2 as the fundamental measurement parameter:

$$d_{00} = \frac{4\lambda f}{\pi D_{in}}$$ (17.2a)

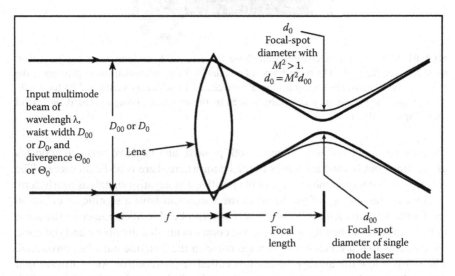

FIGURE 17.9 Curve showing M^2. (Characteristics and equations relating M^2 to the beam-focused spot size.)

$$d_{00} = \frac{M^2 4\lambda f}{\pi D_{in}}$$ (17.2b)

where λ is the wavelength of the laser, f the focal length of the lens, and D_{in} the width of the input beam at the waist.

In measuring and depicting M^2, it is essential that the correct beam width be defined. The ISO standard and beam propagation theory indicate that the second moment is the most relevant beam width measurement in defining M^2. Only the second moment measurement follows the beam propagation laws, so that the future beam size will be predicted by Equations 17.2a and b. Beam width measured by other methods may or may not give the expected width in different parts of the beam path.

M^2 is not a simple measurement to make. It cannot be found by measuring the beam at any single point. Instead a multiple set of measurements must be made as shown in Figure 17.10, in which an artificial waist is generated by passing the laser beam through a lens with known focal length. One additional requirement is to measure the beam width exactly at the focal length of the lens. This gives the divergence of the beam. Other measurements are made near the focal length of a lens to find the width of the beam and the position at the smallest point. The values of beam widths along the propagation axis are fitted to a hyperbolic function using a cubic fit. The coefficients of the cubic fit are then used to solve the M^2 equation in Figure 17.2b. In addition, measurements are made beyond the Rayleigh range of the beam waist to confirm the divergence measurement. With these multiple measurements, one can then calculate the divergence and minimum spot size; then, going backward through Equation 17.2b, one can find the M^2 of the input beam.

The measurement shown in Figure 17.10 can be made in a number of ways. In commercial instruments, shown in Figures 17.11 and 17.12, a detector is placed behind a rotating drum with knife edges, and then the lens is moved in the beam to effectively enable the measurement of the multiple spots without having to move the detector. This instrument works extremely well as long as the motion of the lens is in

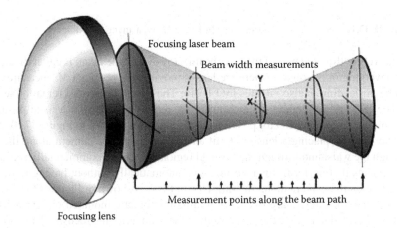

Focusing laser beam

Beam width measurements

Measurement points along the beam path

Focusing lens

FIGURE 17.10 Multiple measurements made for M^2.

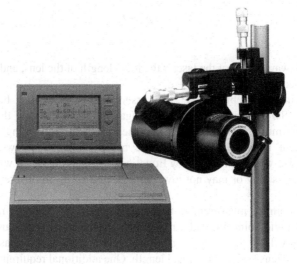

FIGURE 17.11 M^2 measuring instrument and readout based on scanning slit.

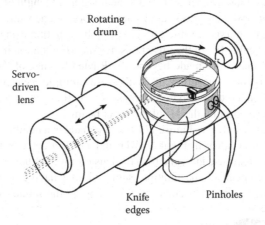

FIGURE 17.12   Schematic diagram of a slit-based M^2 measuring instrument.

a relatively collimated part of the laser beam. However, if the beam is either diverging or converging in the region where the lens is moving, the resultant M^2 measurement can be misleading. It also violates ISO 11146: The lens to laser must remain fixed!

The ISO (1993) method for measuring M^2 requires the lens at a fixed position, then making multiple detector position measurements as shown in Figure 17.10. This can be done by placing a lens on a rail and then moving the camera along the rail through the waist and through the far-field region. There are commercially available instruments that perform this measurement automatically without having to manually position the camera along the rail. One of these is shown in Figure 17.13, where the lens and the camera are fixed, but folding mirrors are mounted on a translation table and moved to change the path length of the beam after the focusing lens. A typical readout of an M^2 measurement is shown in Figure 17.14. In this case, a collimated

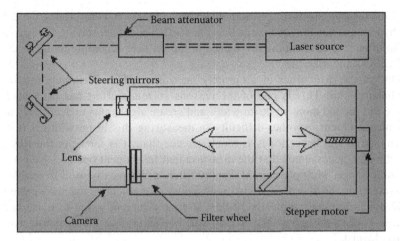

FIGURE 17.13 Schematic diagram of a camera-based instrument with fixed-position lens for measuring M^2.

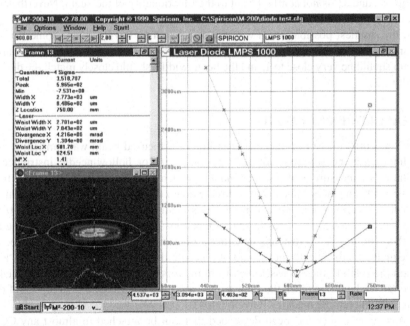

FIGURE 17.14 M^2 measurement display and calculation readout.

laser diode was measured, which gave a much greater divergence in the X axis than in the Y axis. The steep V curve displayed is the X axis of the beam coming to a focus following the lens. The more gradual curve is the focus of the less-divergent Y axis. Notice that while for most of the range the X axis has a wider beam width, at focus the X axis focuses smaller than the Y axis. Also, the X-axis M^2 was 1.46, whereas the Y-axis M^2 was only 1.10. The M^2 reported in the number section is calculated from the measurements of the beam width at the focal length (by fitting the data to the hyperbolic function and extracting the coefficients, then solving the equation), the minimum width, and the divergence in the far field according to the equations in the ISO standard.

One of the difficulties of accurately measuring M^2 is accurately measuring the beam width. This is one of the reasons so much effort has been made to define the second moment beam width and create algorithms to accurately make this measurement. Another difficulty in measuring this beam width is that the irradiance at the beam focus is much greater than it is far from the Rayleigh length. This necessitates that the measurement instrument operates over a wide dynamic range. Multiple neutral density filters are typically used to enable this measurement. An alternative exists with cameras or detectors that have extremely wide dynamic range, typically 12 bits, so that sufficient signal-to-noise ratio is obtained when the irradiance is low, still not saturating the detector near the focused waist.

There are some cases when M^2 is not a significant measure of the functionality of a laser beam. For example, flat-top beams for surface processing typically have a very large M^2, and M^2 is not at all relevant to the functionality of the beam. Nevertheless, for many applications in nonlinear optics, industrial laser processing, and many others, the smallest possible beam with the M^2 closest to 1 is the ideal. Some flat-top beam shapers are designed for an input Gaussian beam; then, the M^2 of the input beam should be very close to 1, and the beam widths should closely match the design width.

MEASUREMENT AT FOCUS

Until recently, measuring the spatial profile of a focused beam was limited to two technologies: the spinning slit and spinning wire or hollow-needle instruments. Rotating slit measurements use a cylinder on which one or more microscopically thin slits have been placed. The beam is directed onto the cylinder, and a single-element detector assembles a major and minor axis beam profile from the intensity–time relationship, which correlates to spatial location. The spinning wire/hollow-needle method (Figure 17.15) depends on either a micron-size pinhole in a hollow tube or an equally small reflector at the end of a needle that passes through the beam, and again the spatial image is assembled from the repeated passes of the wire or needle through the beam, in much the same way as a fax machine assembles a replica of the original on a remote machine (Green, 2004).

A new accessory has been developed that can be attached to almost any CCD camera-based system that performs the attenuation steps in an extremely short distance, so that the array can be positioned at or near the focus of the beam. The system uses standard quartz wedges, which reflect only 4% of the incoming energy and pass the rest through the bulk of the wedge material. By placing additional wedges in proximity to each other, enough attenuation can be achieved by the time the focused spot arrives at the camera array. In this manner, the beam profile of the focused spot as well as just before and just after the focus can be displayed in real time. In addition, for shorter focal length focusing lenses, a negative lens can be attached to translate

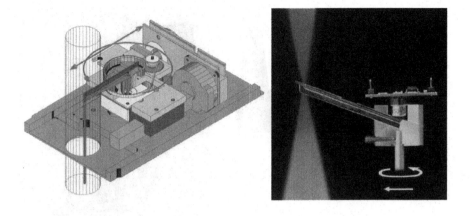

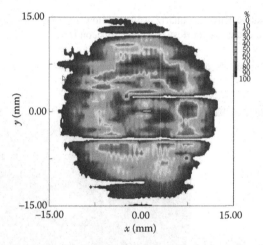

FIGURE 17.15 Spinning wire, hollow-needle, and slit devices all make composite images, taking as much as 10 seconds per image. The gaps in the image occur when a pulsed laser is not emitting. (Courtesy of Primes GmbH, Prometec GmbH.)

the focus further away, so the focus can be imaged on the camera. Figure 17.16 shows this device attached to a CCD camera.

It can be seen from Figure 17.17 that small focused spots can be imaged even with short (fast) focal length lenses. In this case, a 75-mm focal length lens produced a spot of only 65 µm.

LOCATION OF FOCUS AND WHY IT IS CRITICAL TO PROCESSES

The location of the focused spot and its size is critical to every process. Most processes are dependent on the irradiance of the beam squared. Since the irradiance is proportional to the area of the beam, it follows that the irradiance of the beam is proportional to the radius of the beam squared. Thus, the irradiance of the beam at

FIGURE 17.16 Negative lens and quartz wedges allow imaging of focused spots on a charge-coupled device camera. (Courtesy of Ophir-Spiricon Inc.)

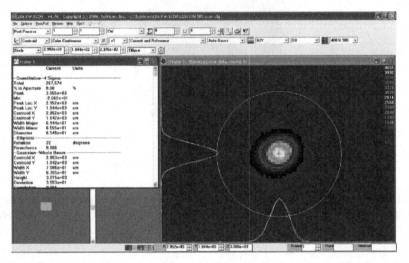

FIGURE 17.17 A 65-μm spot imaged from a 75-mm focal length lens. (Courtesy of Ophir-Spiricon Inc.)

the point of *contact* at the process is proportional to the radius to the fourth power. The beam width then plays a critical role in the size of the processing window. The ability to measure the actual focused beam width and its location can provide the user with new insights into the ability of the beam to produce the proper process ability.

Newer techniques are being developed almost monthly. The scientist and industrial user alike will no doubt wish to contact the major vendors of laser beam-profiling instrumentation to determine the best solution for each application.

REFERENCES

Borghi, R., and Santarsiero, M. (1998). Modal decomposition of partially coherent flat-topped beams produced by multimode lasers. *Opt. Lett.*, 23: 313–315.

Chapple, P. B. (1994). Beam waist and M^2 measurement using a finite slit. *Opt. Eng.*, 33: 2461–2466.

Green, I. (2004). Process monitoring of industrial CO_2 lasers. In *Proceedings of the 2004 Advanced Laser Applications Conference and Expositions*, eds. D. Roessler and N. Uddin, 107–113. ALAC: Ann Arbor, MI.

Herman, R. M., and Wiggins, T. A. (1998). Rayleigh range and the M^2 factor for Bessel–Gauss beams. *Appl. Opt.*, 37: 3398–3400.

International Organization for Standardization. (1993). Test methods for laser beam parameters: Beam widths, divergence angle and beam propagation factor (Document ISO/11146).

Johnston, T. F., Jr. (1990). M-squared concept characterizes beam quality. *Laser Focus World*, 26: 173–183.

Johnston, T. F., Jr. (1998). Beam propagation (M^2) measurement made as easy as it gets; The four-cuts method. *Appl. Opt.*, 37: 4840–4850.

Lawrence, G. N. (1994). Proposed international standard for laser-beam quality falls short. *Laser Focus World*, 30: 109–114.

Roundy, C. B. (1993). Digital imaging produces fast and accurate beam diagnostics. *Laser Focus World*, 29: 117.

Roundy, C. B. (1996). Twelve-bit accuracy with an 8-bit digitizer. *NASA Tech Briefs*, p. 55H.

Roundy, C. B., Slobodzian, G. E., Jensen, K., and Ririe, D. (1993). Digital signal processing of CCD camera signals for laser beam diagnostics applications. *Electro Optics*, 23: 11.

Sasnett, M. W. (1990). *Characterization of Laser Beam Propagation*. Coherent ModeMaster Technical Notes.

Sasnett, M. and Johnston, T. F., Jr. (1991). Beam characterization and measurement of propagation attributes. *SPIE*, 1414, 21–32.

Siegman, A. E. (1990a). Conference on Laser Resonators. Los Angeles, CA: SPIE/OE LASE '90.

Siegman, A. E. (1990b). Conference on Lasers and Electro-Optics. Anaheim, CA: CLEO/IQEC.

Siegman, A. E. (1990c). New developments in laser resonators. *SPIE*, 1224: 2–14.

Siegman, A. E. (1993). Output beam propagation and beam quality from a multimode stable-cavity laser. *IEEE J. Quantum Elec.*, 29: 1212–1217.

Woodward, W. (1990). A new standard for beam quality analysis. *Photon. Spectra*, 24: 139–142.

18 Vibration Control Is Critical*

James Fisher

CONTENTS

INTRODUCTION

When the laser was first experimentally proven in the 1960s, the platforms used were primarily machinery tables or granite slabs that were supported by rigid legs. The more ingenious investigators incorporated truck inner tubes or air suspension components to provide an additional amount of vibration isolation. As the laser has become more complex, and its applications more intricate, the platforms on which they are built have become more advanced in design yet their underlying principles of performance have remained the same. Having a basic understanding of the function and performance of these platforms will not only arm the laser user with knowledge but can also help them make the most efficient use of their budget so that they can assure proper investment is made in adequate laser safety measures in their laboratories.

---

* Portions of this paper include excerpts from the Newport website technical reference and original material prepared by the author.

In this chapter, we provide an overview of the sources of vibration, the key terms used to quantify vibration problems, and provide a detailed review of the common solutions for laser researchers. And finally, we conclude with some helpful hints and safety tips that are essential for all laser researchers.

VIBRATION SOURCES AND LOCATION SELECTION

In most cases, unwanted vibrations can be controlled through proper surveying, planning, and design and implementation of specialized components or equipment. There are generally three approaches used to reduce mechanical excitation of any type of system. The first is lowering the ambient noise level at the source; this includes machinery noise, acoustic noise, and other external vibrations. Locating and understanding the sources and their contributions to the unwanted noise in your system is a critical first step. Figure 18.1 shows typical sources of laboratory vibration and the frequency ranges they typically introduce into the environment. These may or may not disturb the users system depending on the frequency and magnitude of the vibrations and the resonant frequencies of the system and its components.

Where possible, vibration-sensitive systems should be installed on a solid, ground level or below ground levels away from the large machinery such as pumps, generators, or compressors. It is also recommended to locate sensitive systems near load-bearing walls, which generally stiffen the surrounding floor. Wherever

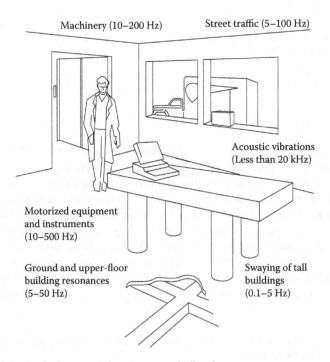

FIGURE 18.1 Typical sources of environmental vibrations.

possible, noise generating machinery should be mounted on elastomeric or rubber pads that attenuate vibrations before they are transmitted to the floor. In some cases, machinery can be contained in acoustic housings to reduce an acoustic noise as well.

In addition to specific room locations and adjacent room disturbances, it is important to consider building location and exterior sources of noise that can affect ambient vibration. Sites near highways and traffic can exhibit very high levels of low-frequency noise. Equipment located on upper floors can be exposed to building sway. Areas near loading docks can be subjected to significant noise from overhead doors or large seismic events caused when trucks *bump* the dock to find the stopping position, often times shaking the entire foundation of the building. Proper building design can reduce structural excitation and isolate the work environment from external seismic noise, so in many cases it is best to design solutions for these during construction or renovation before they become problematic.

REDUCING UNWANTED VIBRATIONS

After an optimal location is selected for the sensitive system, the next step is to understand what level of vibration reduction is necessary and what solutions need to be used to reduce vibrations. Before exploring the various options, it is important to understand a few key vibration control terms and specifications.

VIBRATION ISOLATION

Isolators filter vibrations between structural elements by reducing the transmission of such disturbances. Isolation can be achieved through the use of springs, rubber pads, air-filled bladders, pneumatic self-leveling isolation legs, or active isolation systems using piezoelectric or electrodynamic elements. Each offer varying levels of capability, performance, and of course price. In most cases an isolation performance is measured by a reduction in noise (dB) or a transmissibility plot that shows the isolator's performance over a specified frequency range. A sample transmissibility plot is shown in Figure 18.2. Because isolators prevent vibrations on the floor from reaching the work surface, they generally eliminate 70% of all vibration problems experienced in typical laboratories. However, vibrations that do reach the work surface from either acoustic, surface mounted equipment, or airflow generated sources require effective vibration damping (structural damping within the optical table) to reduce their effect on sensitive systems.

TRANSMISSIBILITY CURVES

Figure 18.2 shows a transmissibility curve of a pneumatic isolator. At 0 Hz, the curve initiates at unity transmissibility (T) = 1 (or 0 dB) vibration transmission. In other words, the isolation leg is essentially rigid at very low frequency, and any vibration amplitudes are exactly transmitted to the tabletop. The curve then begins to rise and peaks at 1–2 Hz. This is the natural frequency of the isolator. Anywhere on the plot where the curve is above unity transmission, the isolator increases the vibration level

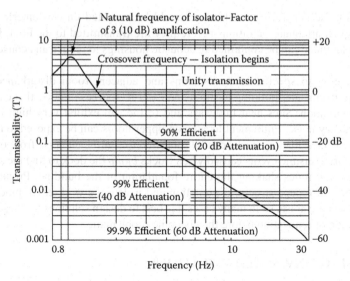

FIGURE 18.2 Typical vibration transmissibility plot indicates performance and resonance of isolator.

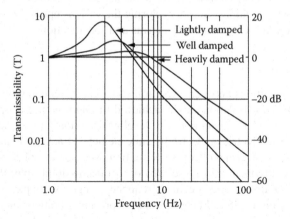

FIGURE 18.3 Transmissibility plots for isolators with different levels of damping.

before it reaches the tabletop. The height of this peak signifies the maximum amplification of the isolator design. Lightly damped isolators will exhibit a tall, sharp peak; heavily damped isolators will exhibit a lower amplitude, more rounded peak as shown in Figure 18.3.

At the natural frequency of most optical table isolators, vibration is amplified by a factor of 3–4 (10–12 dB). As the frequency increases beyond the isolator resonance, it begins to fall sharply. Once the isolator transmissibility curve falls below unity transmission (0 dB), it begins to *isolate*. The isolation crossover point is approximately 1.4 times the isolator resonance frequency. As the curve rapidly descends, the low-pass mechanical filter becomes more and more efficient. By 10 Hz most pneumatic isolators are filtering greater than 90% of the floor vibration. By 100 Hz almost 99% of the vibration is filtered.

What Transmissibility Curves Tell You

Transmissibility curves tell you how much vibration is transmitted at a given frequency and at amplitude levels above the frictional limitations of the isolator. These curves also give some indication of the damping present in the design.

What Transmissibility Curves Do Not Tell You

Transmissibility curves do not give an indication of what floor vibration amplitudes are used to generate the curve. Isolators using bearings or gimbaled pivot surfaces may exhibit significant amounts of friction that must be overcome before the isolator works. This is rarely seen during isolator testing because instrumentation limitations require the system to be shaken at amplitudes above the noise floor of test accelerometers and spectrum analyzers. However, state-of-the-art interferometers, optical inspection, acoustic microscopes, and profiling systems can detect levels far below conventional *vibration measurement systems.* In the nanometer region, isolator friction effects become a real problem. Transmissibility curves also do not give any real information about settling time, repositioning accuracy, and diaphragm life. These must be obtained from the manufacturer or measured in situ on the actual application.

Vibration Damping

Damping dissipates mechanical energy from the system and attenuates vibrations more quickly. For example, when the tuning fork's tips are immersed in water, the vibrations are almost instantly attenuated. Similarly, when a finger touches the resonating mass-beam system lightly, this damping action also rapidly dissipates the vibration energy. Vibration damping is achieved through the use of energy-absorbing dampers, which in their most basic form can consist of alternating layers of solid material and elastomeric material. In more complex and higher performance situations, the use of tuned mass dampers, which address specific frequencies, is preferred. Typically, structural damping is provided by the work surface in the form of an optical table or other rigid structure such as granite or aluminum plate. The differences and advantages of each of these platforms are discussed later in this chapter.

Compliance

Compliance is a measure of the susceptibility of a structure to move as a result of an external force. The greater the compliance (i.e., lower the stiffness), the more easily the structure moves as a result of an applied force. Compliance curves show the displacement amplitude of a point on a body per unit force applied, as a function of frequency. The dynamic performance of an optical table or other work surface is usually characterized with a compliance curve, a log–log plot of the table's dynamic response to random vibration. For nonrigid bodies, a compliance curve shows the structure's resonant frequencies and its maximum amplification at resonance. With other information, compliance curves can also furnish a reliable estimate of how a particular system will perform in your application. Figure 18.4 presents an example

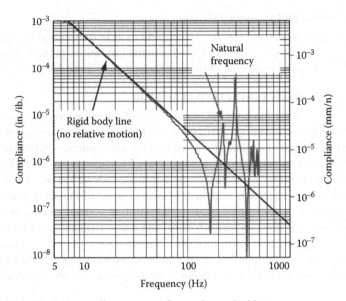

FIGURE 18.4 Typical compliance curve of an undamped tabletop.

of a compliance curve where the main resonant frequencies of the structure are identified. The compliance of a structure is a result of the stiffness-to-weight ratio, level of damping, and general shape.

The compliance curve in Figure 18.4 shows a sample of the positional motion of the structure for a given excitation force and frequency. The curve slopes downward without discontinuities until about 100 Hz. The region where the slope is straight is called the *rigid body line*. Throughout this region, the optical table is rigid and exhibits no relative motion across the surface, which is critical in maintaining beam alignment for laser applications. The position of the curve may change depending on the weight of the structure, but the slope is physically determined. The slope of 40 dB per decade is given kinematically by the relationship between the acceleration (caused by the input force) and the resulting position change. For a given level of excitation, a heavier granite table will exhibit less motion as a monolithic structure in inertial space than a lighter honeycomb structure, but along the straight portion of either curve the relative deflection will remain the same—zero. And except the gross changes in absolute position, it is the elimination of relative motion between components in the setup that is the most critical to the optical researcher.

However, at higher frequencies discontinuities appear along the curve. These peaks represent the amplitudes of the natural modes of the table. The first dominant resonance may be the torsional or bending mode; other modes and harmonics will follow after the dominant modes. The reason honeycomb is preferred over granite in many applications is that the higher stiffness-to-weight ratio produces resonance modes at higher frequencies.

For a given force input, granite may move less as a rigid body at low frequencies when compared with honeycomb, but the natural modes of the honeycomb are usually much higher in frequency. Moving these modes to higher frequencies offers many advantages:

- The rigid body line extends to higher frequencies.
- Less environmental vibration is usually present at higher frequencies.
- A given acceleration (force) produces less displacement (deflection) at higher frequencies.
- Isolation systems are more effective at higher frequencies.
- Optical components on the table typically have resonances in the 100–200 Hz region.

The compliance curve also gives information about the damping built into the structure. Damping reduces the amplitude of the relative motion across the table work surface. If the peaks are very sharp, little damping is present. In general, honeycomb structures have more inherent damping than granite structures. As a result, even though a granite structure may have lower absolute compliance than a honeycomb structure, the damped honeycomb structure generally has less relative motion between points on its surface. Honeycomb tables offer different flavors of damping. Broadband damping techniques are somewhat effective at reducing relative motion across a wide frequency band but in general only three times attenuation at the resonant peaks. In contrast, narrowband damping techniques, also known as tuned mass dampers, can virtually eliminate a natural mode (or modes) altogether and provide 10 times attenuation at the resonant peaks.

What Compliance Curves Tell You

Compliance curves tell you the frequency range over which the optical table acts as an essentially rigid body. The curve also provides the frequency of primary resonance modes and an indication of their relative amplitudes.

What Compliance Curves Do Not Tell You

Compliance curves only give the response at the points measured—not at all points where optical components will be mounted. The most effective measure of worst case deflection for rectangular structures is to place the sensor at the center of the table and/or in the corners. These measure the dominant torsional and bending modes. If the sensors are placed at antinodes (where little motion is present), even a poorly designed structure will look very good. For this reason some manufacturers will not offer compliance curves whereas others are quite liberal in their sensor placement. Compliance curves also do not provide any information about settling time, static deflection, surface damping effects, table flatness, internal stresses, and thermal response characteristics. These measures must be obtained from the manufacturer or measured in situ in the application.

Gaining an understanding of the main sources of vibration and the key terms used to quantify vibration control performance is critical to taking the next step to select the appropriate equipment to reduce unwanted vibrations.

UNDERSTANDING YOUR NEEDS

Determining the amount of vibration reduction necessary for a specific application is ideally done through a measurement analysis of the environment and the actual elements within the application. This would include a vibration site survey, modal analysis, and measurement of the various elements using accelerometers and possibly noncontact vibration analysis using laser Doppler vibrometers. Of course, this is an expensive and timely process reserved for only the most precise and demanding applications such as semiconductor wafer fabrication, interferometry, and holography.

However, any new experiment must take into consideration three distinct issues: first, the lab environment and sources of noise, ideally through a basic site survey; second, the level of protection needed for the experiment; and finally, any potential future changes that may require a higher level of performance.

SITE SURVEYS

A more popular and economical solution for new labs for moderately sensitive applications such as electron microscopy, multiphoton microscopy, and laser micromachining may be to consult with the vibration control supplier to determine if a basic site survey can be performed prior to final system configuration. A basic site survey is useful to determine the general noise level in the lab and verify that the product being purchased will provide the required vibration reduction. These site surveys generally include a multipoint floor measurement, assessment of any water, air, and gas lines to detect heavy flow rates, air handling assessment, and possible acoustic measurements to quantify any need for enclosures or beam protection. Figure 18.5 presents a sample site survey graph at a user lab. The measurement point in this particular lab had approximately 700 μin./s of movement that translates to a VC-B environment. In this environment, it would be very difficult to feel vibrations, and it is likely that less than 40 dB noise would be present.

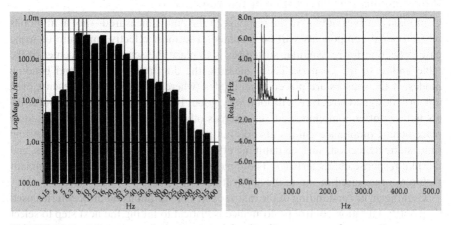

FIGURE 18.5   1/3 octave and power spectral density site survey graphs.

This VC-B environment would permit effective use of standard microscopes to 1000X and other applications with features sizes down to 3 μm with minimal vibration reduction products needed.

The results of any site survey should include the vibration data for review and also a suggested product solution or options to address the measured vibration problems.

CRITERIA CURVES AND COMPARISON

For general purpose, research experiments or studies such as light microscopy, electrophysiology, optical inspection, or precision assembly benches it is likely that a basic pneumatic or elastomeric isolator and medium performance table or breadboard surface is sufficient to meet the vibration reduction requirements. This of course assumes that the environment can be considered a *relatively quiet lab* or as shown in Table 18.1 an Operating Theatre. This table is based on the field data from individual systems and measurements made in facilities both before and after vibration problems had been solved. These basic criteria have been used extensively by leading vibration consultants in the semiconductor manufacturing industry for almost 20 years, and have been extended and refined as the industry has moved to narrower line widths. The figures take into account that equipment used for the most exacting tasks (such as manufacturing semiconductors with smaller device geometries) is stiffer and better isolated. It is, however, important to note that these criteria are for guidance only. For example, like any useful and general rule of thumb, the criterion is reasonably conservative for some specific cases, especially equipment with well-designed built-in vibration control systems. It should also be noted that these criterion curves do not replace the exact measurements provided by high-resolution narrowband spectrum analysis site surveys.

Table 18.2 has been modified to include new ratings developed by Newport to help guide product selection for most applications. The inclusion of *Environment Rating* and *Application Need Rating* serves to help researchers classify what type of environment their lab provides and also what are the needs of their application. Then, by comparing the sum of these two ratings with an optimal score of 10, the researcher can refer to Table 18.2 to determine what type of vibration solutions would help them achieve the level of vibration control needed.

These new ratings provide a guideline to determine the amount of vibration control required for a user's application. To use the ratings, first read through the various descriptions shown in Table 18.1 to categorize your current laboratory environment and then find the respective Environmental Rating. Next, find your Application Need Rating that corresponds to the smallest detail size you will be investigating and add the two numbers together. A score of 10 or less indicates that your environment will meet your application requirements. A score higher than 10 indicates that additional vibration control measures may need to be taken to achieve your goals. Using Table 18.2, you can select the combination of vibration solutions that will bring your score to 10 or below.

TABLE 18.1

Modified Criteria Curve

Criterion Curve/ (Environment Rating)	Max Level[a] μin./s (dB)	Detail Size[b] Achievable/ (Application Need Rating)	Description of Area
Workshop (9)	32,000 (90)	500 μm (1)	Distinctly felt vibration. Significant audible noise present from adjacent machinery, roads, elevators, or overhead doors. Typical of manufacturing areas, workshops, and warehouses. Adequate for heavy manufacturing and assembly
ISO Office (8)	16,000 (84)	250 μm (2)	Noticeable vibration. Noticeable audible noise from machinery, pumps, air handlers, or external disturbances. Typical of centrally located office areas, hallways, or upper floor laboratories. Adequate for basic component assembly stations, basic sample preparation areas, or break rooms
ISO Residential Day (7)	8,000 (74)	75 μm (3)	Barely felt vibration. Low-level audible noise from air handlers. small machinery, water lines, and external road noise. Typical of perimeter offices, laboratories, and buildings in seismic zones. Would be possible to sleep in this environment. Adequate for computer equipment, probe test and precision assembly equipment, lower-power (to 20X) microscopes, scatterometers, and sensitive sample preparation
ISO Operating Theatre (6)	4,000 (72)	25 μm (4)	Very slight vibration felt. Very low-level audible noise from air handlers or lighting. Adequate in most instances for microscopes to 100X and for other equipment and applications of mild sensitivity including optical/visual inspection, multiphoton microscopy, electrophysiology, fluorescence imaging, and optical profilometry
VC-A (5)	2,000 (66)	8 μm (5)	No vibration felt. Minimal audible noise for environmental control equipment. Adequate in most instances for sensitive equipment and applications including optical microscopes to 400X, microbalances, optical balances, proximity and projection aligners, optical trapping, fluid dynamics, or high-resolution laser imaging
VC-B (4)	1,000 (60)	3 μm (6)	No vibration felt and less than 40 dB audible noise. An appropriate standard for optical microscopes to 1000X, inspection and lithography equipment (including steppers) to 3 μm line widths

VC-C (3)	500 (54)	1 µm (7)	No vibration and less than 25 dB audible noise. A good standard for most lithography and inspection equipment to 1 µm detail size
VC-D (2)	250 (48)	0.3 µm (8)	No vibration felt and less than 15 dB audible noise. Suitable in most instances for the most demanding equipment including electron microscopes (TEMs and SEMs) and E-Beam systems, operation to the limits of their capacity
VC-E (1)	125 (42)	0.1 µm (9)	A difficult criterion to achieve in most instances. Assumed to be adequate for the most demanding sensitive systems including long path, laser-based, small target systems, and other systems

TEMs, transmission electron microscopes; SEMs, scanning electron microscopes.

Note: The information given in this table is for guidance only. In most instances, it is recommended that the advice of someone knowledgeable about applications and vibration requirements of the equipment and process be sought.

a As measured in 1/3 octave bands of frequency over the frequency range 8–100 Hz. The dB scale is referred to 1µin./s.

b The detail size refers to the line widths for microelectronics fabrication, the particle (cell) size for medical and pharmaceutical research, and so on. The values given take into account the observation that the vibration requirements of many items depend upon the detail size of the process.

TABLE 18.2
Vibration Reduction Ratings

Support Type	Rigid Frame or Legs	Elastomer (NewDamp™)	Air-Filled Bladder (SLM)	Nonleveling Pneumatic	Auto-Leveling Pneumatic	Active Isolation
Isolation reduction rating	−0.5	−1.0	−1.5	−2.5	−3	−4
Work surface type		Broadband damped work surface		Tune mass damped work surface		Actively damped work surface (SmartTable®)
Damping reduction rating		−1		−2		−4

As rule of thumb, follow these suggestions when selecting the proper solutions:

- Consider choosing a higher Isolation Reduction Rating solution if the unwanted disturbances are originating in the lab floor or are low-frequency vibrations of the entire structure (building).
- Consider choosing a higher Damping Reduction Rating solution if the unwanted disturbances are originating on the work surface, walls, air handling equipment, or acoustics. If acoustics or thermal disturbances are severe enough additional protection such as enclosures or chambers may be required.

For example, a typical research lab could be categorized as a VC-A environment that corresponds to an Environmental Rating of 5.

Next, let us assume that the specimen under investigation has 1–2 μm features that need to be measured or observed. This corresponds to an Application Need Rating of 7.

Add the two numbers, $5 + 7 = 12$, to understand what level of vibration reduction is needed by reviewing Table 18.2.

Based on this information, there are several combinations of products that could help reduce the unwanted vibrations. Using an air-filled bladder, like Newport's SLM isolator, would provide a 1.5 rating reduction and either a broadband (−1) or tuned mass damped table (−2) would deliver the additional reduction to reach an overall rating below 10. There are of course several combinations of products that could provide the needed reduction but the final configuration of products will depend on the source of the unwanted disturbances (floor, walls, air flow, acoustics, machinery, or table mounted equipment).

CONSIDERING FUTURE NEEDS

Generally, laser scientists working within a budget strive to get the maximum performance vibration control solution that their funds will allow. Within this approach are product selection alternatives that should be considered and economical methods to *future-proof* the platform to account for changes in the experiment or environment.

Consider an example where a new lab is constructed with state-of-the-art fixtures, amenities, and storage space. The lab is located in a relatively quiet area so the manager selects vibration equipment that is capable to perform in this environment. However, a year later the facility is required to update its air handling and emergency power equipment and the only place to locate it is in the mechanical closet just on the other side of the new lab. Now the manager is finding that the original platform is not able to isolate the vibrations and acoustic noise coming from the new equipment.

In 2005, when Newport developed the SmartTable, we also began investigating if the active damping system could be used to solve other vibration problems or incorporated into other products. The story of the new lab above was an actual customer situation. They were desperate to find a solution that did not involve completely dismantling their complex experiment and rebuilding it on a new higher performance platform. From this situation, we developed the first upgradeable optical table, the SmartTable UT, and its follow-on the SmartTable UT2. Both products allowed users to *buy into* a base level of vibration damping and, if needed in the future, could purchase additional performance without needing to tear down their experiment. This worked well for those who had made the right investment of the upgradeable table but what about those who need the improved performance but do not have an upgradeable table?

The latest iteration of the SmartTable technology is being called the SmartTable ADD. Seen in Figure 18.6, the SmartTable ADD is an active damping system that can be bolted to nearly any optical table to improve vibration damping performance. For the same price as a quality power meter or other instrument lab managers can save hundreds of hours of time and money by simply adding these active dynamic dampers to their existing experiment and with one press of a button allow the system to auto-tune itself to provide a higher level of damping performance to reduce disturbance affecting their experiment. The difference in performance provided by the SmartTable ADD can be seen in Figure 18.7. You can learn more about Newport's SmartTable ADD at www.newport.com.

FIGURE 18.6   SmartTable ADD damper mounted to an optical table.

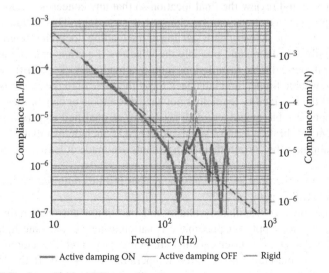

FIGURE 18.7   SmartTable ADD compliance curve.

FINISHING TOUCHES

The final step in putting together an effective vibration isolation system is considering some of the nonperformance issues listed next that could render any elegant product solution useless for your application. Although most of the items seem trivial, they have caused significant problems and delays for many researchers in the past. It is highly recommended that this final list be reviewed with your team and facilities personnel before making a final purchase decision.

1. Verify that the configuration of a doubled optical table or other custom solution is appropriate for your application (size, shape, direction, height, side clearance, bottom clearance, metric or Imperial tapped holes, special feature locations, etc.).

    In most cases, these are customized products built specifically for the customer application and would require significant effort and additional costs to modify if the product is not specified correctly. ALWAYS get design approval from your team, facilities, and major equipment suppliers before making a final purchase decision. This includes a review of the facility layout, the functional and maintenance needs of the major equipment (lasers, scopes, pumps, chambers), and user accessibility (doorway clearance, wall clearance, lab space utilization).

2. Verify that vibration control system(s) can be brought to—and will fit into—the final location.

    Clearly, science had made some significant advancement over these past 20 years but their vast discoveries have not provided a way to get a 10-ft.-long optical table into an 8-ft. room! This seems very basic but it does happen occasionally, even to the most seasoned professionals. Always ask your facilities team, colleagues, or riggers to walk through your installation route and review the final location so that any concerns are addressed and product selections can be changed if necessary. This includes every point along the delivery path starting at the loading dock, the freight elevator, hallways, doors, and possibly the windows if a crane is being used. In general, optical table installations require a loading dock, forklift, wheeled dolly or table flipper, and table jack to set the table onto the supports in the lab. If the site facility team does not have these tools or capabilities contact the optical table manufacturer for a listing of qualified riggers or installers. Most vibration control suppliers DO NOT include installation in their quotations so be certain that arrangements are made at least 2 weeks prior to delivery.

3. Verify that the final location has the proper utilities required for the vibration control system.

    This includes regulated voltage and maximum current requirements, clean/dry air supply for pneumatic isolation systems, adequate lighting if overhead shelves or laser curtains are being used, and ventilation/climate control to maintain the required temperature in the lab. If the location does not have clean/dry air a high-quality air compressor should be used to supply any pneumatic isolators. Dirty or wet air lines can significantly reduce the performance of isolation systems.

4. Carefully inspect the product packaging and follow supplier guidelines for storing and unpacking the product.

Although optical tables are very rigid and stiff structures, they can be damaged by improper handling and unpacking. A precision optical table is no match for a charging forklift! If the product packaging is damaged in any way do not accept the shipment before careful inspection of the contents. Usually, shipping damage is merely cosmetic and would require some touch-up paint, side panel repair or new corner pieces. But, if the damage is caused by dropping the table from a significant height or the fork truck tines penetrate the table core delamination is possible. Always consult the manufacturer the moment the packaging damage occurs and before any shipment is accepted. Shipments are usually insured by the carrier and compensation for damages would be provided by the carrier, not the manufacturer.

Unpacking the product according to the manufacturer's instructions is also very important. High-quality crates are designed with an upper bonnet and base that are separated by removing side screws, which provides maximum protection to the products and make table removal easy. A crowbar should never be used to dismantle this type of crate. Lower cost packing methods use nailed crates that unfortunately require the use of hammers, crowbars, or other devices and extreme care must be taken to not damage the table during unpacking.

When storing an optical table for future use a secure, dry location should be used. Under no circumstances should tables be stored outside or in high-humidity environments or in high-traffic areas where there may be a risk of damage.

5. Product safety is important, human safety is critical.

Under no circumstances should installation of an optical table system involve a dozen people lifting the table while several others move the supports into location. Installation is best left to professional who have the tools, experience, and proper insurance to safely install the system. Also, laboratories in seismic risk zones should have the proper seismic restraints on all system components per the current building codes. Consult facilities and safety personnel to ascertain that all seismic codes are met and large system components are properly restrained.

CLOSING COMMENTS

The material presented in this chapter was intended to provide an adequate level of information to help select an optimum location, product solution, and installation of a vibration isolation system. There is certainly more information available for those wishing to learn more about the design of the components, their performance limitations, and trade-off between performance and price. I would be negligent if I did not mention that a great starting place for this information is the Newport Corp. web site www.newport.com and more specifically at http://www.newport .com/Tutorials/979935/1033/content.aspx.

19 Power and Energy Meters
From Sensors to Displays

Burt Mooney

CONTENTS

From the time the first laser was built, physicians needed a way to measure how much power or energy was coming out of the output coupler. As a result, laser power and energy meters were born.

Because lasers are the good sources of concentrated heat, it was probably assumed that heat sensing methods would best be employed for the measurement of relatively moderate power. The simplest device to measure heat is a thermocouple. Another simple device to measure low-power light is a photodiode. So, some enterprising engineers designed and built such devices. Then they needed an instrument to display the results and give rapid feedback to tweak, align, or adjust the laser for maximum output. Early displays were basically analog meters that had a needle on a dial that went from left to right as the laser power went up.

CALIBRATING DEVICES

This approach was good for tweaking lasers for maximum output, but then calibration became an issue. How could one calibrate these devices to read the correct power? This is still a valid topic today as applications require higher degrees of accuracy and precision.

The National Institute of Standards (NIST) in Boulder, Colorado and Gaithersburg, Maryland have developed gold standards that quantify the uncertainty of products that are used to measure laser power and energy. Manufacturers send their equipment to NIST to have them measure uncertainty values at various wavelengths where they have developed primary standards based on physical constants. In the case of Boulder, they use a water calorimeter and measure the temperature rise of water, a known constant. This calorimeter approach is used to determine the uncertainty of a measured value of laser power at very specific power and/or energy levels.

These products then become the manufacturer's gold standards, which are used to calibrate silver standards and then working standards. Working standards are used on a daily basis to calibrate production volumes of detector heads that are sold to many users of lasers in a broad array of laser applications.

Some manufacturers, such as Gentec-EO, use gold standards to directly calibrate production sensor heads to increase the accuracy, and decrease the uncertainty of measurements traceable to NIST. NIST-traceable spectrophotometers are used to characterize the reflection across a broad spectrum, that is 300–2500 nm. This information can be stored in the E-Prom of the sensor head and used with a smart instrument that can then apply a factor for wavelengths that are not calibrated individually with a laser source.

As lasers progressed and developed other characteristics, like repetitive pulsing, variable pulse widths, high power, more wavelengths, and so on, pyroelectric sensors were developed to measure the energy output whereas variations of form and factor were created to handle the widening array of lasers being developed for new and exciting applications. Measuring precise laser power and/or energy became an important issue in the world that now included semiconductor manufacturing, uVia drilling, dermatology, and ophthalmic applications such as LASIK vision correction.

In the following sections, we will review the basic sensor types—thermopile, pyroelectric, and silicon photodiode—and cover the instruments that are used to display the power and/or energy of the lasers. We will explore what parameters one must look at to determine the best detector head for a particular laser and then match it up with a readout. This chapter will give some historical perspective on the evolution of power and energy meters and what the future holds.

DETERMINING SENSOR TYPE

One of the first things you need to do to select the correct sensor head is look at the laser parameter. Determine whether the laser is continuous wave (CW) or pulsed. If the laser is CW, then just finding out the minimum and maxim power that is expected tells you whether you will need to use a photodiode- or thermopile-type sensor head to measure it. Of course, the wavelength of the laser may be the first consideration as photodiodes have a limited wavelength range. Thermopile sensor heads have a very broad wavelength response.

If your laser is pulsed, then you may want to consider a pyroelectric sensor to measure pulse-to-pulse energy in joules. In addition, they typically allow you to convert joules to average power in watts and measure the pulse repetition frequency when used with a display instrument that supports such functionality.

If you are just measuring the output of the laser to tweak it up or adjust optics, a thermopile (series of thermocouples) power meter works just fine. They are less expensive than pyroelectric energy sensors and meters, and their associated processing electronics are much simpler than pyros. If you want to know the stability of the laser on a pulse-to-pulse basis, a pyroelectric or silicon sensor head allows you to measure every pulse up to 250 kHz pulse repetition frequency.

Let us take a closer look at the details of thermopile sensors, pyroelectric sensors, and photodiodes.

THERMOPILES

Thermopile theory and parameters:

- Generate voltage when there is a temperature difference at junction between two dissimilar metals, an array of thermocouples.
- A thermopile is an array of thermocouples.
- Broadband spectral response: deep ultraviolet (DUV) to far-infrared (IR).
- Dynamic range is from 50 µW to 25 kW.
- Used to measure average laser power, such as for CW and repetitively pulsed sources.
- Can be used in *single-pulse* mode in which a single pulse of energy is measured. There is a time delay between pulses of ~3–5 seconds whereas the algorithm and electronics integrate the pulse and the sensor head cools sufficiently between pulses.
- Diameters of the sensing area range from 10 to 210 mm.

There are two basic types of thermopiles:

Axial flow
- Heat flows along the axis; these have long time constants, from tens of seconds to minutes.

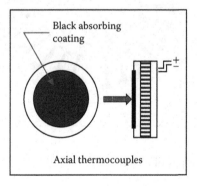

Radial flow
- Heat flows from the center to the edges, shorter time constants (in seconds) unless probe has very large area.
- With radial flow thermopiles DC voltage is generated when heat flows from hot to cold junctions between two dissimilar metals.

- Most companies today use thermopiles that are radial flow (primarily because of their faster speed of response).
- Disks are made of aluminum for powers to 300 W and copper for kilowatts sensor heads.

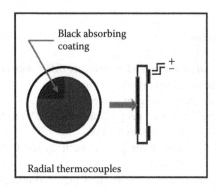

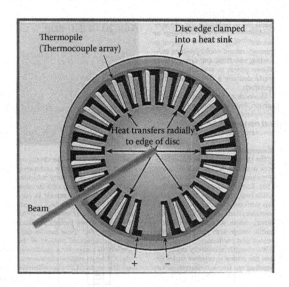

There are two basic types of absorbers:

Surface absorbers
The laser light is absorbed in the front surface of the sensor.
- Optical black paint: 500 W/cm^2, 50 mJ/cm^2 at 10 nanoseconds
- High-temperature ceramic: up to 45 kW/cm^2, 600 mJ/cm^2 at 10 nanoseconds
- Typically used for CW lasers and some pulsed lasers

Volume absorbers

The laser light is absorbed in a bulk material and then conducted to the metal disk. Typical lasers include Q-switched relative high energy Nd:YAG, Ruby, Alexandrite, and Erbium lasers with energy >100 mJ.

- Colored glass: good for high peak power, 10 J/cm^2 at 1064 nm, but low average power threshold at 50 W/cm^2.
- Combination of glass/ceramic: good for high peak power and moderate average power.
- All volume absorbers are typically used for high-energy short pulse width lasers.
- Diffusers: They will diffuse and scatter the light before reaching absorbing surface.

Thermopile sensors heads are cooled in one of three ways:

Air cooled

Water cooled

Fan cooled

Generally, air-cooled sensors have a larger physical size and have fins to use convection to get rid of the heat. If one needs a smaller physical size, then either a sensor with a built-in fan or a sensor with water channels to get rid of the heat would be considered.

When choosing a particular model of thermopile sensor, one needs to consider various laser parameters and match one up that meets those parameters:

- Maximum power
- Minimum power
- Beam spot size
- Does one need a surface absorber or volume absorber?
- Are there any size constraints?

PYROELECTRICS

Pyroelectric sensors are suitable for measuring pulse-to-pulse energy from nearly any type of pulsed laser. They are spectrally broadband, from x-ray to millimeters, so measuring lasers from DUV (i.e., 157 nm) to far-IR (i.e., 10.6 μm) to THz is not a problem. They can measure single pulses and repetitive pulses to hundreds of kilohertz.

Pyroelectric theory and parameters:

- Respond to the *rate of change of temperature*, the source must be pulsed, chopped, or modulated.
- When designed as a joulemeter sensor, they essentially act like capacitors in that they *integrate* pulses and produce an AC signal whose peak voltage is proportional to the energy in the pulse.

- Very fast, up to 250 kHz per pulse, possible at 1 MHz and beyond in the future.
- Broadband DUV to far-IR.
- Dynamic range from sub microjoules to joules.
- A sensor made from a pyroelectric material is an AC thermal detector.
- Ideal for measuring chopped or pulsed light sources (lasers).
- The current output from the sensor is proportional to the rate of change of temperature.
- Broadest band sensor known, x-ray to millimeters.
- All must be surface absorbers, as any type of volume absorber would add heat dissipation time making them virtually useless as fast temporal sensors.
- Four decades of range from a single detector.
- 1–100 mm diameter active areas.

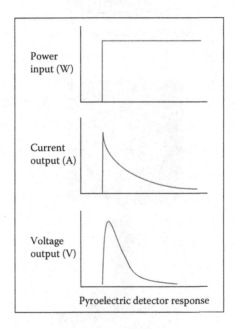

Pyroelectric detector response

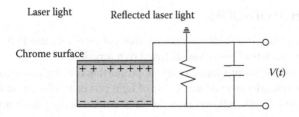

50 mm² pyroelectric sensor head

95 mm diameter pyroelectric sensor head

SILICON PHOTODIODES

A silicon photodiode is a solid-state device that converts incident light into an electric current. It consists of a shallow diffused p–n junction.

A photodiode sensor produces a current proportional to light intensity and has a high degree of linearity over a large range of light power levels—from fractions of a nanowatt to about 2 mW. Above that light level, corresponding to a current of about

1 mA, the electron density in the photodiode becomes too great and its efficiency is reduced causing saturation and a lower reading. Some manufacturers have sensors with a built-in filter that reduces the light level on the detector and allows measurement up to 30 mW without saturation. In addition, another removable filter allows measurement to 300 mW or all the way up to 3 W.

Photodiode theory and parameters:

- Spectral range from 190 to 1100 nm for silicon.
- Dynamic range from sub nanowatts to milliwatts, 3 W with filters.
- A response time of 50 milliseconds for power and average power measurements.
- Can measure CW power, and average power of repetitive pulses.
- Can be designed to measure energy per pulse up to 250 kHz (with Gentec-EO Mach6).
- Diameters of sensors are typically 5–10 mm.

DISPLAYS

When power meters were first developed, they were typically used with just one type of sensor. There was one type of display that you could plug into one type of sensor: pyroelectric, thermopile, or photodiode. This was well before the advent of microprocessors and E-PROMs. One had to use dials and screws to adjust the gain and correction factors for each head.

Initially most displays were analog meters, which worked for tweaking or tuning lasers. To read the actual laser power, one then had to look at the ticks on the analog dial to take the reading. As readouts evolved, so did the way they displayed their readings.

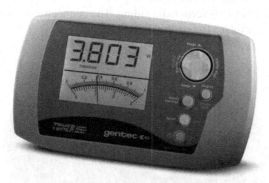

Analog/digital laser power meter display monitor

With the coming digital and computer electronics transformation, displays got a lot smarter. Almost overnight, displays had small computer chips and heads had E-PROMs. A display could read the contents of the E-PROM of each individual sensor head, know what type it was, and what dynamic range it had so that it would set range limits. One could even program each individual head's discrete calibration points and obtain a complete spectral curve. This revolutionized the world of power and energy meters. Meters and sensors became plug-and-play. All three sensor types could be used with one display. This not only made life much easier for users of power and energy meters, but also reduced the cost of ownership as one meter could be used with multiple heads.

With LCDs and, later on, dot matrix displays, one could more easily read laser power, although many people still liked to use analog meters for tweaking and tuning because of their fast response and easy trend displays, up or down. Some manufacturers started making instruments with both analog and digital displays, and some simulated an analog meter movement with a dot matrix display (see photo of analog/digital monitor).

Riding on the coattails of the display technology evolution, with the proliferation of color screen devices like PDAs, GPSs, and cell phones, the power and energy meter manufacturers can now afford to put color screens on their displays producing vivid vibrant displays that can be altered for color depending on the type of laser used and the glasses needed. For this technology the future has arrived. Some are now like dedicated mini touch screen hand-held tablets (see photo of Maestro).

As part of the computer revolution, the ability to record data became easier. With older analog displays, some had analog outputs that could be connected to an analog recorder that then plotted on paper a record of the meter's recordings. Then along came direct computer connectivity. The first was a serial interface, RS-232, and then a parallel interface, IEEE-488 (GPIB). Computer programs were created to collect and analyze the data on a computer.

Some instruments, such as Gentec-EO's Maestro, have a built-in USB flash drive that record as much data as the flash drive can hold, in addition to RS-232, analog output, and the first of its kind, an Ethernet port.

Today, RS-232 and GPIB seem to be fading. The connector of choice is now USB. Who knows what it evolves to into the future.

Color touch screen display monitor

PC OPTIONS

Sometimes one just wants to control the sensor and record the output data via a PC. They do not need a meter with an analog needle of digital display. Everything would be handled remotely from the PC. And while you are at it, can you make them more than one channel, and a version that I can connect to Ethernet and therefore read if from nearly everywhere in the world? Well, that is here now, with USB interface boxes like the P-LINK below, that are available in 1, 2, or 4 channels. And the S-Link below has either a USB or a USB and Ethernet connection in 1 or 2 channels.

USB interface box

INTO THE FUTURE

Recently introduced instruments into the marketplace are *built-in USB* sensor heads. These completely do away with the meter and even the PC USB interface box. They have the conditioning electronics all imbedded in the sensor head cable that terminates into a standard USB 2 connector. This simplifies everything; no extra box to plug into, that then connects to the PC. Just plug the sensor head cable, which has USB 2 connector on the end, straight into your PC or laptop. The included software then recognizes it and guides us through all the measurement parameters.

Who knows what other developments may occur in the future? I, for one, am looking for the unobtainium and indestructium materials that will be able to stand up to the most powerful lasers ever developed with nary a burn mark or fracture or pit in the absorbing surface. The data will be instantly analyzed and transmitted anywhere in the world in picoseconds and displayed on some kind of four-dimensional interactive display. The possibilities are endless.

Maybe there will be an application that works with Google Glasses, which collect the data wirelessly from the sensor head and allows the user to look to the side, up, or down, for measurement readings while adjustments to the laser are made.

20 Laser Shutters

David C. Woodruff

CONTENTS

DEFINITION OF THE LASER SHUTTER

The laser shutter is an optical and mechanical physical apparatus mounted at or near the output of a laser. It allows a beam to pass through unobstructed when commanded to open and safely terminates the specular beam with minimal backscatter when commanded to close. Closure takes place at the end of an operation or if the system experiences a safety breach. Well-engineered units provide a fail-safe closure on power loss or control signal loss. The beam is terminated by optical absorption inside the enclosure. A lightweight moving mirror system is used to steer the beam to the absorber, or beam dump, when in the closed position. The mirror and absorber optical properties must be rated for the demands of the laser irradiance. This separates a laser shutter from imaging-type shutters in cameras and vision systems and from rotary choppers. High reflectivity, high-damage-threshold mirrors, and high-absorption, ultra-low outgassing beam dumps are required in most laser shutters.

The goal of 100% throughput and 100% termination for open and closed states sets the laser safety shutter apart from any of the modulation-type devices, including electro-optic, acousto-optic, and liquid crystals. Rapid closing speed is a desired feature, attainable through careful design. A position-sensing circuit, using various switch types, is a common feature added to the laser shutter to independently monitor the state of the aperture regardless of the open/close command signal and power.

ROLE OF THE LASER SHUTTER

The primary role of the laser shutter is safety. This falls in two categories: general safety of personnel and living organisms and economical safety protection of materials and lost time.

Of utmost importance is the protection of the laser operator, staff, and visitors in any part of the potential exposure zone. Second, research animals, living tissue cultures, and microbes under irradiation and study must have protection in the case of an irradiation problem. Irreversible damage or destruction of living organisms usually carries much more than just a material loss, especially in advanced biological research.

Exposure of personnel or other living organisms can come from many sources. The primary source is usually the laser beam itself or excessive scatter after incidence on other items in the beam path. Additional concerns come from the laser beam producing photochemical-induced toxic vapors, toxins in tissues, thermal burns, cell nucleus damage, and ignition of materials resulting in fire and potentially toxic smoke.

A laser shutter needs to react quickly to any signal commanding it to close so threats mentioned do not have significant time to develop. As the laser power levels increase, closing speed of the shutter becomes more important. A well-designed laser shutter can often close more quickly than an electrical power-down command to the actual laser can perform. Stored charge in capacitors and energy in inductors in the laser power supplies can sometimes delay the shutdown, whereas a typical shutter may close in under 10 ms.

The economic impact of a safety breach usually results in lost time or materials in production processing environments and research applications. If something goes wrong in a long train of elements, such as an optic burn, then the laser safety shutter will quickly block the laser and protect the rest of the system. Although there are many means to protect personnel, such as enclosures, curtains, and the like, the laser shutter is usually the final element to protect the system from component failure escalation and the loss of valuable processed materials and tools. The shutter allows the laser to keep running with full thermal equilibrium and stable modes, permitting beam exit on command. This sharply cuts time lost waiting for laser stabilization after a safety breach has occurred. The breach is investigated and acknowledged and exposure is restored without the long wait, usually measured in minutes; a long time in processing systems. Typical types of safety breaches seen in the economic sense are $x–y$ table movement errors, processed material surface irregularities, and foreign objects entering the work area, such as material remnants or slugs. In each case, the system would detect a safety problem and close the shutter; the operator would quickly rectify the problem and the process would continue. A minute of time may have been consumed.

Another safety aspect of the laser shutter is the use in extremely important applications in which multiple shutters are used redundantly. Three shutters may be placed in series, opened one by one, to facilitate critical operations such as pyroinitiation of rocket propellants, explosives, or one-shot state-of-the-art research. Here, the economic impact can be quite severe, and the mechanical laser shutter is a layer of safety along with many sensor systems and controls.

REQUIRED FEATURES AND PERFORMANCE

Because the laser shutter needs to absorb all of the energy entering it when closed and needs to reduce any backscatter to a very low level, the optical and thermal systems in the design must scale with the laser power. Fluence and power density become important issues to avoid mirror damage and absorber damage. The shutter needs to be a drop-in, clean element that does not generate significant particles or outgassing contamination. Usually, there are constraints on the shutter switching speed, aperture, physical envelope size, and available control power (electrical or pneumatic). A designated lifetime, usually measured in cycles open/closed, is often specified. Most designs will experience some mechanical wear, fatigue, and particulates over time. Because the laser is nearby and optics trains typically follow the shutter, particulates and outgassing are a major design criterion.

When closed, the shutter sees the full optical load and will experience a temperature rise. The internal beam dump absorber must be designed with excellent thermal conductivity in addition to optical absorption. This heat can then be conducted out to a convenient surface for cooling. Conduction and water convection are the most popular methods. Forced-air cooling is usually not preferred because of air gradients and turbulence around the optical zone. Water cooling is usually implemented above about 50 W. Most systems will absorb power levels lower than this into their system mass. Designs should be for the worst case, where the beam is continuously dumped in the shutter for long periods.

Position sensor systems include mechanical microswitches, opto-interrupters, and magnetic reed switches. The independent audit function serves to validate that the command to open actually occurred, and the shutter is not at some point between open and closed. Potential obstructions in the aperture are more likely to occur as the aperture size increases. Many systems use dual switches in both open and closed states for the added security of redundancy.

USER SAFETY APPLICATIONS

The most popular use of laser safety shutters is in original equipment manufacturer (OEM) capital equipment. These manufacturers use the shutter primarily for safety shutdown purposes and occasionally for safe processing control and low cycle rates, about 1–2 Hz. Because the shutter is providing dual duty for safety and processing, the demands on the design for high cycle lifetime with minimum wear or change are imperative. Products designed for the OEM market tend to be the most robust by trading off some speed and packaging miniaturization for long life or in some cases practically unlimited lifetime. An OEM laser shutter will be linked to the machine's main information-processing board for control, and the position sensors are used to monitor the aperture state and decide when to perform the next operation. Panel switches on the instrument usually interlock with the shutter to protect any user from irradiation when opening the instrument while power is on.

Researchers are another primary category of users implementing the laser shutter on their optical table, breadboard, or chassis. It becomes just one of many elements, but is usually the first one, adjacent to the laser. These applications usually implement a control box, commercially supplied by laser shutter manufacturers, that provides interlocking features for laboratory door switches, curtain switches, or pressure mats. Light-emitting diode (LED) displays give the researcher the status of the shutter with a simple glance.

Processing applications have primarily economic concerns or concerns for safety of their precious materials, so the shutter is safeguarding overexposure. This can be in welding, cutting, or photochemical processes such as lithography. Any detection of overexposure can be a direct signal to close the shutter.

TECHNOLOGIES IMPLEMENTED

Because every application has a different set of features emphasized, the designer can tailor the technical features for the job. For commercial manufacturers, the design is more complicated because they must try to implement as many desirable features as possible to cover a wide range of applications. Currently, there are only a handful of manufacturers offering laser shutters with a majority of the desired features designed in one device.

Mirrors

The primary element, and the first in the optical path, is a lightweight mirror capable of handling high continuous wave (CW) power and surviving the pulse threat at the wavelength range of interest. Because most commercial lasers range from 157 nm to about 11 μm in wavelength, a move to use broadband mirrors cannot be successful at high powers. Usually, metal mirrors are used for CW at lower powers in the ultraviolet (UV)/visible range and for higher powers in the infrared. For pulsed laser applications, the mirror nearly always is mandated to glass substrate, dielectric coated mirrors.

How the mirror is mounted is very important regarding light leakage through the coating. As power levels increase, leakage of light for a 99.5% reflective coating can become significant. Nearly all designs call for a mirror bracket, usually made of metal, to hold the mirror and to absorb, scatter, or redirect the leakage via specular reflection.

Metal mirrors can sometimes be used as a monolithic bracket/mirror. The application of metal mirrors often requires using a high incidence angle to spread the energy over the surface. Damage thresholds are typically an order of magnitude less than a dielectric mirror. Attaching a mirror to a bracket requires attention to shock and vibration so that movement of the mirror in the bracket does not generate particles or allow the mirror to chip and fracture. Spring-loaded concepts and elastomeric bonding are popular. The mirror bracket needs to stay reasonably cool to avoid damage, and also because it is adjacent to the beam when it is pulled out of the way. If it is very hot, temperature gradients occur in the air where the beam passes. This can introduce wave front distortion. Designs often mandate that the mirror bracket be able to withstand a complete failure of the coated optic, or full irradiance of the beam, for a selected period of time. Typically, this is on the order of several seconds, but only in systems that monitor smoke, surface temperature of the mirror bracket, or excessive scatter with independent sensors tied to the safety system. In such a case, the safety system would shut down the power to the laser. When using a moving dielectric mirror, one must maintain the angle of incidence within a few degrees on the coating during all movement. Some relief on this constraint is provided with metal mirrors, but for polarized sources and high angles of incidence one should be aware of the principal angle (analogous to the Brewster angle) of the metal and corresponding absorption for one polarization state. Polished, coated, ferromagnetic metal flexure mirrors represent the lightest weight, fastest moving high-power mirrors. The flexure is pulled by an electromagnet. Although dielectric coatings can be applied to a polished metal surface, the damage threshold is usually much lower than glass due to the surface roughness.

One should also design the moving mirror edge to roll across the beam so that only the coated surface or transparent substrate edge is irradiated. The use of *window frame* optic brackets is not advisable except at low-power levels. It is possible to capture any optic and leave a path for the extended section of the optic to interrupt the beam and have a metal bracket section, slightly within the edge of the coating, to block the leakage through the coating.

The use of a lightweight, high-integrity mirror allows one to direct the energy to a stationary absorber or beam dump. It is very difficult to cool an absorber that is moving in and out of the beam with any appreciable speed. Thermal cooling umbilical straps, hoses, and the like have significant mass. For this reason, moving mirrors are used almost exclusively. The faster we can close, the less damage we do when an unsafe condition has been detected.

MIRROR MOVEMENT

Once we have a mounted mirror choice, a system for moving the mirror must be designed. Electromagnetic is by far the most popular approach, with pneumatic systems having special niches. For an electromagnetic system, we need to have the mirror system attached to some ferromagnetic mass to which we can attract an electromagnet. Some simple solutions are readily available, such as partial rotation motors and solenoids. More advanced systems design custom magnetic geometries and commutating ferromagnetic mirror assemblies. Common design approaches to mirror movement are motor shafts, hinges, levers, flexures, pistons (solenoid and pneumatic), and linear slides. For highest reliability, designs with the fewest moving parts are preferred.

Rotating shaft devices, such as motors and galvanometers, have low torsional friction. This is due to bearings. A design emphasis in most bearings is point contact (balls) or line contact (rollers). Both exhibit extremely low cross-sectional area for thermal flow. They are usually lubricated. Sleeve-type bearings are only slightly better for thermal flow and are almost always lubricated or impregnated. If we mount a mirror on a mirror bracket, then attach this to the rotor of such a device, the leakage light for dielectric mirrors or absorption in a metal mirror will manifest itself as heat flowing into the shaft and rotor, warming the lubrication. For higher laser power levels, this approach can lead to outgassing. Enough outgassing on nearby optics is a potential failure mechanism and safety concern. Design emphasis should be placed on getting residual heat out of the moving mirror element. Hinges are another example of a heat flow roadblock. Flexure devices can carry heat down their length to the fixed end. Careful design can allow any of the above techniques to work if intimate contact of the mirror bracket is made with a large mass in the closed position. This lets leakage light heat to take a preferred path out some simple mechanical rest plate.

To arrive at some sort of fail-safe design, independent of gravity, most designs will use some spring loading that the electromagnet or pneumatic system must work against to open the shutter. Then, this stored spring energy should be capable of closing the shutter if the control power is lost in any way. Here, friction can play a role, and designs that show no change over time in required closing time are highly preferred. Bearing wear, lubrication loss or change, gummed rollers, or slides all result in frictional changes, potentially being capable of dominating the stored spring force to close. Stiction of the moving mirror with any surface must be carefully reviewed. This is often a concern with polymer materials and life or temperature changes. A failure of the moving mirror to release from an open position stop due to stiction can render the safety shutter useless. The devices used to open and hold open the mirror in general will dissipate heat, and this heat must be managed. Unless the heat is very small, a thermal path should be designed to allow the joule heating from a magnetic winding to be removed via conduction to a convenient surface for cooling. The fail-safe design is the most popular because many laser systems can be oriented in many directions with respect to gravity. They are also used in space and aerospace applications, where high shock/vibration conditions can exceed gravity forces, requiring a preset, known spring force to be built into the design.

Pneumatic systems are popular when used in high magnetic field environments or if any electrical discharge could initiate a safety concern. Sometimes, a magnetic field from another optical device, such as a faraday rotator, can influence a laser shutter using an electromagnet system. A magnetic shield, in the form of a steel aperture plate, is usually sufficient to reroute the field and allow the laser shutter to operate as intended. Another method is to locate the shutter further from the device or vice versa.

Absorbing the Beam

Now that we can move the laser beam with the mirror on demand, we need to focus on the stationary absorber. We want this optical absorber to be stationary so that we can attach it and carry away heat developed as the beam is absorbed. Volume absorbers, such as crystals or doped glass, have limited use at higher power levels because of low thermal conductivity. Surface absorbers in general have the best high-power capability. Water cooling can always be added to their mounting plane. The base material must have significant thermal conductivity, usually dictating copper or aluminum for economic reasons. The focus now becomes how to make this element compact within practical limits, highly absorbing, surviving the pulse threat, and not outgassing or generating particulates over time. We must give special attention to geometry, move away from the concept of a black surface and think about a black path. Trying to achieve all of the desired features with a single incidence absorption is not possible for most lasers, especially as powers increase. We would like to spread the power through the surfaces of the absorber so that we can choose geometries that use high incidence angles and fold the partially reflected or forward-scattered beam as it bounces through the absorption path. The idea here is to achieve somewhat of a progressive absorption flux waveguide. The first incidence does not need to absorb everything; even 50% or less can be effective as long as we continue absorbing and using forward scattering surfaces for each bounce. Clever designs can fit progressive absorbers into the smallest of volumes. We want to avoid surfaces with high potential to backscatter to the mirror and back to the user. This usually favors high angle of incidence paths in the absorber.

With the geometry designed, now the absorber surface morphology and material are chosen. The material is usually a plating. But, because surface morphology, or microscopic roughness, is more important, we need to choose a coating, plating, or evaporated finish that can be manipulated to produce a desired surface morphology. There are not a great deal of choices, but if carefully researched for the band or wavelength of interest, one can find materials that furnish both good atomic or molecular absorption from the material and the potential to be modified, usually by chemical or irradiation techniques, for ideal surface morphology. This ideal morphology would be tall, conical pillars, on the order of tens of wavelengths. Such a surface may only be possible in an attenuated region of the absorber or for lower power levels. High-energy pulsed lasers can destroy such surfaces because the absorbed energy cannot flow as heat fast enough down the pillars to eliminate melting or vaporization. Plating materials should be thin to get the absorbed heat to the main substrate, which was chosen for high thermal conductivity and convenient mounting. Even the highest melting point refractory metals can exhibit damage and oxide smoke when exposed to pulsed lasers, so geometry, surface morphology, and choice of atomic absorption coating are the design hierarchy to avoid such damage.

Cleanliness

The sophistication of a modern laser shutter requires it to be treated as an optical instrument. Particulate generation is given attention for sensitive applications, such as in semiconductor manufacturing. Outgassing properties are sensitive issues to all parties using the shutter with other adjacent optical elements or in sealed instrument environments. The amount of design effort given to these issues can highly influence costs, and a good balance of compromise is usually found for most applications. With the popularity of UV sources for industrial processing, many challenges are present. Replacing convenient materials in the shutter construction with photochemically insensitive ones is a major effort. Clean shutter construction, and designing such that subsequent cleaning can be performed by a user, is a desirable feature of design.

PROPER USE OF A LASER SHUTTER

A first step in using a laser shutter is to mount the device rigidly, with some alignment capability and with good thermal flow to a larger mass. Typically, any shutter dumping over about 5 W should have significant thermal design in the bracket. Post and base configurations are very poor due to near-zero thermal cross section. The thermal bracket should extend, with full cross-sectional area, from the shutter mounting plane all the way to the *infinite mass*, commonly identified as an optical table or chassis. A water-cooled chiller plate is also viewed as an infinite mass. Because we can experience software control bugs and operator errors, it is recommended to design the long-term thermal capacity of the mounting arrangement for worst-case conditions. Most safety shutters see the energy for only a short period before laser power shutdown, but if the shutter does both safety and processing, it could see long-term closed cycles and require worst-case cooling. Without proper cooling, many shutter designs will lose efficiency in their electromagnetic systems and may open slowly or not at all. The issue of air temperature gradients also emerges without proper cooling. Actual forced-air cooling is not very popular because of flowing air around optics, particulate redistribution, and turbulence air gradients.

If the shutter mirror and absorber are accessible, the operator should have a maintenance cycle to blow off dust with clean gas or clean them with standard optics solvents. Many designs are closed, requiring clean precautions from initial use throughout the product lifetime other than dry gas blow-offs.

Nearly all laser shutter designs are mechanical. The motion of the mirror system can generate shock and vibration and even some recoil on impact, commonly known as *bounce*. Mechanical design considerations can eliminate or reduce this action, with added compromises, or it can be addressed with advanced electrical circuitry. Using overdamped or critical damping, the accelerations near stopping points can be gradual, with low shock. Recoil bounce can be eliminated. Depending on the sophistication of the electrical drive and the shutter design, the user should be cautioned that the bounce could expose an edge of the beam, essentially extending the total time to open/close. If position sensors are integral to the product, then the user's circuitry used to process the sensor signal needs to wait until the mirror has ceased bouncing and settled. Failure to build in settling *wait* periods for sensors can generate an unsafe condition, either economically in processing or physically for operator safety, even though the time frame may be milliseconds or tens of milliseconds.

CONTROL BASICS

Drivers or controllers for the shutter deliver power to create the desired motion, open and closed. They typically accept a low-level signal, such as transistor–transistor logic, and convert this to a current waveform or release of air pressure sent to the shutter. Most arrangements seek some degree of speed and will send a more powerful impulse to open the shutter, then drop to a lower level to hold it open long term. Electrical circuits used are similar to those implemented for other electromechanical systems, such as motors, valves, and solenoids.

Units providing damped motion for the shutter are similar to circuits used in galvanometer drives. These driving techniques can be tailored to achieve low vibration, high speed, or high electrical efficiency, provided the shutter design is intended for the particular feature. The driver and shutter are a mated system, and for highest extraction of a single- or multiple-performance feature, must be viewed as a system from the initial design phase. They function much like a speaker and amplifier matched pair.

The controllers that are commercially available offer many safety features to allow convenient stand-alone use or connection with door interlocking and warning lights. Most offer an interlock connection, a simple make/break contact to run through any series of door, curtain, or pressure pad switches. An armed condition indicator allows the user to know whether the interlocks are closed and the system is ready for actuation. Reset functions allow the user to reset the system after a safety breach. A breach maybe from a door interruption or restoration, the user acknowledgment of this situation is by using a reset function to arm the controller of the shutter. From that point, either external signals from computer systems or manual toggle switches can be used to actuate the shutter open. OEM systems are usually fully automated, with microprocessors handling the decision making and reading position sensor information. Numerous displays are used to indicate status of the shutter, with LEDs the most common. A stand-alone controller instrument for a research application measures about the size of a common brick, whereas an OEM, with limited space, will design the circuitry on their system printed circuit board to be as small as 2×2 in.

GUIDELINE FOR DESIGN OR COMMERCIAL PURCHASE

Whether deciding to embark on a custom design, developing a statement of work for an outside engineering contractor, or making a commercial purchase, a hierarchy should be used to rapidly resolve compromises and extract the maximum benefit of the convolved features. For safety shutters, the recommended progression usually follows this path:

1. Determine physical package envelope constraints; this is your working space.
2. Definitely know the full beam size; aperture selection is later.
3. Get CW maximum power, wavelength, and polarization information.
4. Pulsed laser parameters, pulse width, repetition rate, energy, wavelength, and polarization must be determined.
5. Aperture determination depends on alignment, dithering, and potential diffraction rings.
6. Switching speed is determined; magnetic/pneumatic designs use more space for speed.
7. Controller circuit decision; whether to design or commercially purchase and the available power.
8. Determine if position sensors are required or not.

These items cover most applications. In special cases for clean room use or space applications, more design or selection criteria steps are involved.

21 Free-Electron Laser Safety Challenges

Focus on the Jefferson Lab Free-Electron Laser

Patty Hunt and Stephen Benson

CONTENTS

INTRODUCTION

The free-electron laser (FEL) technology dates back to the 1970s. When compared with conventional lasers, the significant difference lies with the laser medium. In the FEL, the laser medium is no longer a chemical, such as a gas or a crystalline structure, but an electron beam in an oscillatory magnetic field. Current FELs cover a wavelength range from millimeters to angstroms. The advantages of the FEL over conventional lasers include wavelength tunability and high powers that can be obtained in areas of the infrared and ultraviolet spectrum that have not been achieved using conventional lasers.

Applications that have benefited from use of FELs include biology, physics, medicine, and materials science. In the field of materials science, FELs can offer substantial cost and capability advantages, including advantages for high-volume materials processing over traditional manufacturing tools. FELs are being developed to process plastics, synthetic fibers, advanced materials, and metals as well as components for electronics, microtechnology, and nanotechnology. Prospective products include durable yet attractive polymer fabrics for clothing and carpeting; cheap, easily recyclable beverage and food packaging; corrosion-resistant metals with increased toughness; mechanical and optical components with precisely micromachined features; microcircuitry; and electronics for use in harsh conditions.

The price paid to achieve FEL-quality laser light is high. The conventional tabletop "black box" that houses the laser medium of a conventional laser must be replaced by an intricate facility that houses an electron beam accelerator and a magnetic field–producing device that causes the electrons to emit coherent light. The costs include the cost of construction, maintenance, and staffing. For this reason, most FELs in the United States are operated as federal or academic-funded user facilities.

In such facilities, the electron beam is usually created and accelerated using a large and complicated linear accelerator (linac). The beam of electrons is then sent through a periodic magnetic field known as a wiggler, where the light is produced. The light can be either recycled in a laser resonator or amplified to saturation in a single pass through a long wiggler. The wavelength of the light is a function of the electron beam energy and the strength and wavelength of the wiggler. Because of this, the FEL is very broadly tunable. Under the right circumstances, the accelerated electrons in the wiggler develop density modulations with an optical spacing (this phenomenon is referred to as microbunching). Over the length of the wiggler (or after multiple passes of the light through the wiggler), the electrons in the microbunches begin to oscillate in step (coherently), thereby giving rise to light with properties characteristic of conventional lasers. Because the electron microbunches can be less than a picosecond in length, the light generated comes in ultrashort pulses that can be used for strobe-like investigations of extremely rapid processes.

This chapter discussion is based on the FEL Facility at Jefferson Lab (JLab). At the JLab facility, the laser medium is isolated from the laser light laboratory. With effective isolation, the safety considerations can also be addressed separately; therefore this chapter will primarily address the safety considerations of the FEL laser laboratory and its interface with the laser medium (Figure 21.1).

FIGURE 21.1 Wiggler (arrow) and section of accelerator (metal cylindrical can in background).

JEFFERSON LAB FREE-ELECTRON LASER FACILITY

DESCRIPTION OF THE FACILITY

The JLab FEL Facility is located at the Thomas Jefferson National Accelerator Facility in Newport News, VA. The program and this user facility derive from the primary mission of JLab, namely, nuclear physics research and the world's first large superconducting accelerator for generating continuous multibillion-volt beams of electrons, called the Continuous Electron Beam Accelerator Facility.

The FEL specifications are outlined in Table 21.1. A comparison of the FEL specifications with other light sources is presented in Figure 21.2.

OVERVIEW OF THE JLAB FEL COMPONENTS

The basic components of the FEL at JLab include the lasing medium, the beam transport and delivery system, and the laser laboratories.

Lasing Medium

The electron beam accelerator and wiggler that serve as the laser medium are located in the basement of the facility, referred to as the vault. Figure 21.3 depicts a schematic view of the device. The JLab FEL is based on an energy-recovered linac. Electrons are released from the injector and are accelerated in a superconducting linac. After emerging from this linac, the electrons pass into a laser cavity that has a wiggler near its center. This wiggler causes the electrons to oscillate and emit light, which is captured in the cavity and used to induce new electrons to emit even more light.

TABLE 21.1
FEL Specifications

Wavelength range (infrared)	0.9–10 μm
Energy/pulse	120 μJ
Pulse repetition frequency	Up to 75 MHz
Pulse length	200–17,000 fs FWHM*
Maximum average power	>10 kW
Wavelength range (UV/VIS)	250–1,000 nm
Energy/pulse	20 μJ
Pulse repetition frequency	Up to 75 MHz
Pulse length	100–1,700 fs FWHM
Maximum average power	>1 kW

*FWHM, full width at half maximum.

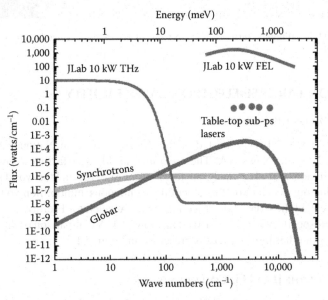

FIGURE 21.2 Jefferson lab free-electron laser (JLab FEL) power versus conventional sources.

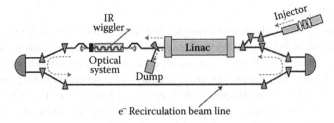

FIGURE 21.3 Schematic of a simple FEL beam accelerator and wiggler.

After exiting the optical cavity, the electrons then travel around the beam transport loop and back into the linac. There they give up most of their energy to a new batch of electrons, making the process highly efficient.

Laser Beam Transport and Delivery System

The laser beam delivery system transports the laser beam from the FEL to a series of laser laboratories. The beam is shared among the laboratories. Required components include the initial beam delivery pipe, optics control room (OCR), laser lab beam delivery pipe, mirror cassettes, and beam dump. All operations of the accelerator and laser laboratories are managed from a central control room.

Initial Beam Delivery Pipe

The laser transport system is generally a stainless steel vacuum line. This allows the beam to be transported with minimal losses even at wavelengths at which the air is strongly absorbing. The vacuum system must be properly designed with pressure relief valves and well-shielded high-voltage lines for the vacuum pumps.

Optics Control Room

From the initial beam delivery pipe, the laser enters the OCR, where part of the beam is picked off for diagnostics, and the remainder is prepared for delivery to the laser laboratories (Figure 21.4). This is the first location where laser light is accessible on an optical table.

Laser Lab Beam Delivery Pipe

The laser lab beam delivery pipe is a stainless pipe under vacuum that transports laser light from the OCR to the laser laboratories.

FIGURE 21.4 Laser diagnostics table in optics control room.

Mirror Cassettes

Mirror cassettes are remote-controlled mirror systems that reflect the beam from the beam delivery pipe into the laser laboratory.

Beam Dump

The beam dump is located at the end of the laser lab beam delivery pipe. It is a water-cooled, copper cone coated with a low-reflectivity coating that can absorb up to 50 kW of laser light.

The control room (Figure 21.5) provides an area where all operations of the accelerator and laser laboratories are computer controlled.

Laser Laboratories

The second floor of the FEL houses the laser laboratories and radio frequency (RF) power supplies for the beam electron accelerator (Figure 21.6). Laser laboratory capabilities are provided to permit testing of various materials. Since the targets vary with each experiment, maximum flexibility in configuration of the facilities is desired. In each lab, control of the laser must be provided. Required light characteristics for each lab that can be manipulated include polarization, wavelength control, power control, pulse length, rastering, and focal spot waist and position.

FIGURE 21.5 Control room.

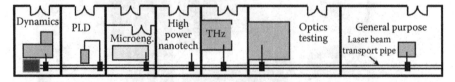

FIGURE 21.6 Laser laboratories on second floor. PLD is pulsed laser deposition. The microengineering lab operates in the ultraviolet. The laser delivery transport enters from the right and ends in the high-power dump (dark gray box). Gray areas are hutches. Small black boxes on the laser beam transport pipe are mirror cassettes.

Advantages of the JLab FEL

Advantages associated with the use of the JLab FEL are similar to those associated with most FELs. Since the gain medium only exists in the wiggler for a few nanoseconds and then leaves at the speed of light, there is no waste heat issue for FELs. Because of this they are, in principle, capable of producing very high average power. This power is available at any desired wavelength, and the wavelength can be continuously tuned using either the wiggler strength or the electron beam energy. The gain medium (the electron beam) is smaller than the optical mode (the laser beam). This leads to a repeated spatial filtering that results in an output beam with excellent optical mode quality. In FEL oscillators especially, the optical mode is a pure TEM_{00} mode. Because the gain medium exists in a vacuum, the laser is capable of operating at wavelengths where there are no transparent materials. In addition, if the experiment is mounted in a vacuum chamber, the light never needs to go through a window. This eliminates the need for a transparent window at the laser wavelength. Most FELs take advantage of high repetition rate picosecond or subpicosecond electron pulses as their gain media. The output is then a high repetition rate series of ultrashort, high-energy micropulses. This time structure is very effective for materials modification and processing.

Applications for the JLab FEL

Currently, the JLab FEL Facility has seven laser laboratories. Experiments currently under study have applications in the manufacturing, nanotechnology (Figures 21.7 and 21.8), and medicine. To secure beam time at the FEL, an application is

NASA/JLab nanotube synthesis — research to production

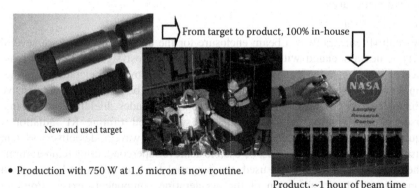

New and used target

From target to product, 100% in-house

Product, ~1 hour of beam time

- Production with 750 W at 1.6 micron is now routine.
- Production rate of 2-6 g/hour of as-grown, high quality, "research grade" raw material is already cost competitive in $400/g market.
- Nanotube diameter is strong function of laser parameters, suggesting the possibility of "designer" tubes (selectable diameter and perhaps chirality).
- Experimental trends indicate improved gross and net yield with soon-to-be-available shorter FEL wavelengths and higher power (no scale-up issues).

FIGURE 21.7 Nanotechnology experiment. (From Mike Smith, NASA LaRC.)

Differential heating of fatty tissue — News coverage

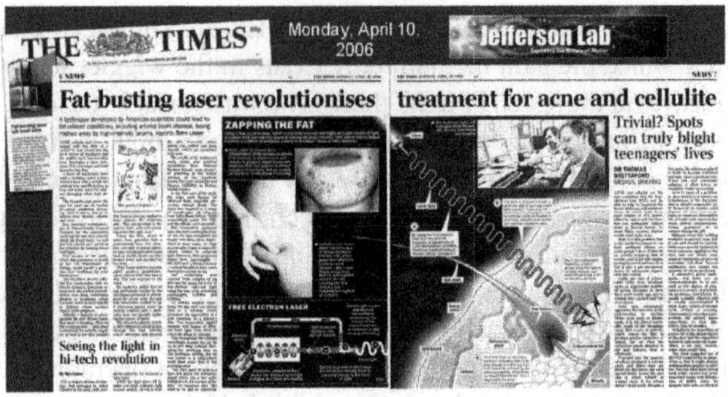
THE TIMES

Monday, April 10, 2006

Jefferson Lab

Fat-busting laser revolutionises treatment for acne and cellulite

ZAPPING THE FAT

Trivial? Spots can truly blight teenagers' lives

FREE ELECTRON LASER

Seeing the light in hi-tech revolution

FIGURE 21.8 News coverage of FEL experiment in medical application research.

submitted and reviewed by committee. When an experiment is approved, actual experimental time is allotted at no cost to the user. In addition to the FEL, users can and do bring ancillary equipment such as magnets, furnaces, and other Class 4 lasers.

FEL Infrastructure

The required infrastructure for the JLab FEL includes physical structures and management systems for the laser medium, delivery system, and laser laboratories.

Physical Structures

Accelerator Vault

The vault (Figure 21.9) is a beam enclosure for the accelerator; there are several safety issues associated with the accelerator vault. Most significant is the existence of ionizing radiation. Whenever an electron beam is accelerated in a vacuum, bremsstrahlung (known as braking radiation) can be produced. An extensive radiation control program is required. The program includes shielding, monitoring, personnel entry control interlock system, and specialized training. RF radiation is required to accelerate the electrons, so an RF leak-tight waveguide delivery system must be used. Liquid helium must be used to achieve superconducting temperatures. Wherever the liquid helium is used, the potential for heat influx and subsequent evaporation and pressurization of the accelerating components exists. Pressure relief devices must be present to mitigate the explosion potential associated with overpressurization, and a program to manage oxygen-deficiency hazards (ODHs) associated with unexpected venting of liquid helium must be in place. The program must address requirements such as ODH assessments and controls, oxygen monitors, local exhaust, and training.

FIGURE 21.9 Cryomodule in the vault. Note the radio frequency (RF) waveguide coming from ceiling penetrations and terminating into the cryomodule (stainless-steel cylinder).

FIGURE 21.10 Electronic racks above penetrations on the second floor of the facility.

RF Gallery

The electronics (Figure 21.10) that drive the accelerator must be located outside the vault due to the potential for damage to electronics by ionizing radiation. The electronics and RF power supplies are located on the second floor of the facility. Cabling is fed down to the vault through penetrations. The most significant safety issue here is associated with exposure to ionizing radiation associated with the penetrations. A radiation safety program mitigates this issue. Electrical safety issues are also associated with the penetrations: Maintenance performed on such systems requires a lockout, tag-out program to control electrical hazards that present simultaneously both in the racks and in the vault. Another significant hazard is the potential for helium release to migrate from the vault up through the cabling penetrations to create oxygen deficiency, and this must be controlled by the ODH program.

Control Room

The control room (Figure 21.5) is located on the second floor of the facility. Operation of the accelerator is managed here. Control of the laser states and delivery of the laser beam to the laser labs is also managed here.

Optics Control Room

The OCR receives the laser beam from the optical transport system (Figure 21.4). It is from the OCR that the beam is delivered into the laser laboratories. A laser safety system now becomes necessary to control exposure to the laser beam. The OCR is accessed from the FEL Control Room. There are two purposes for the OCR. The first is to enable and disable the FEL. The second is to provide diagnostics to ensure that the correct beam will be received by the laser lab.

Laser Laboratories

All the light-tight laser laboratories receive the FEL beam through the laser lab beam delivery pipe. A beam dump is located at the end of the last laser lab. A mechanism to reflect the beam into each laboratory is needed; that mechanism is called a mirror cassette. A laser safety system is necessary in all laser laboratories.

Management Structures

The management structure for an FEL system is complex. The JLab FEL Line Management is responsible for the safe operation and maintenance of the facility; for planning, review, and acceptance of user experiments; and for providing vision for the future applications of the FEL. Key positions that exist include

- Associate Director
- FEL Facility Manager
- Safety Officer
- Laser Safety Officer

An optimal organizational structure for the FEL includes the following groups:

1. *Operations Group*: The operations group operates the control room. This group is responsible for the operation of the accelerator, the laser delivery system, and the state of laser operations in each laser lab.
2. *Accelerator Systems Management*: At least seven subgroups are required for adequate management of the accelerator.
   a. *Safety Systems Group*: It is responsible for the control of prompt radiation associated with the accelerator. Since the laser medium is a high-energy electron beam, it must be situated in a radiation-shielded

enclosure. Radiation controls must be in place to prevent any inadvertent radiation exposure. A personnel safety system must be installed to ensure that no one is in the electron beam accelerator enclosure during beam operations. Systems that this group installs and maintains include the controlled access system for the vault, run/safe crash boxes (Figure 21.11), controls to maintain the accelerator in beam permit and high-power permit, and the oxygen-deficiency detector and alarm system that must exist in the facility due to the use of liquid helium to support the superconducting state of the accelerator.

b. *Instrumentation and Controls Group*: This group is responsible for all equipment used to start, optimize, and operate the accelerator and the laser beam.

c. *RF Group*: The electron beam is accelerated by RF power. Associated klystrons (50 kW for the FEL accelerator; Figure 21.12) and power supplies are managed by this group.

d. *Injector Group*: The injector is the source of the electrons for the accelerator. This group must maintain the photocathode electron gun (Figure 21.13), along with its high-voltage power supply and vacuum system.

e. *Cryogenic Group*: This group maintains the liquid helium supply required for the superconducting accelerator.

f. *Vacuum Group*: This group maintains the vacuum system required for the accelerator and beam transport system.

g. *Facility Management*: The mechanical engineering group for utilities ensures that all nonbeam safety controls are functional. Examples include chemical fume hoods and fire protection system.

FIGURE 21.11 Run-safe box—part of the personnel safety system in the vault.

FIGURE 21.12 One of the klystrons used to produce RF for the accelerator. Klystrons are located in the RF gallery (second floor). The power supply is in an adjacent electronics rack (see Figure 21.10).

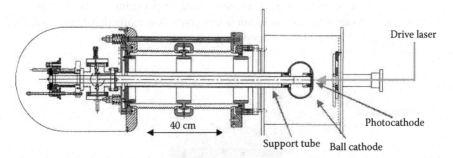

FIGURE 21.13 Schematic of the electron gun; note that a Class 4 drive laser is used to interact with the cesiated photocathode to produce free electrons.

LASER SAFETY

The laser safety system for an FEL is complex. Key positions necessary to maintain and inspect this system are the following:

- *Laser Safety Officer*: Provides independent oversight to the laser safety program.
- *Laser System Supervisor*: Normally a physicist with a doctoral degree, the laser system supervisor is responsible for the overall operational safety of the FEL laser beam.

- *Laser Interlocks Manager*: Normally this position is held by an electrical engineer who is responsible for maintaining the laser interlocks associated with the FEL and any other Class 3B or 4 laser used in the facility.
- *FEL Laser Operator*: An FEL employee that has full control of the beam delivery to each lab.

LASER SAFETY HAZARDS AND CONTROLS SPECIFIC TO THE FEL

There are hazards that must be considered with the FEL. These hazards are related to the control of laser light to multiple laser laboratories, extremely high power, the presence of harmonics and tunability, and the presence of a user population that is generally transient.

Control of Laser Light to Multiple Laser Laboratories

The FEL is located in the accelerator vault. Once the laser light is generated, a complex system of shutters becomes necessary.

Shutter Configuration

The FEL laser beam needs seven or eight pieces of equipment to be moved to transport the laser light into the lab. The first two are a pair of FEL shutters located in the FEL optical cavity in the vault. Inserting either of these shutters greatly increases the losses in the FEL cavity and shuts off the laser. The third is a high-power shutter that sends the beam into a diagnostic power meter. This must be withdrawn to send the laser light into the optical transport. The fourth is a collimator mirror cassette that has an open position allowing experiments to be carried out in the vault. If the mirror cassette is moved to a position with a mirror, the beam is sent upstairs into the OCR. In the OCR is another higher-power shutter that, when opened, allows beam into the laser laboratories. This is the fifth shutter. The sixth shutter is a mirror cassette that must be off its home position to send the beam into a specific lab. In the home position the light is passed onto the next lab. The seventh shutter is the specific lab shutter, which must be withdrawn to send beam into the lab. In laser labs with a hutch there is one more shutter that allows the beam going into the hutch to be shut off (the hutch bypass shutter). The hutch bypass shutter is tied to the lab shutter so that one always has two shutters between the user and the beam. The user only has control of the seventh and eighth shutters.

Shutter Configuration Control

The FEL shutter is interlocked to the vacuum for the beam delivery lines, so the FEL cannot be operated with the beamline in an incomplete (especially open) state. All optical transport vacuum pipes are labeled as such and are not to be opened without permission from the optical transport system owner. Any attempt to open the transport line results in a vacuum loss, which will shut off the laser.

All the laser optics and the transport optics are remotely controllable via a Programmable Logic Controller–based control system. The FEL operator usually controls these systems from the FEL control room and has sole control over the first six shutters. Only the FEL operator can provide the enabling voltage for the lab shutter. The user must be able to lock out the lab shutter, however, if it is necessary to prevent the laser from entering the lab. This may be accomplished by locking out the FEL shutter in the lab. Only the user can operate the hutch bypass shutter, though the FEL operator can monitor its state.

High Power

The FELs are capable of producing very high power. There is potential for devastating injuries associated with both eye and skin exposure. Small specular reflections from beamline elements can be as large as Class 4 by themselves. If the laser is operated in the ultraviolet, the user will generally want to keep the beam enclosed due to the severe skin hazard. When the laser is operating at kilowatt levels, the nominal hazard zone for even dermal exposure from scattered light can be quite large. These conditions necessitate establishment of an exclusionary area for operations at full power.

Configuration

The established states for the laboratories are listed in Table 21.2 and described here:

Open: When all status lights are off, anyone can enter. The lab contains no Class 3B or Class 4 laser hazards. Many nonlaser hazards may still be present.

Hutch Mode: During hutch mode, a light-tight hutch is used to contain the laser. Personnel may be present in the laser lab during hutch mode. The hutch must have engineering controls to keep personnel from entering or opening the hutch when laser light is present.

When the lab is in hutch mode, a green (laser users only) lamp is illuminated on the door, and any laser user approved for the lab can enter after presenting an electronic radio frequency identification card (RFID card or smart card) to the card reader next to the door. The maglocks and the door interlocks are bypassed for 15 seconds after the card is accepted. If the door remains open after this delay, the lab interlock system will crash. If an approved user does not have a working smart card, they may ring the doorbell next to the lab door and have someone inside the lab let them in using the Exit button on the inside of the door. Users needing to leave the lab may also press the Exit button to leave the lab. Again, the maglocks and door interlocks are bypassed for 15 seconds when this is done.

TABLE 21.2

Summary of the Allowed Access States for the Laser Labs

Light on Door	Access State	Laser Protection	Other Requirements
None	Open	None required	None
Approved users only (green)	Hutch	Class 1 laser enclosure	Hutches interlocked/ floor mats
Goggles required (yellow)	Alignment	Laser safety eyewear required	Harmonic blocking filter inserted
No access (red)	Exclusionary	Sweep procedure must be followed	Check beam paths before sweep

Alignment Mode: When the yellow light (goggles required; see Figure 21.14) is illuminated on the door, the lab is in alignment mode. In this mode, any approved laser user may enter the lab using a smart card or after being let in by another approved user. It is essential to wear laser safety eyewear when the yellow light is on. The maglocks and door interlocks may be temporarily bypassed using the smart card reader or the Exit button when the lab is in hutch mode. There are three submodes in alignment mode:

Local Laser Mode: This mode allows for the use of ancillary conventional lasers not related to the FEL. It is selected using the local laser/FEL mode key. When in this mode, the lab shutter and the mirror cassette are locked out so that FEL light cannot be delivered to the lab. The lab will stay in alignment mode even when the accelerator is running continuous electron beam. Only local lasers may run, and laser safety eyewear is required at all times. The interlock system will allow only this mode to engage if the mirror cassette is in the home position.

FEL Mode: The FEL light can be delivered to the lab as long as the accelerator is in a limited duty–cycle state, which reduces the average power so that laser safety eyewear can be specified.

HeNe Alignment Mode: It is possible to operate the system with an upstream ultraviewer Class 3R helium–neon (HeNe) laser. The upstream HeNe laser can then be used to align an experiment. An FEL operator must activate HeNe alignment mode using a key in the OCR.

Exclusionary Mode: When the red (no access) light is illuminated on the door, no access is permitted. The lab is in exclusionary mode. If a smart card is presented to the card reader, the maglocks and door interlocks are not bypassed. The Exit buttons crash the lab in this mode.

Configuration Control

The operating mode of each individual lab is set by the FEL operator using the keys in the laser labs and the OCR. The state of the labs can be monitored from the FEL control room. The control system must make it clear at a glance where the FEL beam is being sent at any time. The controls used include an interlock system, an administrative procedure to activate the interlock system, and a communication system between laser operator and users.

FIGURE 21.14 (Left) Example of a small hutch in a laser lab with one interlocked cover removed. (Right) Entry status lights when in hutch mode.

The interlock system is adapted to each laboratory. The interlocks are designed in accordance with *American National Standard for Safe Use of Lasers* (ANSI Z.136.1) specified by American National Standards Institute (ANSI). In addition to those requirements, the interlock system must have the following:

Redundant Interlocks: Two shutters or insertable mirrors must be in place between any personnel and the full-power FEL beam at all times and between any unprotected personnel and alignment mode or full-power beam.

Failsafe Laser Shutters: The shutter that turns off the FEL must be failsafe in that the loss of either power or air must close the shutter. The power for the shutter must come from the interlock system.

Provision for Separate Class 3B or 4 Lasers: Lasers are interlocked to the lab interlock system so that they cannot be operated unless the lab is secured. If the lab drops to an open state, the lasers must be disabled from lasing or must drop to a Class 1 state.

Interlock Beam Stops to Water Flow: All laser beam stops must have temperature interlocks that shut the laser beam off if the external beam stop temperature exceeds 60 °C.

FEL must be interlocked to transport vacuum integrity. If the pressure in the optical beam delivery pipe exceeds 10 mm Hg, the FEL must be disabled so that the light cannot be transported into the laser labs. All vacuum pipes must be labeled so that they are not disconnected during operation.

Administrative Interaction with Interlock System: Prior to engaging the interlock system, the laboratory must be cleared of all personnel. An administrative sweep of the room ensures that no personnel remain. The sweep path (Figure 21.15) is designed to incorporate physical interaction of the sweep personnel with the interlock system (Figure 21.16) in the form of sweep buttons that are located at several key points in the laboratory. This at least ensures that the sweeper has fully traversed the lab. At the end of the sweep, a complete sweep button must then be depressed, and the sweeper exits the lab. Warning lights both inside and outside the room and an alarm clearly audible throughout the inside of the room under normal operating conditions must activate.

Communication System: The door for each lab has three status lights indicating whether the room is in hutch mode (green), alignment mode (yellow), or exclusionary mode (red) or open (no lights illuminated). The laser operator in the control must be in continuous communication with the laser user for FEL operations.

The operating lasers and wavelengths must be selected when the lab is made up. The permissive buttons are selected at the entrance to the laser lab. The buttons are the interface to a programmable logic controller. The controller will not permit use of two lasers with conflicting eyewear requirements (Figure 21.17). This ensures the user to select appropriate eyewear before entry.

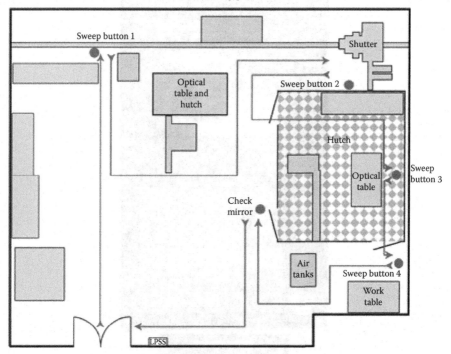

FIGURE 21.15 Schematic of sweep path in one of the laser labs at JLab. The Laser Personnel Safety System (LPSS) interlock chassis shown in Figure 21.16 is to the right of the door. A typical sweep button is shown on the right.

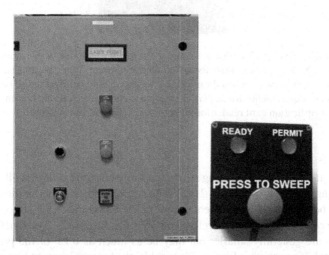

FIGURE 21.16 (Left) LPSS interlock chassis with crash button and sweep initiate button. (Right) Sweep button. When the "Ready" light is lit, one can press the button to arm the box.

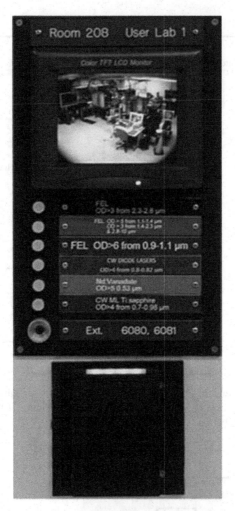

FIGURE 21.17 State-of-the-laboratory display for a laser lab (User Lab 1) located outside entrance door. Note real-time video display of interior of room. White buttons have laser system identified next to them; when illuminated, the system is to be expected to be lasing, and safety eyewear requirements are displayed adjacent to each button. At the bottom is the radio frequency identification card reader for the room electronic access system.

Harmonics

The FELs are unique in producing strong, partially coherent beams at all harmonics of the lasing wavelength (Figure 21.18). The second and third harmonics can be as large as a percent of the fundamental power. The harmonics have almost complete transverse coherence and are therefore considered as laser light. The fourth and fifth harmonics are generally smaller but may be as much as 0.1% of the power of the fundamental. Since the fundamental is tunable, the harmonic might be at any wavelength in the visible.

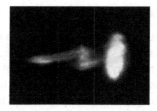

FIGURE 21.18 Fifth, sixth, and seventh harmonics during 3 μm FEL operation dispersed in a prism.

When operating at near-infrared wavelengths, the user must either block out all harmonics or use "virtual" laser-protective eyewear that takes advantage of virtual reality goggles developed for gaming tied to small cameras on the goggles. An example of a harmonic blocking filter is a gallium arsenide plate that stops all light with wavelengths shorter than 800 nm.

Tunability

The wideband tunability spans many damaging regions of the spectrum. It is imperative that the state of the laser with respect to the fundamental wavelength selected is detectable and controlled.

Cultural Differences of User Community

Because of the high cost of overhead associated with the electron beam accelerator laser medium, FELs are generally operated as user facilities; their operation is similar to synchrotron light sources. The infrastructure support group must maintain vigilance over the maintenance of the engineering controls and ensure strict compliance with administrative controls. This vigilance results in a tightly structured and documented safety program. This mode of operation is quite different from a typical single user in a university lab. The users must adapt to a structured and documented safety regime.

ELEMENTS OF THE LASER SAFETY PROGRAM UNIQUE TO THE FEL

The FEL Facility requires engineering and administrative controls to ensure the safety of the user and the infrastructure staff. In addition to the elements of a laser safety program specified by the ANSI *American National Standard for Safe Use of Lasers* (ANSI Z136.1), the laser safety program requires an approval process for experiments to be performed at the facility and a robust laser safety interlock system coupled to administrative controls.

Laser Safety Training

There are two courses required to become a laser user. The first is given by the Laser Safety Officer and must meet ANSI standard requirements. The student must pass a written test. The second set of training is FEL specific. It consists of training on the facility in general and on the specific lab and experiment to be used. The user must pass a written test specific to the laser laboratory in which they will be working.

FEL EXPERIMENT REVIEW PROCESS

An experiment consists of four phases: proposal, setup, run time, and decommissioning.

During the proposal phase, the safety requirements are identified by the user lead scientist using the Experiment Safety Approval Form (ESAF; see Appendix). This form allows for a preliminary hazard analysis: It identifies the target material, optical setup, required optics and beam dumps, and lists any chemicals and ancillary equipment that may be required for the experiment. The form also allows for identification of significant environmental aspects that would be associated with the experiment, such as discharges to air or sewer and waste generation. The ESAF is reviewed by the Technical Advisory Committee (TAC) appointed by the facility manager. The team consists of the laser safety officer, the laser system supervisor, and the safety officer. Other safety expertise may be required, such as the facility fire protection engineer or industrial safety representative.

During the setup and before runtime, a physical inspection is conducted by the TAC and any other required safety experts. Examples of issues that may need correction include the following (for a complete list see the Task Hazard Analysis form in the Appendix, form 3130-T_3):

- Ancillary equipment such as magnets may not have all interlocks and electrical barriers installed. Such equipment must be inspected to ensure that its installation meets electrical and fire safety codes.
- Users may have brought respiratory protection to the facility. Evidence that the user home institution has an Occupational Safety and Health Administration compliant respiratory program, that the respirator is necessary for the experiment, and that the respirator user is trained and fit tested may be required. In most cases, substitution of facility engineering controls will be required.
- Any Class 3B and Class 4 lasers brought by the user must meet Center for Devices and Radiological Health requirements; users may have "fabricated" diode lasers from laser components that do not meet these requirements.

During an experiment, the laser operator must maintain close communication with the laser user during FEL beam delivery. During and after decommissioning, the TAC must inspect to ensure that the laser lab is left in a safe state.

CONTROLS SUMMARY

The laser safety interlock system is robust. In addition to the system, administrative controls are required such as

- Laser Operational Safety Procedures in place for each laboratory.
- Procedures for each experiment must be reviewed and approved.
- The laser operator must be present in the Control Room during FEL operations. The FEL operator is a JLab employee trained to operate the electron beam accelerator and deliver beam through the OCR. The operator is an independent observer of the experiment.
- Permission to use the laser can only be issued by the FEL operator, who has available video camera monitors of each laboratory.
- Routine safety oversight must be provided. The user must be supervised by the FEL operator to ensure that administrative controls are followed.

FUTURE OF THE FEL

New facilities designed specifically to produce X-rays are under construction. It is expected that additional operational and safety requirements will evolve along with this capability.

APPENDIX

Appendix

Jefferson Lab ®Thomas Jefferson National Accelerator Facility	**FEL Experiment Safety Approval Form** (See ES&H Manual Chapter 3130 Appendix T1 FEL Experiment Safety Approval Form – Instructions)	

FEL Experiment ID:		(Assigned by FEL Basic Research Program Manager)

This form documents your experiment. The Lead Scientist completes ALL numbered questions (write "not applicable" or "none" when: appropriate) . If your experiment changes before the form expires, you must notify the FEL Facility Manager. Most changes are easily accommodated and should not result in significant delay.

1. Lab Number:		
2. Expected Start Date:	(Once approved, this form is valid for two years.)	
3. Experiment Title:		
4. Document Owner(s) (Lead Scientist):		4a. Contact Information:

5. List all Experimenters who will be working at the FEL:

First & Last Name (Print)	Affiliation:	Phone:	E-Mail:

6. Name of People who completed this form :

First & Last Name (Print)	Affiliation:	Phone:	E-Mail:

(When form is complete submit it FEL Basic Research Program Manager for review and approvals)

This part to be completed by Jefferson Lab

SUPPLEMENTAL TECHNICAL VALIDATIONS

Subject Matter Expert Review and Acceptance

Hazard Reviewed (per ES&H Manual 2410-T1):	Print	Signature	Date:
[Enter Hazard]			
[Enter Hazard]			
[Enter Hazard]			

APPROVALS

	Print	Signature	Date:
FEL Basic Research Program Manager			
FEL Facility Manager:			
FEL Laser System Supervisor:			

Jefferson Lab	FEL Experiment Safety Approval Form
Thomas Jefferson National Accelerator Facility	

FEL Experiment ID: _____

(Assigned by FEL Basic Research Program Manager)

Document History:

Revision:	Reason for revision or update:	Serial number of superseded document

Distribution: Original: FEL Control Room, **Copies:** Lead Scientist, author(s), affected area, Division Safety Officer, ESH&Q Document Control, FEL ESH & Q Liaison, Area Safety Warden
After expiration: Forward original and log sheet of trained personnel to ESH & Q Document Control.

7. Experiment Overview

Provide a brief description of your planned activities. Include the approximate duration of the program.
[Start Typing Here]

Does this experiment involve modification to the basic optical beam delivery system?		YES		NO
If YES describe the Modifications.				

[Start Typing Here]

8. Task Hazard Analysis

Instructions:	Answer the following questions. When answers indicate a hazard may exist – document the resolution(s) and hazard mitigation techniques.			
General Conditions	**Keywords**	**Yes**	**No**	**Resolutions**
Will chemicals be used? Note: such use must meet the appropriate MSDS requirements including Personal Protective Equipment (PPE).	acids, flammable gases and solvents, heavy metals (lead, etc.), respirator, gloves, aprons, face shield, safety glasses, working with flammables			
Will you create dust, welding arcs, heat, excessive noise, RF or x-rays?	welding, grinding, painting, x-rays, respirator, gloves, RF, lasers, chemicals, epoxies			
Are there any fire or explosive hazards associated with the work?	painting, welding, grinding, brazing, mixing chemicals, battery charging			
Could the work create headaches, breathing problems, or dizziness from odors, etc.?	Motor exhaust, painting, ozone, solvents, acids, bases, chemicals, portable heaters			
Will compressed or liquefied gasses be used?	cryogenics, nitrogen, helium, argon, carbon monoxide			

This document is controlled as an on line file. It may be printed but the print copy is not a controlled document. It is the User's responsibility to ensure that the document is the same revision as the current on line file. this copy was printed on 4/12/2013.

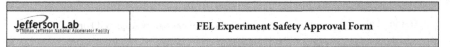

| Jefferson Lab | FEL Experiment Safety Approval Form |
| Thomas Jefferson National Accelerator Facility | |

FEL Experiment ID: _____

(Assigned by FEL Basic Research Program Manager)

Instructions: Answer the following questions. When answers indicate a hazard may exist – document the resolution(s) and hazard mitigation techniques.				
General Conditions	**Keywords**	**Yes**	**No**	**Resolutions**
Does the task require work in areas or with materials subject to temperature extremes?	welding, soldering, brazing, cryogenics, resistive heating			
Does the work involve the use of hoists or robotics?	manlifts, subcontractors, rentals, slings, rigging			
Will powered hand tools be used?	drills, saws, PPE, GFCI, power activated tools			
Does the work involve the risk of electrical shock or other forms of hazardous energy?	LOTO, compressed gases, power supplies, pressure, cryogenics			
Does the task involve lifting, pulling, pushing, or carrying heavy objects, or repetitive motion?	Posture, back injury, twisting			
Does the task involve work with pressurized or vacuum vessels?	resistive heaters, GFCI, pressure relief, tanks, containers			
Does the task require any permits?	welding, grinding, open flame soldering			
Does the task require specialized training?	Respirator			
Will waste products require special handling or disposal requirements?	chemicals, by products, discharges to sanitary sewer or air			
Any other hazards we may have overlooked with this list?				

9. Experimental Details

List all materials (and quantities) to be used in your experiment. List Target Material first and include all chemicals, gases, sample materials, etc.
[Start Typing Here]

Describe any airborne contaminants that may be produced. Include the expected composition/decomposition; the method of exhaust; fixture description; and expected interaction with the FEL beam.
[Start Typing Here]

Describe the beam stop construction and its ability to handle power.
[Start Typing Here]

This document is controlled as an on line file. It may be printed but the print copy is not a controlled document. It is the user's responsibility ensure that the document is the same revision as the current on line file. This copy was printed on 4/12/2013.

Page
3 of 6

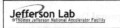Jefferson Lab _{Thomas Jefferson National Accelerator Facility}	**FEL Experiment Safety Approval Form**

FEL Experiment ID: _____

(Assigned by FEL Basic Research Program Manager)

List Personal Protective Equipment Required.
[Start Typing Here]

Additional Precautions (e.g. posting requirements, process restrictions, equipment limitations, laser beam containment, interlocks)
[Start Typing Here]

10. Additional Laser Usage

Describe any additional lasers to be used in the experiment. (Use of additional Class 3b and above lasers will require a separate, additional Laser Operational Safety Procedure (LOSP).)
[Start Typing Here]

11. Outline Experiment Procedure

Layout of equipment and room (e.g. a brief description of any special requirements, overhead floor plan.)
[Start Typing Here]

General Experiment Procedures (Please be concise. Provide sufficient information to illustrate what will be done, who will do it and where procedures will occur. You may refer to the LOSP for the particular lab, its hardware, and procedures. However a specific Test Plan will be filed on a separate form after technical and safety approval and scheduling have been assigned.
[Start Typing Here]

Residual Hazards (Contaminants, Disposal, Safe Disassembly,...)
[Start Typing Here]

Any Other Safety Considerations
[Start Typing Here]

This document is controlled as an on line file. It may be printed but the print copy is not a controlled document. It is the user's responsibility to ensure that the document is the same revision as the current on line file. This copy was printed on 4/12/2013.

Page
4 of 6

Jefferson Lab Thomas Jefferson National Accelerator Facility	**FEL Experiment Safety Approval Form**

FEL Experiment ID: _____

(Assigned by FEL Basic Research Program Manager)

12. Regulatory Requirements

Regulatory Requirements	
☐ Yes ☐ No	Does the proposed experiment utilize viruses, viable bacteria, or material presenting a biological hazard at the FEL? Certain biological hazards require notification to agencies outside Jefferson Lab.
☐ Yes ☐ No	Does the proposed experiment require any radioactive materials or radiation producing equipment?
☐ Yes ☐ No	Does the proposed experiment require any industrial chemicals to be brought or shipped to Jefferson Lab? All chemicals must include a MSDS for each material shipped.
☐ Yes ☐ No	Does the proposed experiment create any chemical hazards?

13. Environmental Management Information
(See EMP–04 Project/Activity/Experiment Environmental Review)

Is this a Water-Based Project?	☐ YES ☐ NO	
If YES provide details:		
Source of the Water and estimated quantity.		
How is water to be discharged or disposed of:		
Sanitary Sewer		
Special Sanitary Sewer Discharge		
Surface Water		

Will the Experiment Generate Waste?	☐ YES ☐ NO	
If YES list all wastes including anticipated quantities and disposal approach for each type.		
Anticipated Air Emissions		
Other Waste Water		
Hazardous Waste		
Solid Waste (landfill or recycling)		
Power/Natural Resource Consumption Expected		

Jefferson Lab	**FEL Experiment Safety Approval Form**
Thomas Jefferson National Accelerator Facility	

FEL Experiment ID: _____

(Assigned by FEL Basic Research Program Manager)

14. Decommissioning/Shutdown Procedure (if necessary):

How will you ensure the lab is left in a safe and clean state after the experiment? Provide guidelines or process steps which outline closeout actions. Think about what needs to be done and plan enough time to do it:

☐ Hazardous material to be removed from the lab.	☐ User provided equipment to be removed.	☐ Lab to be left in a clean and orderly state.

[Start Typing Here]

1.0   Revision Summary

Revision 1 – 11/23/10 – Updated to reflect current laboratory operations.

ISSUING AUTHORITY	TECHNICAL POINT-OF-CONTACT	APPROVAL DATE	EFFECTIVE DATE	EXPIRATION DATE	REV.
ESH&Q Division	George Neil	11/23/10	11/23/10	11/23/15	1

This document is controlled as an on line file. It may be printed but the print copy is not a controlled document. It is the User's responsibility to ensure that the document is the same revision as the current on line file. this copy was printed on 4/12/2013.

22 Tunable External Cavity Diode Lasers

Diana Warren

CONTENTS

INTRODUCTION

Tunable external cavity diode lasers (ECDLs) are useful in many applications, including atomic and molecular laser spectroscopy, laser cooling, atomic clocks, environmental sensing, phase-shifting interferometry, coherent optical telecommunications, and exciting new fields such as optical microresonators and quantum computing. Aside from tunability, these applications require continuous-wave, single-mode, and narrow-linewidth laser sources. Semiconductor diode lasers are attractive because of their low cost, although not an ideal choice since they typically operate with several longitudinal modes lasing simultaneously, leading to low coherence and therefore large linewidths. However, controlled optical feedback into the laser diode with external optics can extract highly coherent light and narrow linewidths from a semiconductor-based laser. Figure 22.1 shows a simple ECDL schematic. The laser cavity is now external to the diode itself and defined by the back facet of the laser diode and output coupler. By putting an antireflection (AR) coating on the front facet of the laser diode, the diode becomes purely a gain element. This is essential to the performance of tunable ECDLs because it suppresses mode competition and self-lasing because of internal diode reflections.

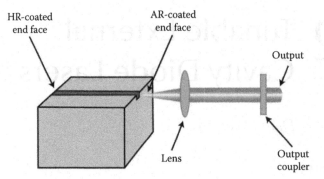

FIGURE 22.1 Illustration of a semiconductor diode within an external cavity. The high-reflectivity (HR)-coated face and output coupler define the laser cavity. The antireflection coating turns the laser diode into a gain medium.

LITTMAN–METCALF AND LITTROW EXTERNAL CAVITY DIODE LASERS

Two typical configurations enabling tuning across the diode gain band are the Littrow and Littman–Metcalf designs. Both use a grating to provide optical feedback into the diode chip, as shown in Figure 22.2. The output from the diode is collimated and directed onto a diffraction grating. In the Littrow design, the mode is selected by rotating the diffraction grating, whereas in the Littman–Metcalf design, an additional mirror in the laser cavity reflects the first-order diffraction off the grating to provide the feedback. The specular reflection or zero-order diffraction off the grating serves as the output beam of the laser. There are advantages and disadvantages to both the Littman–Metcalf and Littrow cavity designs. In general, the Littrow design results in higher-output laser power; however, advances in chip manufacturing technology and optical coatings have led to higher-power Littman–Metcalf ECDLs. The optical feedback in a Littrow is much stronger, and high-quality AR coatings, which can be challenging, on the diode facet are not critical as they are for the Littman–Metcalf design. The Littman–Metcalf ECDL achieves mode-hop-free tuning ranges in the tens to hundreds of nanometers and the Littrow ECDL typically has less than a tenth of a nanometer mode-hop-free tuning. See the following section for further description of mode-hop-free tuning.

SOME EXTERNAL CAVITY DIODE LASER CHARACTERISTICS

Mode-Hop-Free Tuning

A very important specification for ECDLs is the mode-hop-free tuning range. In many applications, it is desirable to have a single-mode laser source that can tune continuously over tens of gigahertz and even tens of nanometers. In a Littman–Metcalf cavity, there are two competing wavelength selection constraints, the mirror-grating angle and the laser-cavity length. The grating configuration (groove spacing, incident, and diffracted angles of the laser beam) and laser-cavity length define a discrete set of possible wavelengths that can lase in the cavity. Rotation of the end mirror causes both parameters to change. Therefore, it is important for

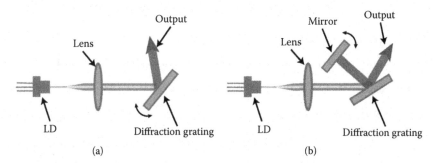

FIGURE 22.2   (a) Littrow and (b) Littman–Metcalf laser cavities.

an appropriately selected pivot point of rotation to synchronize the two, so that the cavity length remains the same number of half-wavelengths of the feedback frequency while the mirror is rotated. Thus mode-hop-free tuning is achieved. When this condition is not met, the lasing wavelength will periodically hop from one mode to the next.

Linewidth and Frequency Stability

The short-term frequency stability of the laser is referred to as linewidth. Once single-mode operation is established by the optics in the external cavity, the linewidth of the laser can be affected by acoustic coupling and cavity temperature variations, each of which can change the cavity length. It is also affected by electrical noise coupling, which causes changes in the index of refraction of the diode and in the piezo length (also affecting the cavity length). Since various noise contributions occur on different timescales (thermal > acoustic > electrical), the length of time over which a linewidth measurement is taken is important. Representative linewidths are measured over milliseconds.

A common way to measure linewidth is by using a heterodyne beating signal between two lasers, one being the laser under test and the other one either a similar laser or a laser with known linewidth.[1] The linewidth is obtained with a high-speed photodetector followed by an electrical spectrum analyzer (ESA). The frequency difference between the two beating lasers is tuned to be within the photodetector bandwidth, and the ESA bandwidth must be sufficient to cover the beating frequency. The beams from the two lasers are overlapped well enough before incidence onto the photodetector to get high mixing efficiency. The result is a *beat note* (see Figure 22.3) and the width of the peak at 3 dB from the maximum height is the linewidth.

Longer time wavelength drifts (over seconds) of narrow-linewidth tunable lasers occur because of floor vibrations, small temperature drifts, and even acoustic noise from people talking in the vicinity of the laser. The *free-running* linewidth, or short-term stability of the laser, is often not adequate for many applications without active stabilization of the laser frequency. The wavelength can be stabilized by using feedback control to the tuning element. To do this, an optical reference such as a gas cell or high-finesse optical cavity is used to provide feedback to the piezoelectric transducer (piezo) that controls the grating or tuning arm to shift the laser wavelength so

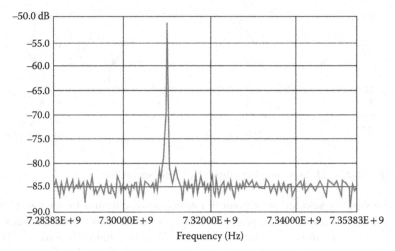

FIGURE 22.3 Heterodyne beat note of two New Focus TLB-6712 Velocity lasers. The linewidth is measured to be 200 kHz. The integration time is 50 milliseconds.

that it always stays fixed on the stable resonance line shape of the reference, regardless of possible external disturbances. Precision spectroscopy and the manipulation of atomic and molecular systems have directly benefited from the resulting improvement in laser stability over the past several decades. For example, atomic clocks based on optical transitions require extremely stable laser sources to accurately probe the subhertz linewidths available in laser-cooled samples.[2] Interferometric measurements such as the search for gravitational waves (e.g., LIGO2) also critically depend on the availability of narrow-linewidth laser systems with extremely low frequency and amplitude noise.

TAPERED AMPLIFIERS

The typical laser output power of an ECDL ranges from 10 to 100+ mW, depending on several factors, including the capability of the laser diode, external cavity optics and configuration, and drive electronics. Oftentimes, higher laser power than can be provided by an ECDL is desired. For this purpose, tapered waveguide semiconductor chips can be used to amplify ECDLs, providing up to 2+ W of power. This is achieved in a master oscillator power amplifier setup in which the ECDL output (seed laser) is coupled into the tapered waveguide for single-pass amplification. The tapered gain region is electrically pumped and as the input propagates from the narrow end to the wide end of the taper, the beam expands to fill the cross section of the device.[3] Both facets of the tapered chip are AR coated to prevent the tapered chip from self-lasing. The seeding requirements are ample input power to surpass the lasing threshold of the tapered chip, and matching seed and tapered chip wavelength. The wavelength of the input seed laser and other characteristics, such as single longitudinal mode and linewidth, are preserved. These systems are especially useful to provide sufficient power for laser cooling and harmonic generation (Figure 22.4).

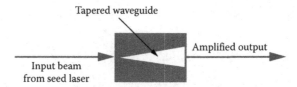

FIGURE 22.4 Schematic of the input and output of a tapered waveguide semiconductor amplifier and gain region.

SOME APPLICATIONS

Microresonator Biodetection

A relatively new device using widely mode-hop-free tunable ECDLs is microresonator for applications in biodetection, optomechanics, optical frequency combs, optical switching, third harmonic generation, and more.[4] The unique properties of microresonators, including high-quality factors, micron-scale sizes, and narrow linewidth, lead to interesting dynamics when coupled with laser light. One type of microresonator is whispering gallery mode (WGM) microtoroids, approximately 100 μm diameter silica disks elevated on a pedestal and fabricated by photolithography and etching techniques.[5] In these devices, light is evanescently coupled into the resonator through a tapered optical fiber (see Figure 22.5). The wavelength of the laser coupled into the tapered fiber is swept and the transmission is collected by a photodetector and viewed on an oscilloscope. At the resonant frequency, there will be a dip in transmission since much of that wavelength has been coupled into the toroid. Because of the small sizes of the resonators, resonant wavelengths are discreet and a function of the toroid diameter. The exact resonant frequency of a microresonator cannot be known a priori because fabrication techniques inevitably lead to an ensemble of sizes. This highlights the importance of widely tunable lasers (tens of nanometers) that are mode hop free to ensure the exact resonant frequency can be delivered by the laser source.

Whispering gallery mode toroids can be used for sensitive biodetection. There are two main types of biodetection: fluorescence-based and label free. WGM biodetection uses the latter, which is advantageous because target molecules are not altered by a fluorescent dye and therefore are in their natural forms.[7] In these techniques, the toroid is functionalized by adhering biospecific ligands to the surface. Since the resonance frequency of a microtoroid is sensitive to environmental conditions, manifesting itself by a shift in resonant frequency, when the biomolecule attaches to the ligand that shifts, is observed. Since shifts can be very small, on the order of femtometers, the narrow linewidth of ECDLs is critical to resolve the spectrum.

Quantum Computing

Optical Lattices: Laser-trapped atoms comprise a useful platform for quantum computing.[8] Interference of counter-propagating laser beams can form a light field with periodic potentials. Through laser cooling and repumping optical excitations, certain atoms in a localized quantum state can be prepared. The periodic potentials in the light field form what is called optical lattice traps, which are populated with single atoms as the lattice potential depth is increased through various techniques. Each atom can be thought of as occupying a qubit state and can be manipulated (Figure 22.6).

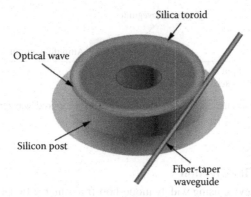

Silica toroid

Optical wave

Silicon post

Fiber-taper
waveguide

FIGURE 22.5 Illustration of a microtoroid ring. The fiber-taper waveguide is put close to the toroid ring and the resonant frequency is evanescently coupled into the toroid. (From Vahala, K.J., *Nature*, 424, 839–846, 2003.[6])

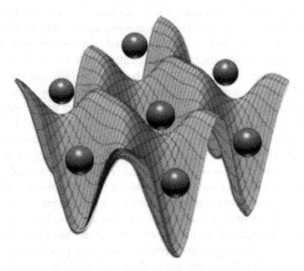

FIGURE 22.6 Schematic of atoms confined to potential wells in an optical lattice.

The phenomenon of interest is the quantum interactions between these atoms.[9] The coherent quantum behavior of these trapped atoms has also improved the accuracy of atomic clocks in the past few years.[10]

Nitrogen Vacancy Centers: Another platform for quantum computing is nitrogen-vacancy (NV) color centers in diamond, a natural diamond defect site.[11] In these systems, a carbon atom in the diamond lattice is substituted by a nitrogen atom and accompanied by a vacancy in an adjacent lattice site (Figure 22.7). The strong optical absorption and long coherence time of the spin states make NV centers ideal for quantum information processing.[12] While the resonant wavelength of the NV centers is known to be 637 nm, experiments often use different environmental conditions such as magnetic fields that shift this frequency. Therefore, mode-hop-free widely tunable lasers are valuable in this application.

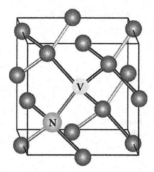

FIGURE 22.7 Figure of a diamond lattice. The dark gray balls represent carbon atoms. Nitrogen substitutes a carbon atom and is always accompanied by a vacancy on an adjacent site.

Phase-Shifting Interferometry

Determining the surface quality of a critical component, such as an optic or semiconductor surface, is crucial to high-precision manufacturing.[13] Traditional interferometry methods consist of an interferometer in which fringe patterns (interferogram) can be analyzed to provide the surface flatness of the measured sample. While analyzing interferograms is an extremely powerful technique, there are a few disadvantages. Finding the location of the fringe centers, which ultimately limits the accuracy of the technique, is a difficult task, and any intensity variations across the interferogram or sensitivity variations in the photodetector introduces errors. Another drawback is that data are obtained only along the fringe centers and not on the regularly spaced grid that many analysis routines demand. Lastly, since a few widely spaced fringes can be measured more accurately than many closely spaced ones, there is a trade-off between resolution and the number of data points. Phase-shifting interferometry overcomes these limitations. First, phase-shifting interferometry does not rely on finding the fringe centers. Second, measurements can be taken at every element in the photodetector array directly yielding optical path-length differences on a regularly spaced grid. The concept behind phase-shifting interferometry is to apply a time-varying phase shift between the reference and test wavefronts. This can be achieved, for example, by mounting the reference optic on a linear transducer, such as a piezoelectric crystal. While many phase-shifting interferometers use HeNe lasers as the source, using an external-cavity tunable diode laser provides significant advantages, including the capability of choosing the exact wavelength to match the operating wavelength of the optics, of particular value when the optics are coated and may not be reflective at 633 nm. Second, the time-varying phase shift can be achieved by unbalancing the two arms and varying the laser wavelength, eliminating the need for a linear actuator to translate the reference optic (Figure 22.8).[14]

CLOSING REMARKS

Founded in 1990 with the mission of providing Simply Better™ Photonics Tools, New Focus has built a portfolio of high-performance products that includes tunable lasers, optoelectronics, high-resolution actuators, stable optomechanics, vacuum and

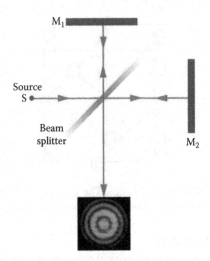

FIGURE 22.8 Typical schematic of an interferometer.

ultraclean solutions, and original equipment manufacturer-engineered solutions. The portfolio of tunable ECDLs was first introduced in 1992 and New Focus is now a world leader of tunable lasers for test and measurement and research. Today, New Focus is a proud member of Newport Corporation's family of brands. The information provided in this chapter is sourced from peer-reviewed journal articles, published Newport application and technical notes, internal tests, and discussions with scientists and engineers across various fields. The information is meant to provide a general overview of ECDLs and some useful applications to spark an interest in the reader and prompt further investigation. More information about New Focus tunable lasers can be found at the company's Web site (www.newport.com).

REFERENCES

1. F.L. Walls and D.W. Allan, "Measurements of frequency stability," *Proceedings of the IEEE* 74, No. 1 (1986).
2. B.C. Young, F.C. Cruz, W.M. Itano, and J.C. Bergquist, "Visible lasers with subhertz linewidths," *Physical Review Letters* 82, 3799, 1999.
3. J.N. Walpole, "Semiconductor amplifiers and lasers with tapered gain access," *Optical and Quantum Electronics* 28, 623–645, 1996.
4. A.B. Matsko, A.A. Savchenkov, D. Strekalov, V.S. Ilchenko, and L. Maleki, "Review of applications of whispering-gallery mode resonators in photonics and nonlinear optics," *IPN Progress Report 42-162*, 1–51, 2005.
5. D.K. Armani, T.J. Kippenberg, S.M. Spillane, and K.J. Vahala, "Ultra-high-*Q* toroid microcavity on a chip," *Nature* 421, 925–928, 2003.
6. K.J. Vahala, "Optical microcavities," *Nature* 424, 839–846, 2003.
7. X. Fan, I. M. White, S. I. Shopova, H. Zhu, J. D. Suter, and Y. Sun, "Sensitive optical biosensors for unlabeled targets: A review," *Analytica Chimica Acta* 620, 8–26, 2008.
8. P.S. Jessen, D.L. Haycock, G. Klose, G.A. Smith, I.H. Deutsch, and G.K. Brennen, "Quantum control and information processing in optical lattices," *Quantum Information and Computation* 1, Special Issue, 20–32, 2001.

9. J.J. García-Ripoll and J.I. Cirac, "Quantum computation with cold bosonic atoms in an optical lattice," *Philosophical Transactions of the Royal Society of London A: Physical, Mathematical and Engineering Sciences* 361, 1537–1548, 2003.

10. Jun Ye, *Science* 331, 1043, 2011.

11. L. Robledo, H. Bernien, I. van Weperen, and R. Hanson, "Control and coherence of the optical transition of single defect centers in diamond," *Physical Review Letters* 105, 177403, 2010.

12. A. Batalov, V. Jacques, F. Kaiser, P. Siyushev, P. Neumann, L.J. Rogers, R.L. McMurtrie, N.B. Manson, F. Jelezko, J. Wrachtrup, "Low temperature studies of the excited-state structure of negatively charged nitrogen-vacancy color centers in diamond," *Physical Review Letters* 102, 195506, (2009).

13. Leslie L. Deck, *Proceedings of SPIE* 4451.

14. Newport Corporation, "*Phase-Shifting Interferometry for Determining Optical Surface Quality*," application note.

23 Practical Ideas

Ken Barat

This chapter is a collection of beam control and other laser-related practical solutions. All of these have been or are being used in laser-use areas today. Regardless of how good an idea they may be, the human element can defeat them. This is why an appreciation and genuine desire for laser safety is required on the behalf of the laser user.

The best safety systems are the simple ones. Simple can be depended on and is usually inexpensive. The rest of this chapter is images of simple and real laser safety solutions. They may be useful in a setting you know or give you an idea of how you can do things.

There are many such examples in my earlier text *Laser Safety Management*. Well, here you go. See Figures 23.1 through 23.47.

FIGURE 23.1 Totally enclosed optics, beam tube, and perimeter guard at edge of table.

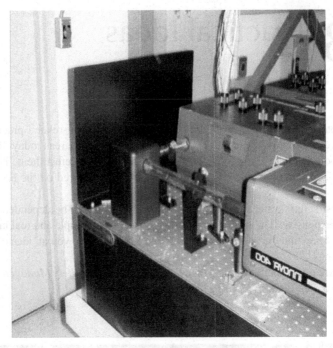

FIGURE 23.2 Clear beam tube, perimeter guard blocking beams from exiting doorway. Putting tape on clear tubes would block diffuse green light from pump laser.

FIGURE 23.3 Total enclosure, similar to Figure 23.1.

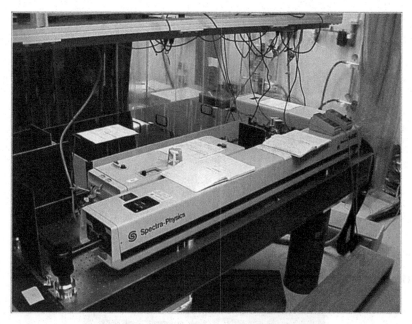

FIGURE 23.4 Good laser setup, stools to step on, beams contained, and so on.

FIGURE 23.5 Optical table entirely enclosed with viewing windows (see Figure 23.10).

FIGURE 23.6   Inside view of Figure 23.5 optical table.

FIGURE 23.7   Optical table with sliding panels and interior light; light could be replaced.

FIGURE 23.8   Labeled beam tube, better is a solid color.

FIGURE 23.9   Set up for beam tube leaving table; shutter drops if tube is not present or falls out.

FIGURE 23.10   Collection of individual beam blocks, homemade; Newport has a series of inexpensive models.

FIGURE 23.11   Labeled cap on chamber, security strap keeps it in place.

FIGURE 23.12　View of vertical beam block.

FIGURE 23.13　View of label on vertical block.

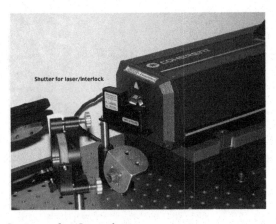

FIGURE 23.14　One type of nmLaser shutter.

FIGURE 23.15   Laser size is not related to power; beam going into fiber.

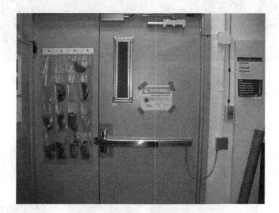

FIGURE 23.16   Eyewear holder on inside of lab door, designed as over door shoe holder.

FIGURE 23.17   Interior view of lab, half curtain blocks beams from exiting doorway and others from seeing in.

FIGURE 23.18 Laser curtain around optical table; do not want floor to ceiling unless needed to block out light.

FIGURE 23.19 Perimeter guard at edge of work bench protects people nearby (see Figure 23.20).

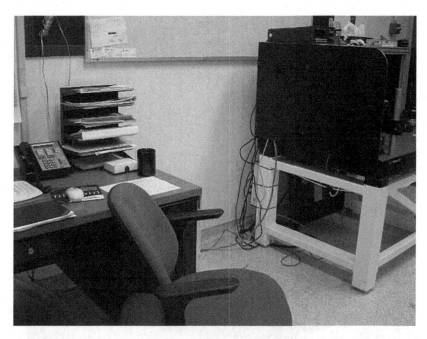

FIGURE 23.20   Desk that needed protection.

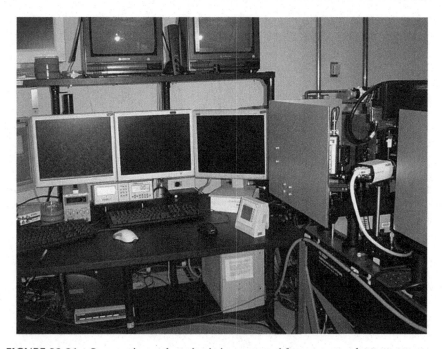

FIGURE 23.21   Once again, work station being protected from any stray beams.

FIGURE 23.22   Curtain on rail, making room within room.

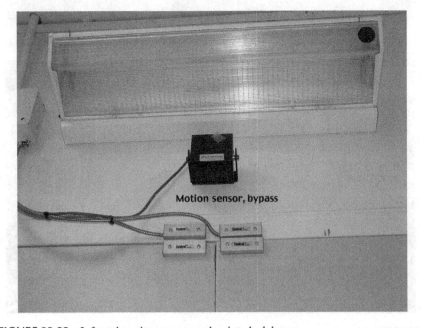

FIGURE 23.23   Infrared motion sensor used as interlock bypass.

FIGURE 23.24   Plastic bridge to protect wires and reduce trip hazard, can connect to others.

FIGURE 23.25   Really large optics.

FIGURE 23.26   Vertical bread board, saves room.

FIGURE 23.27   Illuminates sign; best if illuminated by light-emitting diodes and low voltage.

FIGURE 23.28   Digital scroll sign is programmable.

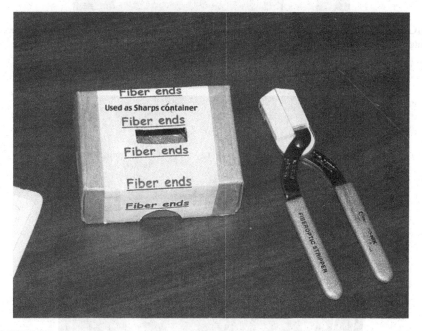

FIGURE 23.29   Homemade sharps container; if you use commercial sharps container, make sure to cover biohazard symbol.

FIGURE 23.30   Interlocked door, window covered, posted, illuminated sign and magnetic interlock, Emergency button, key pad. Illuminated sign too high to be useful.

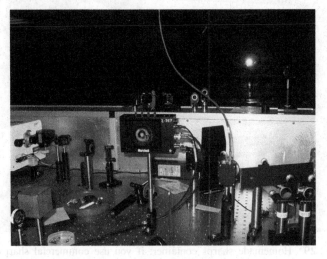

FIGURE 23.31   Homemade infrared viewer using webcam.

FIGURE 23.32 Using $3 bike flasher for temporary sign.

FIGURE 23.33 Another crosswalk way protection device.

FIGURE 23.34   Enclosure panel; poor design, needs great deal of space to lift up.

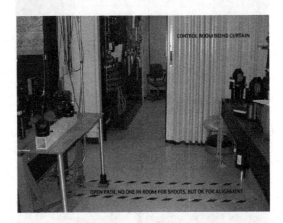

FIGURE 23.35   Tape on walkway only indication of possible beam path; a poor approach.

FIGURE 23.36   Workstation in need of laser operator protection.

FIGURE 23.37 Yet another workstation in need of help; all beams headed back to workstation.

FIGURE 23.38 Water hoses over electrical wires; trouble is waiting.

FIGURE 23.39 All open set up, but after considered by user (see Figure 23.40).

FIGURE 23.40   Users took a compartmentalization approach and are very happy.

FIGURE 23.41   Pipe insulation padding on hanging shelf.

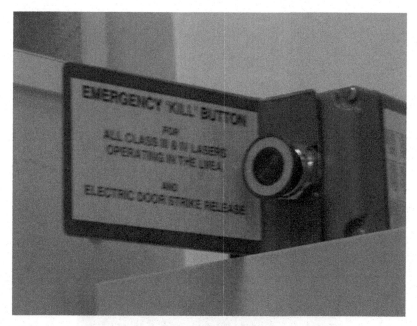

FIGURE 23.42   Sign saying what button is for; a great idea.

FIGURE 23.43   Liquid-crystal display programmable warning sign.

FIGURE 23.44   Electric access lock, keeps unauthorized out, not connected to laser.

FIGURE 23.45   Using locked slide to give user in room control of beam coming from floor above.

FIGURE 23.46 White packing foam can be good beam block for low-power visible beams.

FIGURE 23.47 Multiremote viewing cameras displayed on one screen.

24 User Facilities

Ken Barat

CONTENTS

In the research community, the term *user facility* usually refers to highly sophisticated experimental facilities that are available to researchers from universities, industry, and other government laboratories. In the United States, they are usually funded or managed by the Department of Energy (DOE) or National Science Foundation. Few examples include Advance Photon Source, National High Magnetic Field Lab, linac coherent light source, and Jefferson Lab Free Electron Laser. Their purpose is to advance national research and development, and minimizing unnecessary duplication of effort (usually from an equipment perspective), promoting beneficial scientific interactions, and making the most effective use of costly and, in many cases, unique equipment.

Some institutions have internal user facilities, which by definition are available to in-house staff only. Most commonly as stated, user facilities are government-run and funded facilities. Therefore, access follows the following general policy; access to user facilities is a twofold process. Prospective users submit research proposals directly to the facility of interest. The acceptance of proposals depends on scientific merit, suitability of the facility for the project, selection of a collaborator at the user facility. Concurrently, a contractual agreement implies between the user and the user facility institution. This *user facility agreement*, which can be either proprietary or nonproprietary, stipulates the terms and conditions (including disposition of intellectual property) for the project. Some facilities are available for nonproprietary research at no cost, whereas some facilities must recoup the actual costs incurred for staff and equipment time.

Obtaining laser safety or any type of safety at such facilities is a challenge that one cannot overlook. Because it is difficult for short-term users (which is the majority of user facility users) to buy into the stricter safety culture found at most user facilities. Especially, when they are open to a worldwide user population.

This challenge is recognized by ANSI Z136.8 and some guidance is included in the standard. Training of users can present a real issue, as most are on-site for only a short time, may be a day or two before their assigned beam time, if not arriving the day of their beam time. These facilities are most commonly open 24/7 that adds to the safety challenge. From the user point of view, they want to spend their time doing science, not taking training, and other administrative tasks. So the successful user facility tries to get as much as possible accomplished online before they arrive; regardless, some duties must be done on-site, such as experimental setup and review.

The Advanced Light Source (ALS) is a synchrotron facility operated by Lawrence Berkeley National Laboratory on behalf of the DOE. Requirements to acquire and work there will be used as a typical example of user facility practices.

The Advanced Light Source (or ALS) is located in Berkeley, California. The original building, situated in the East Bay Hills overlooking San Francisco Bay, was completed in 1942. Designed by Arthur Brown, Jr. (designer of the Coit Tower in San Francisco), the domed structure was built to house Berkeley Lab's namesake E. Orlando Lawrence's 184-inch cyclotron, an advanced version of his first cyclotron for which he received the 1939 Nobel Prize in Physics. Today, the expanded building houses the ALS, a third-generation synchrotron and national user facility that attracts scientists around the world. The ALS has over 35 beamlines, where samples may be illuminated with x-ray, ultraviolet, or infrared light to explore the structure and electronic properties of materials. The ALS operates as a national user facility, and is open to researchers worldwide to submit proposals for research. There is considerable laser use at the ALS.

The basic steps to obtain beam time at the ALS are the following:

1. Apply for beam time: Submit a new proposal or a beam time request using an existing active proposal.
2. Establish a user agreement: The user's institute must have a signed agreement with Berkeley Lab before you may do work. Safety documentation must be completed and reviewed before the beam time. Experiments involving any biological material or radioactive material require more extensive review.
3. Register with the user office: New and returning users need to register with the user office 2 weeks before arriving at the ALS. Users arriving out of regular office hours must either have a valid Berkeley Lab ID badge or have completed registration to be granted access to the ALS.
4. Compete online safety training: All users are required to complete their online training before they will be granted unescorted access to the ALS.
5. Prepare for your visit: Logistics, which includes obtaining a user account or email, and shipping and receiving equipment, samples, gases, and

hazardous materials. Information on how to reach the ALS, parking, and accommodations may be found in the sidebar to the left.

6. While you are here: On arrival, complete on-site registration (photo ID required, foreign nationals will need complete visa and passport documentation). Information about facilities is available to the users at the ALS and Berkeley Lab (cafeteria, machine shops, storage, etc.).

7. User feedback survey: Please complete a survey after your visit. The results are used to make improvements at the ALS, and are reported to the DOE each year.

8. Report publications, awards, talks: All publications of work done in whole, or in part, at the ALS need to include the standard ALS acknowledgment and be entered in the ALS Publications database. Awards, invited lectures, theses, patents, books, and book chapters can also be reported here.

FACILITY SAFETY CULTURE

All user facilities should work toward establishing a viable safety culture, in particular when it applies to laser safety. The key to any safety is having documented expectations in the form of principles and procedures.

1. Guiding principles: The aim of user policy at the ALS is to provide a framework for establishing a challenging yet congenial environment where talented scientists from different backgrounds can work together in pursuit of the new scientific opportunities presented by the availability of this innovative facility. The user policy must address a variety of user needs and sensitivities. On the one hand, the qualified researcher with little financial backing needs the assurance of adequate access to the facility. On the other hand, the qualified groups that make a large commitment of time and resources need some assurance of an equitable return on their investment. A national facility should be accessible to all qualified researchers and at the ALS, there are three modes of access: as a general user, a member of an approved program, or as a member of a participating research team. All proposals for research to be conducted at the ALS are evaluated based on the criteria endorsed by the International Union of Pure and Applied Physics.

2. Safety: Safety is a critical function for all research performed at the ALS. All users and staff are expected to know and follow applicable safety requirements at all times. For information about user and experiment safety, please see the safety for users Web pages.

HOW DO WE ENSURE LASER SAFETY?

USER SAFETY OVERVIEW

The ALS experiment coordinators are available to support you through this process. Please contact them at any stage if you have questions or need more information.

PRIOR TO YOUR ARRIVAL AT THE ADVANCED LIGHT SOURCE

1. Complete or update an ESS: If you did not submit a General User Proposal, you must submit an ESS one month prior to arrival at the ALS.
2. Biological, radioactive, hazardous, and electrical materials, and lasers: If your experiment involves the use of any of the above materials—no matter how small the quantities are or how innocuous the sample may be—additional authorization may be required. Please submit your ESS early and clearly identify your materials. Our staff will assess the hazards and contact you about any necessary supplementary documentation.
3. ALS Chemistry Lab access: If you need access to the ALS Chemistry Lab, you will need to complete training and documentation.
4. Experiment coordinator will contact you: If using lasers, biological, radioactive, hazardous, or electrical materials, you can expect to be contacted 2–3 weeks prior to your arrival to confirm experiment details, including a list of participants and to establish requirements. For nonhazardous materials, you will be contacted 1 week prior to the experiment start date.

ON ARRIVAL AT THE ADVANCED LIGHT SOURCE

5. Locate your ESS: On the outer wall of the ALS at the appropriate beamline sector board, you will find a paper copy of the ESS plus any other required documentation in the plastic ESS holder.
6. Check the ESS for accuracy and have all participants sign on final page: The ESS also needs to be signed by subject-matter experts, depending on the hazards, and finally by the beamline scientist to authorize the experiment. No work should be done until the ESS is authorized.
7. Changes to an experiment: At this stage, any changes to the experiment, including sample materials, hardware, or personnel must be documented on a Functional Change Form. Please notify the beamline scientist or experiment setup coordination any time you need to make a change.
8. User Experiment Form (UEF): The UEF (to be placed in front of the ESS) also needs to be completed and signed by the participants, and initialed by the beamline scientist. This form provides *real time* experiment information, that is, dates and times when the work is being performed, participant contacts, and possible experiment hazards.
9. Final experiment approval: Once all documentation is signed, and the ESS and UEF are posted, the experiment may proceed.

DURING THE EXPERIMENT

10. Hazards: All work must be carried out following the hazard controls identified in the ESS and with appropriate personal protective equipment for the type of work and for the area where the work is being performed.

THE SAFETY OFFICER CHALLENGE

The real challenge is getting short time users to buy into ones safety approach, regardless if anyone is watching.

I say watching, for many times experimental work will be performed off hours when the majority or all the safety staff are home worrying about what may be happening. I have encountered many a user whose home institution's safety background or culture can be summed up in one word, *Duck*. As long as no one has been hurt, everything must be running well. Of course, this view is only fooling one self, the home institution, and only surprises everyone when an accident does happen.

THE SOLUTION

Is the expectation and living a solid laser safety culture, where users speak out to peers who are not working in a safe manner and more importantly those comments are accepted and welcomed.

For repeat users, it is easier to get their buy in on safety, for they learn it as rules of the road for operating at the facility.

EXPERIMENTAL SAFETY SHEET AND ON-SITE INSPECTION

This is a tool the ALS has found useful is the ESS and a *kick the tires* inspection of all experiments before they go operational. Such an inspection can find items missed or not covered on design reviews, such as reach in problems, compatibility of equipment with electrical systems, the Nationally Recognized Testing Laboratory approval, mechanical problems.

EXPERIMENT SAFETY SHEET: PROCEDURES AND DOCUMENTATION

An ALS ESS is required for all researches performed on the ALS floor.

The ESS defines the work to be done, including the materials, equipment, and people coming to the ALS, and identifies hazards and appropriate controls. It is authorized for one year.

An ESS is created automatically when a General User Proposal is submitted, or it can be submitted separately.

In either case, an Experiment ID (ALS-xxxx) is generated, which can be used to reference both the proposal and the ESS. A UEF also needs to be completed on arrival for each visit to the ALS.

Structural biology beamlines have an umbrella ESS that covers the materials and equipment for the majority of experiments on these beamlines. Users on these beamlines should not complete their own ESS; however, on arrival at the ALS, they must sign the umbrella ESS and complete a UEF for each visit.

The ESS input form is very similar to the proposal input form and is intended to be self-explanatory, but the following points might be useful:

- Experiment ID: If you already have an active General User Proposal for this experiment, please use the Experiment ID in your beam time allocation letter. Otherwise leave blank.

- Technical description: There should be sufficient details about the procedures to be used to allow our reviewers to understand the type of work to be performed and determine the associated hazards.
- Principal investigator/Experimental lead: Both these people will receive emails confirming the submission. It is generally expected that the experimental lead will be presented at the ALS during the experiment.
- Experimenter's page: Please include the names of all experimenters you expect to work at the ALS.
- Hazard checklist: Select all relevant sample types, equipment types, and procedures. Lasers, biological, nano, and radioactive materials require additional information. Be sure to check these boxes so that you are asked for the additional information. Biological materials include any material of biological origin, hazardous or not.
- Sample materials and chemicals: Please enter sample materials and all chemicals.
- Biological, nano, and radioactive materials: Should be entered on subsequent pages which will appear only if you checked the boxes on the previous page.
- Vacuum chambers and endstation apparatus: This only needs completing if you are bringing your own vacuum chamber or vacuum equipment.
- Gas cells and gas systems: Enter details of any gas cells/systems you are bringing to the ALS.
- Radiation hazards: Tell us about radiation-producing equipment you are bringing to ALS.

SUMMING UP

Safety at any user facility is a challenge, but for laser safety it can easily be a greater concern, all because laser users have direct access to the laser and beam path, as well as 24/7 access. Also the user might feel little responsibility to the user facility. Therefore, their actions have little effect to them. This is not always the case; users have been barred from facilities due to reckless safety performance. The author knows of at least three such cases. The information on the ALS is taken from their Web site and the author's personal experience at the facility. I know their safety staff works very hard to accommodate users and make sure a safe work environment exists for all users, staff, and visitors.

OVERLOOKED ITEM

Safety procedures at a user facility need to also apply to in-house staff. If one exempts them from required training or paperwork, it first undermines the importance of those procedures. Second, it can lead to a situation where an injury occurs whose cause would have been found it the operation was being performed by an outside user. This author has run across such incidents during his career.

25 Laser Disposal

Ken Barat

CONTENTS

INTRODUCTION

Laser technology is continuously advancing and in some settings, such as universities and research institutions, lasers become outdated or applications change such that particular lasers are no longer needed. Typically, these lasers go to storage, but space is very valuable and a time comes when these laser systems go to salvage. The question arises; can they just be thrown out in routine trash? If one is unaware of the individual components of lasers, the answer seems a simple yes; while, in fact, a number of laser components require careful consideration when it comes to disposal. Hazards and components vary by laser. As an example, the majority of lasers manufactured before July 1, 2006 that use electricity as their source of energy contain lead. In this case, if disposed of improperly, lead may seep into the environment. However, high power gas laser plasma tubes may contain carcinogens. This chapter aims at giving the reader some suggestions and options. The topic has been partially addressed in Z136.1 and Z136.8, in a number of presentations on the Laser Safety Officer (LSO) Workshop series and a laser disposal guide developed by the laser safety program at Lawrence Berkeley National Laboratory (while I was LSO there).

At this time, two groups have a responsibility to see that the laser systems are handled in the proper manner; the user (present owner) and the institution's property management/salvage department.

DISTRIBUTION OF RESPONSIBILITY

QUESTIONS FOR USER

1. Have you contacted Property Management to ask which forms need to be filled out?
2. If you are disposing of a dye or excimer laser, have you flushed out the chemicals in the pump containers, tubing, and inner cavity? For instructions, please see the Dye Lasers or Excimer Lasers sections.
3. Do you have the user manual for the laser? If so, send it for disposal with the laser.
4. Did you contact the LSO to see if he can find a new home for your laser?
5. Remember to remove the laser from your inventory record.

USER RESPONSIBILITIES

When a laser is getting ready to be released for surplus, the user should find the laser system user manual and send it along with the laser. Although commercial lasers should have several labels on them, the most important for surplus are the manufacturer label and the logo label. From the manufacturer label, one obtains the model and serial numbers. The logo label contains wavelength and output data and might indicate the general laser name (i.e., argon, HeNe). Before sending the laser to surplus, the user shall place a sign on the laser indicating its optical laser medium (i.e., Nd:YAG, dye, argon). This will be a great help to the property disposal group especially if the user manual cannot be found.

QUESTIONS FOR HAZARDOUS WASTE GENERATOR ASSISTANT (IF SUCH A POSITION EXISTS AT YOUR INSTITUTION)

1. If it is a dye or excimer laser, ensure the user has flushed out the chemicals in the pump containers, tubing, and inner cavity, and disposed of correctly or saved for you.
2. Majority of power supplies built before 2000 with capacitors will contain capacitor oil; see that the oil is drained and placed in the proper container.
3. Any other type of laser has no special precautions required for transportation.

QUESTIONS FOR SURPLUS RECEIVER

1. Have you received the proper paperwork?
2. Is the laser type identified, so the proper disposal steps can be taken?

GENERAL APPROACHES TO LASER DISPOSAL

There are several alternatives to laser disposal that may be considered. Some laser manufactures have a cradle-to-grave service and will accept old lasers for recycling value as a service to the user. Surplus personnel are advised to contact the laser manufacture and ask if the laser system may be returned for disposal, refurbishment, usable components, or possible equivalent to new repair. However, not all manufacturers have this option and some require the user to pay for shipping.

It is also possible to transfer the laser to donate it to a local university's engineering or physics department. It is necessary to ensure that the laser system complies with all applicable safety instructions for operation and maintenance and that the receiving department has a viable laser safety program. A *Limit of Liability* document will also need to be generated. If it is a Class 3B or 4 laser or laser system that will be transferred, the LSO will need to be contacted. He or she may contact the designated institution to see if they have an LSO and the steps that will be taken to use the laser in a safe manner.

If it is determined that those two options are not viable, the laser will be disposed. The following is an excerpt on general laser disposal:

> Lasers, which are to be surplus by the institution, are not to go to public auction. This is due to concerns over the misuse of the laser system; one example is lasers being used to expose commercial aircraft pilots while in flight. The laser must be rendered unusable.
>
> Knowing the manufacturer and model number, a call to the manufacturer is prudent to check on any possible hazardous material components.

Power Supplies

- Simple action is to cut off the plug and as much as possible of the AC source cord.
- Lasers that use electricity as their main source of energy and manufactured before July 1, 2006 have lead in the printed circuit boards. Dispose of them as electronic waste (e-waste).
- Many laser systems utilize a high voltage capacitor system hence an electrical shock hazard is a real possibility. Standard capacitor safety needs to be followed; for example, use of ground hooks, and so on.

Optics

- For optics, remove the optics and place them in a ziplock bag. Label bag with laser manufacturer information and then send it to the LSO.

Various lasers require different actions, the following are examples.

Dye Lasers

Active Concern

These lasers use a liquid medium. This medium comprises organic dyes and solvents, all of which must be considered carcinogenic or mutagenic.

Disposal

Wear personal protective equipment. This includes goggles, particle masks, chemical-resistant apron, and gloves. Rinse pump containers, tubing, and inner cavity several times with methanol and then with water until internal circulating fluid appears to be clear. All washing must be considered hazardous waste and dealt with as such.

Once done, cut power cord, remove optics. Place the optics in a ziplock bag, label it, and send it to the LSO.

Common Dyes and Solvents

- Coumarin
- Rhodamine
- Exalite
- Stilbene
- Oxazine
- Dimethyl sulfoxide

Excimer Lasers

Active Concerns

Excimer lasers use a combination of halogen and noble gas. Each has its own risks depending on the quantity and concentration. It also includes an internal electrical system that is required to make the two family of gases form a dimer.

Common Excimer Lasers

- Argon fluoride
- Hydrogen fluoride
- Krypton fluoride
- Xenon fluoride

Disposal

User should have flushed out the resonator before sending to surplus as well as any premix chambers. Wear gloves and safety glasses. Remove chamber and crack open in a well-vented area. Remove any circuit boards and dispose of as e-waste. Remove power cord.

Hazardous Materials to Consider

- Argon
- Fluoride
- Krypton
- Xenon

DIODE/SEMICONDUCTOR LASERS

Active Concerns

From a size perspective 95% of these laser units are made up of heat sink, current controls, and so on. The actual laser diode is smaller than a paper clip. Some diode laser systems may be part of a diode fiber system.

Common Diode Lasers

- Gallium aluminum arsenide
- Gallium arsenide
- Indium gallium aluminum phosphide laser

Disposal

For individual diode units or arrays, simply break the housing unit taking care to wear protective eyewear. Only if multiple units are received at one time should they be sent to the Hazardous Waste Group (if such a system is present within your institution). They will be treated as e-waste. For fiber optic systems, while wearing safety glasses and protective eyewear, the fiber should be cut near the diode end and fiber segment be put in a sharps container.

Hazardous Materials to Consider

- Gallium arsenide (hazardous waste)

DIODE/TELECOMMUNICATIONS LASER SYSTEMS

Active Concerns

In a majority of cases, these lasers contain optical fibers, which are used to deliver laser radiation. A majority of injuries come from handling the fibers. Treat them as potentially *sharp* components.

Common Telecommunication Lasers

- Gallium aluminum arsenide
- Gallium arsenide
- Indium gallium aluminum phosphide laser
- Ytterbium-doped fiber

Disposal

Fibers are to be treated as sharp optical components and need to go into sharp, disposable container. Safety glasses and gloves are to be worn when handling (cut or removes from diode on laser box). Diode to be treated as e-waste, as stated earlier.

Hazardous Materials to Consider

- Silicon

GAS LASERS

Active Concern

Many of these will have a glass or metal plasma tube. The tube can contain beryllium (Be) oxide in a ceramic form, safe for handling unless broken up, in which case it will then generate Be dust (Be dust will have to be handled as hazardous waste). Many gas lasers also contain extensive windings of copper (i.e., Spectra Physics model 2016–2080). Argon ion lasers can also have an internal electrical capacitor.

Common Types of Gas Lasers

- Argon laser
- Argon ion laser
- Carbon dioxide laser
- Copper vapor laser (rarely found, contain copper core)
- Helium–cadmium laser
- Helium–neon laser
- Krypton laser
- Nitrogen laser
- Excimer laser (see Excimer Lasers section)

Disposal

Remove power cord from unit. Crack glass tubes in a well-vented area (cracking of one end is sufficient). Metal plasma tubes should be sent to Hazardous Waste Group for disposal. Common examples of this are the Spectra Physics laser models 160, 163, 177, 183D, 185, and 185F. Any internal optics are to be removed, placed in a ziplock bag, labeled, and sent to the LSO for possible reuse by other laser users.

Hazardous Materials to Consider

- Argon
- Carbon dioxide
- Cadmium

- Neon
- Krypton
- Nitrogen
- Beryllium oxide dust

SOLID-STATE AND ROD LASERS

Active Concern

Some of these are on export control list. Circuit boards may contain lead.

Common Solid-State Lasers

- Alexandrite laser
- Erbium:Glass
- Ho:YAG
- Nd:YAG laser
- Nd:YLF
- Nd:YVO4
- Ruby laser
- Ti:Sapphire
- Yb:YAG
- Ruby

Disposal

Remove power cord, circuit boards (e-waste), and any obvious optics or crystal rods. Place the optics in a ziplock bag, label the bag, and send it to the LSO. Wear protective eyewear and safety gloves. Small quantity of gallium arsenide can be found in diodes.

Of the laser systems noted earlier, most contain Rod lasers modules that generally contain xenon or krypton arc *Flash Lamps* that serve as the excitation mechanism for the laser medium. Keeping in view the nature of these components, extreme care needs to be taken when handling the flash lamps as they can potentially explode and/or break upon removal. Treat like glass bulb waste.

Hazardous Materials to Consider

- Alexandrite
- YAG
- Gallium arsenide

BIBLIOGRAPHY

Marshall D. "Laser Disposal, How and Where," Lawrence Berkeley National Laboratory, 2010. Available at: http://www-afrd.lbl.gov/lsow/LSOW_PDF/7_2_Marshall.pdf.

Glossary

aberration: Deviation from what is normal; in optics, defects of a lens system that cause its image to deviate from normal.

absorption: Transformation of radiant energy to a different form of energy by interaction with matter. Or, the loss of light as it passes through a material.

accessible emission limit (AEL): The maximum accessible emission level permitted within a particular class.

accessible optical radiation: Optical radiation to which the human eye or skin may be exposed for the condition (operation, maintenance, or service) specified.

active medium: A medium in which stimulated emission will take place at a given wavelength.

alpha max (α_{max}): The angular limit beyond which extended source MPEs (maximum permissible exposures) for a given exposure duration are expressed as a constant radiance or integrated radiance. This value is defined as 100 mrad.

alpha min (α_{min}): *See* limiting angular subtense.

aperture: An opening or window through which radiation passes.

aphakic: Term describing an eye in which the crystalline lens is absent.

apparent visual angle: The angular subtense of the source as calculated from source size and distance from the eye. It is not the beam divergence of the source.

attenuation: The decrease in the radiant flux as it passes through an absorbing or scattering medium.

authorized personnel: Individuals approved by management to install, operate, or service laser equipment.

average power: The total energy in an exposure or emission divided by the duration of the exposure or emission.

aversion response: Closure of the eyelid, eye movement, pupillary constriction, or movement of the head to avoid an exposure to a noxious stimulant or bright light. In this standard, the aversion response to an exposure from a bright laser source is assumed to occur within 0.25 seconds, including the blink reflex time.

beam: A collection of rays characterized by direction, diameter (or dimensions), and divergence (or convergence).

beam diameter: The distance between diametrically opposed points in that cross section of a beam where the power per unit area is $1/e$ (0.368) times that of the peak power per unit area.

beam divergence (ϕ): *See* divergence.

beam expander: An optical device that increases beam diameter while decreasing beam divergence (spread). Its simplest form consists of two lenses, the first to diverge the beam and the second to recollimate it. Also called an *upcollimator*.

beam splitter: An optical device that uses controlled reflection to produce two beams from a single incident beam.

blink reflex: The blink reflex is the involuntary closure of the eyes as a result of stimulation by an external event, such as an irritation of the cornea or conjunctiva, a bright flash, the rapid approach of an object, or an auditory stimulus or with facial movements. The ocular aversion response may include a blink reflex.

Brewster windows: The transmissive end (or both ends) of the laser tube, made of transparent optical material and set at Brewster's angle in gas lasers to achieve zero reflective loss for one axis of plane polarized light. Nonstandard on industrial lasers, but some polarizing element must be used if a polarized output is desired.

calorimeter: A device for measuring the total amount of energy absorbed from a source of electromagnetic radiation.

carcinogen: An agent potentially capable of causing cancer.

coagulation: The process of congealing by an increase in viscosity characterized by a condensation of material from a liquid to a gelatinous or solid state.

coherent: A light beam is said to be coherent when the electric vector at any point in it is related to that at any other point by a definite, continuous function.

collateral radiation: Any electromagnetic radiation, except laser radiation, emitted by a laser or laser system that is physically necessary for its operation.

collecting optics: Lenses or optical instruments having magnification and thereby producing an increase in energy or power density. Such devices may include telescopes, binoculars, microscopes, or loupes.

collimated beam: Effectively, a *parallel* beam of light with very low divergence or convergence.

conjunctival discharge (of the eye): Increased secretion of mucus from the surface of the eyeball.

continuous wave (CW): The output of a laser that is operated in a continuous rather than a pulsed mode. In this standard, a laser operating with a continuous output for a period of 0.25 seconds or longer is regarded as a CW laser.

controlled area: An area where the occupancy and activity of those within are subject to control and supervision for the purpose of protection from radiation hazards.

cornea: The transparent outer coat of the human eye that covers the iris and the crystalline lens. The cornea is the main refracting element of the eye.

critical frequency: The pulse-repetition frequency above which the laser output is considered CW. For a 10-second exposure to a small source, the critical frequency is 55 kHz for wavelengths between 0.4 and 1.05 μm and 20 kHz for wavelengths between 1.05 and 1.4 μm.

cryogenics: The branch of physical science dealing with very low temperatures.

denaturation: Functional modification of the properties of protein by structural alteration via heat or photochemical processes.

depigmentation: The removal of the pigment of melanin granules from human tissues.

dermatology: A branch of medical science that deals with the skin, its structure, functions, and diseases.

dichroic filter: Filter that allows selective transmission of colors of desired wavelengths.

diffuse reflection: Change of the spatial distribution of a beam of radiation when it is reflected in many directions by a surface or by a medium.

diopter: A measure of the power of a lens, defined as $1/f_0$, where f_0 is the focal length of the lens in meters.

divergence (ϕ): For the purposes of this standard, divergence is taken as the plane angle projection of the cone that includes $1 - 1/e$ (i.e., 63.2%) of the total radiant energy or power. The value is expressed in radians or milliradians.

duty cycle: Ratio of total *on* duration to total exposure duration for a repetitively pulsed laser.

effective energy (Q_{eff}): Energy, in joules, through the applicable measurement aperture.

effective power (ΦF_{eff}): Power, in watts, through the applicable measurement aperture.

electromagnetic radiation: The flow of energy consisting of orthogonally vibrating electric and magnetic fields lying transverse to the direction of propagation. X-ray, ultraviolet, visible, infrared, and radio waves occupy various portions of the electromagnetic spectrum and differ only in frequency, wavelength, and photon energy.

embedded laser: An enclosed laser with an assigned class number higher than the inherent capability of the laser system in which it is incorporated, where the system's lower classification is appropriate because of the engineering features limiting accessible emission.

enclosed laser: A laser that is contained within a protective housing of itself or of the laser or laser system in which it is incorporated. Opening or removing of the protective housing provides additional access to laser radiation above the applicable MPE than possible with the protective housing in place (an embedded laser is an example of one type of enclosed laser).

endoscope: An instrument utilized for the examination of the interior of a canal or hollow organ.

energy: The capacity for doing work. Energy content is commonly used to characterize the output from pulsed lasers and is generally expressed in joules (J).

epidemiology: A branch of medical science that deals with the incidence, distribution, and control of disease in a population.

epithelium (of the cornea): The layer of cells forming the outer epidermis of the cornea.

erythema: Redness of the skin because of congestion of the capillaries.

extended source: A source of optical radiation with an angular subtense at the cornea larger than α_{min}.

eye-safe laser: A Class 1 laser product. Because of the frequent misuse of the term *eye-safe wavelength* to mean *retina safe* (e.g., at 1.5–1.6 µm) and *eye-safe laser* to refer to lasers emitting outside the retinal hazard region in this spectral region, the term *eye safe* can be a misnomer. Hence, the use of eye-safe laser is discouraged.

fail-safe interlock: An interlock where the failure of a single mechanical or electrical component of the interlock will cause the system to go into, or remain in, a safe mode.

femtoseconds: 10^{-15} seconds. 1 femtoseconds = 0.000,000,000,000,001 seconds.

F-number: The focal length of lens divided by its usable diameter. In the case of a laser, the usable diameter is the diameter of the laser beam or a smaller aperture that restricts a laser beam.

fiber optics: A system of flexible quartz or glass fibers that uses total internal reflection to pass light through thousands of glancing (total internal) reflections.

flashlamp: A tube typically filled with krypton or xenon. Produces a high-intensity white light in short-duration pulses.

fluorescence: The emission of light of a particular wavelength resulting from absorption of energy, typically from light of shorter wavelengths.

focal length: The distance, measured in centimeters, from the secondary nodal point of a lens to the secondary focal point. For a thin lens imaging a distant source, the focal length is the distance between the lens and the focal point.

focal point: The point toward which radiation converges or from which radiation diverges or appears to diverge.

fundus: *See* ocular fundus.

funduscopic: Examination of the fundus (rear) of the eye.

gain: Amplification.

gas laser: A type of laser in which the laser action takes place in a gas medium.

gated pulse: A discontinuous burst of laser light made by timing (gating) a CW output—usually in fractions of a second.

Gaussian curve: Statistical curve showing a peak with normal even distribution on either side. Maybe either a sharp peak with steep sides or a blunt peak with shallower sides. Used to show power distribution in a beam. The concept is important in controlling the geometry of the laser impact.

ground state: Lowest energy level of an atom.

half-power point: The value on either the leading or trailing edge of a laser pulse at which the power is one-half of its maximum value.

heat sink: A substance or device used to dissipate or absorb unwanted heat energy.

Helium–neon (HeNe) laser: A laser in which the active medium is a mixture of helium and neon. Its wavelength is usually in the visible range. Used widely for alignment, recording, printing, and measuring.

hertz (Hz): The unit that expresses the frequency of a periodic oscillation in cycles per second.

image: The optical reproduction of an object; produced by a lens or mirror. A typical positive lens converges rays to form a *real* image that can be photographed. A negative lens spreads rays to form a *virtual* image that cannot be projected.

incident light: A ray of light that falls on the surface of a lens or any other object. The *angle of incidence* is the angle made by the ray perpendicular (normal) to the surface.

infrared (IR): The region of the electromagnetic spectrum between the long-wavelength extreme of the visible spectrum (about 0.7 μm) and the shortest microwaves (about 1 mm).

infrared (IR) radiation: Electromagnetic radiation with wavelengths that lie within the range 0.7 μm to 1 mm.

installation: Placement and connection of laser equipment at the appropriate site to enable intended operation.

integrated radiance: The integral of the radiance over the exposure duration; expressed in joules per square centimeter per steradian ($J \cdot cm^{-2} \cdot sr^{-1}$).

intrabeam viewing: The viewing condition by which the eye is exposed to all or part of a laser beam.

ionizing radiation: Electromagnetic radiation having a sufficiently large photon energy to directly ionize atomic or molecular systems with a single quantum event.

iris: The circular pigmented membrane that lies behind the cornea of the human eye. The iris is perforated by the pupil.

irradiance: Radiant power incident per unit area on a surface; expressed in watts per square centimeter ($W \cdot cm^{-2}$). Synonym: power density.

joule (J): A unit of energy. 1 joule = 1 watt·second.

KTP (potassium titanyl phosphate): A crystal used to change the wavelength of an Nd:YAG laser from 1060 nm (infrared) to 532 nm (green).

Lambertian surface: An ideal surface with emitted or reflected radiance that is independent of the viewing angle.

laser: A device that produces radiant energy predominantly by stimulated emission. Laser radiation may be highly coherent temporally, spatially, or both. An acronym for light amplification by stimulated emission of radiation.

laser barrier: A device used to block or attenuate incident direct or to diffuse laser radiation. Laser barriers are frequently used during times of service to the laser system when it is desirable to establish a boundary for a temporary (or permanent) laser-controlled area.

laser diode: A laser employing a forward-biased semiconductor junction as the active medium. Synonyms: injection laser; semiconductor laser.

laser pointer: A Class 2 or Class 3A laser product that is usually handheld that emits a low-divergence visible beam of less than 5 mW and is intended for designating specific objects or images during discussions, lectures, or presentations as well as for the aiming of firearms or other visual targeting practice.

laser safety officer (LSO): One who has authority to monitor and enforce the control of laser hazards and effect the knowledgeable evaluation and control of laser hazards.

laser system: An assembly of electrical, mechanical, and optical components that includes a laser.

lesion: An abnormal change in the structure of an organ or part because of injury or disease.

limiting angular subtense (α_{min}): The apparent visual angle that divides small-source viewing from extended-source viewing, α_{min} is defined as 1.5 mrad.

limiting aperture diameter (D_f): The diameter of a circle over which irradiance or radiant exposure is averaged for purposes of hazard evaluation and classification from.

limiting cone angle (γ): The cone angle through which radiance or integrated radiance is averaged when photochemical effects are considered in hazard evaluation and laser classification.

limiting exposure duration (T_{max}): An exposure duration that is specifically limited by the design or intended use.

macula: The small, uniquely pigmented, specialized area of the retina of the eye that, in normal individuals, is predominantly employed for acute central vision (i.e., area of best visual acuity).

magnified viewing: Viewing a small object through an optic that increases the apparent object size. This type of optical system can make a diverging laser beam more hazardous (e.g., using a magnifying optic to view an optical fiber with a laser beam emitted).

maintenance: Performance of those adjustments or procedures (specified in user information provided by the manufacturer with the laser or laser system) that are to be performed by the user to ensure the intended performance of the product.

maximum permissible exposure (MPE): The level of laser radiation to which a person may be exposed without hazardous effect or adverse biological changes in the eye or skin.

measurement aperture: The aperture used for classification of a laser to determine the effective power or energy that is compared to the AEL for each class.

meter: A unit of length in the International System of Units currently defined as the length of a path traversed in vacuum by light during a period of 1/299792458 seconds. Typically, the meter is subdivided into the following units: centimeter (cm) = 10^{-2} m; millimeter (mm) = 10^{-3} m; micrometer (μm) = 10^{-6} m; nanometer (nm) = 10^{-9} m.

minimum viewing distance: The minimum distance at which the eye can produce a focused image of a diffuse source, usually assumed to be 10 cm.

mode: A term used to describe how the power of a laser beam is geometrically distributed across the cross section of the beam. Also used to describe the operating style of a laser such as continuous or pulsed.

mode locked: A method of producing laser pulses in which short pulses (~10^{-12} seconds) are produced and emitted in bursts or a continuous train.

modulation: The ability to superimpose an external signal on the output beam of the laser as a control.

monochromatic light: Theoretically, light consisting of just one wavelength. No light is absolutely single frequency since it will have some bandwidth. Lasers provide the narrowest of bandwidths that can be achieved.

multimode: Laser emission at several closely spaced frequencies.

nanometer (nm): A unit of length in the International System of Units equal to one-billionth of a meter. A measure of length, with 1 nm equal to 10^{-9} m; the usual measure of light wavelengths. Visible light ranges from about 400 nm in the purple to about 760 nm in the deep red.

nanosecond (ns): One-billionth (10^{-9}) of a second. Longer than a picosecond or femtosecond but shorter than a microsecond. Associated with Q-switched lasers.

Nd:glass laser: A solid-state laser of neodymium:glass offering high power in short pulses. An Nd-doped glass rod used as a laser medium to produce 1064-nm light.

Nd:YAG (neodymium:yttrium aluminum garnet) laser: A synthetic crystal used as a laser medium to produce 1064-nm light.

near-field imaging: A solid-state laser imaging technique offering control of spot size and hole geometry, adjustable working distance, uniform energy distribution, and a wide range of spot sizes.

NEMA: Acronym for National Electrical Manufacturers' Association, a group that defines and recommends safety standards for electrical equipment.

neodymium (Nd): The rare earth element that is the active element in Nd:YAG lasers and Nd:glass lasers.

noise: Unwanted minor currents or voltages in an electrical system.

nominal hazard zone (NHZ): The space within which the level of the direct, reflected, or scattered radiation during normal operation exceeds the applicable MPE. Exposure levels beyond the boundary of the NHZ are below the appropriate MPE level.

nominal ocular hazard distance (NOHD): The distance along the axis of the unobstructed beam from a laser, fiber end, or connector to the human eye beyond which the irradiance or radiant exposure, during installation or service, is not expected to exceed the appropriate MPE.

non-beam hazard: A class of hazards that result from factors other than direct human exposure to a laser beam.

ocular fundus: The interior posterior surface of the eye (the retina) as seen on ophthalmoscopic examination.

operation: The performance of the laser or laser system over the full range of its intended functions (normal operation).

ophthalmoscope: An instrument for examining the interior of the eye.

optical cavity (resonator): Space between the laser mirrors where lasing action occurs.

optical density (OD): A logarithmic expression for the attenuation produced by an attenuating medium, such as an eye protection filter. The logarithm to the base 10 of the reciprocal of the transmittance: $OD = \log_{10}(E_i/E_t)$, where OD is optical density, E_i incident beam irradiance (W/cm^2) worst-case exposure, and E_t transmitted beam irradiance (MPE limit in W/cm^2).

optical fiber: A filament of quartz or other optical material capable of transmitting light along its length by multiple internal reflections and emitting it at the end.

optical pumping: The excitation of the lasing medium by the application of light rather than electrical discharge.

optical radiation: Ultraviolet, visible, and infrared radiation (0.35–1.4 nm) that falls in the region of transmittance of the human eye.

optically aided viewing: Viewing with a telescopic or magnifying optic. Under certain circumstances, viewing with an optical aide can increase the hazard from a laser beam. *See* telescopic viewing; magnified viewing.

optically pumped lasers: A type of laser that derives energy from another light source such as a xenon or krypton flashlamp or other laser source.

output coupler: Partially reflective mirror in laser cavity that allows emission of laser light.

output power: The energy per second measured in watts emitted from the laser in the form of coherent light.

photochemical effect: An effect (e.g., biological effect) produced by a chemical action brought about by the absorption of photons by molecules that directly alters the molecule. For example, one photon of sufficient energy can alter a single molecule. Such effects are generally important in the shorter visible and ultraviolet regions of the optical spectrum. The threshold radiant exposure is constant over a wide range of exposure durations (the *Bunsen–Roscoe law*).

photophobia: An unusual intolerance of light. Also, an aversion to light usually caused by physical discomfort on exposure to light.

photosensitizers: Substances that increase the sensitivity of a material to irradiation by electromagnetic energy.

pigment epithelium (of the retina): The layer of cells that contains brown or black pigment granules next to and behind the rods and cones.

plasma radiation: Black body radiation generated by luminescence of matter in a laser-generated plume.

point source: No longer used. *See* small source.

power: The rate at which energy is emitted, transferred, or received. Unit: watts (joules per second).

protective housing: An enclosure surrounding the laser or laser system that prevents access to laser radiation above the applicable MPE level. The aperture through which the useful beam is emitted is not part of the protective housing. The protective housing may enclose associated optics and a workstation and limits access to other associated radiant energy emissions and to electrical hazards associated with components and terminals.

pulse duration: The length of a laser pulse, usually measured as the time interval between the half-power points on the leading and trailing edges of the pulse.

pulse-repetition frequency (PRF): The number of pulses occurring per second; expressed in hertz.

pulsed laser: A laser that delivers its energy in the form of a single pulse or a train of pulses. In this standard, the duration of a pulse is less than 0.25 seconds.

pump: To excite the lasing medium. *See* optical pumping; pumping.

pumped medium: Energized laser medium.

pumping: Addition of energy (thermal, electrical, or optical) into the atomic population of the laser medium; necessary to produce a state of population inversion.

pupil: The variable aperture in the iris through which light travels to the interior of the eye.

Q-switch: A device for producing very short (~10–250 nanoseconds), intense laser pulses by enhancing the storage and dumping of electronic energy in and out of the lasing medium, respectively.

Q-switched laser: A laser that emits short (~10–250 nanoseconds), high-power pulses by means of a Q-switch.

radian (rad): A unit of angular measure equal to the angle subtended at the center of a circle by an arc with a length that is equal to the radius of the circle. 1 radian ~57.3°; 2π radians = 360°.

radiance: Radiant flux or power output per unit solid angle per unit area expressed in watts per centimeter squared per steradian ($W \cdot cm^{-2} \cdot sr^{-1}$).

radiant energy: Energy emitted, transferred, or received in the form of radiation. Unit: joules (J).

radiant exposure: Surface density of the radiant energy received; expressed in units of joules per centimeter squared ($J \cdot cm^{-2}$).

radiant flux: Power emitted, transferred, or received in the form of radiation. Unit: watts (W). Also called *radiant power*.

radiant intensity: Quotient of the radiant flux leaving a source and propagated into an element of solid angle containing the direction, by the element of solid angle. Radiant intensity is expressed in units of watts per steradian ($W \cdot sr^{-1}$).

radiant power: Power emitted, transferred, or received in the form of radiation; expressed in watts (W). Synonym: radiant flux.

radiometry: A branch of science that deals with the measurement of radiation.

Rayleigh scattering: Scattering of radiation in the course of its passage through a medium containing particles with sizes that are small compared with the wavelength of the radiation.

reflectance: The ratio of total reflected radiant power to total incident power. Also called *reflectivity*.

reflection: Deviation of radiation following incidence on a surface.

refraction: The bending of a beam of light in transmission through an interface between two dissimilar media or in a medium with a refractive index that is a continuous function of position (graded index medium).

refractive index (of a medium): Denoted by n, the ratio of the velocity of light in vacuum to the phase velocity in the medium. Synonym: index of refraction.

repetitive pulse laser: A laser with multiple pulses of radiant energy occurring in a sequence.

resonator: The mirrors (or reflectors) making up the laser cavity, including the laser rod or tube. The mirrors reflect light back and forth to build up amplification.

retina: The sensory membrane that receives the incident image formed by the cornea and lens of the human eye. The retina lines the inside of the eye.

retinal hazard region: Optical radiation with wavelengths between 0.4 and 1.4 µm, for which the principal hazard is usually to the retina.

rotating lens: A beam delivery lens designed to move in a circle and thus rotate the laser beam around a circle.

ruby: The first laser type; a crystal of sapphire (aluminum oxide) containing trace amounts of chromium oxide.

safety latch: A mechanical device designed to slow direct entry to a controlled area.

scanning laser: A laser having a time-varying direction, origin, or pattern of propagation with respect to a stationary frame of reference.

scintillation: The rapid changes in irradiance levels in a cross section of a laser beam.

secured enclosure: An enclosure to which casual access is impeded by an appropriate means, such as a door secured by a magnetically or electrically operated lock or latch, or by fasteners that need a tool to remove.

service: The performance of those procedures or adjustments described in the manufacturer's service instructions that may affect any aspect of the performance of the laser or laser system.

shall: The word *shall* is to be understood as mandatory.

should: The word *should* is to be understood as advisory.

small source: In this document, a source with an angular subtense at the cornea equal to or less than alpha min (α_{min}), that is, less than or equal to 1.5 mrad. This includes all sources formerly referred to as *point sources* and meeting small-source viewing (formerly called point source or intrabeam viewing) conditions.

small-source viewing: The viewing condition by which the angular subtense of the source α_{min} is equal to or less than the limiting angular subtense α_{min}.

solid angle: The three-dimensional angular spread at the vertex of a cone measured by the area intercepted by the cone on a unit sphere with a center that is the vertex of the cone. Solid angle is expressed in steradians (sr).

source: A laser or a laser-illuminated reflecting surface.

spectator: An individual who wishes to observe or watch a laser or laser system in operation and who may lack the appropriate laser safety training.

specular reflection: A mirror-like reflection.

standard operating procedure (SOP): Formal written description of the safety and administrative procedures to be followed in performing a specific task.

steradian (sr): The unit of measure for a solid angle. There are 4π steradians about any point in space.

stromal haze (of the cornea): Cloudiness in the connective tissue or main body of the cornea.

surface exfoliation (of the cornea): A stripping or peeling off of the surface layer of cells from the cornea.

synergism: A condition in which the combined effect is greater than the sum of the effects of individual contributors.

T_1**:** The exposure duration (time) at which MPEs based on thermal injury are replaced by MPEs based on photochemical injury to the retina.

T_2**:** The exposure duration (time) beyond which extended-source MPEs based on thermal injury are expressed as a constant irradiance.

telescopic viewing: Viewing an object from a long distance to increase its visual size. These systems generally collect light through a large aperture, magnifying hazards from large-beam, collimated lasers.

TEM: Abbreviation for transverse electromagnetic modes. Used to designate the cross-sectional shape of the beam. The radial distribution of intensity across a beam as it exits the optical cavity.

TEM$_{00}$: The lowest-order mode possible with a bell-shaped (Gaussian) distribution of light across the laser beam.

thermal effect: An effect brought about by the temperature elevation of a substance (e.g., biological tissue). Photocoagulation of proteins resulting in a thermal burn is an example. The threshold radiant exposure is dependent on the duration of exposure and heat transfer from the heated area.

threshold limit (TL): The term is applied to laser protective eyewear filters, protective windows, and barriers. The TL is an expression of the *resistance factor* for beam penetration of a laser protective device. This is generally related by the TL of the protective device (expressed in $W \cdot cm^{-2}$ or $J \cdot cm^{-2}$). It is the maximum average irradiance (or radiant exposure) at a given beam diameter for which a laser protective device (e.g., filter, window, barrier) provides adequate beam resistance. Thus, laser exposures delivered on the protective device at or below the TL will limit beam penetration to levels at or below the applicable MPE.

T_{max}: *See* limiting exposure duration.

t_{min}: For a pulsed laser, the maximum duration for which the MPE is the same as the MPE for a 1-nanosecond exposure. For thermal biological effects, this corresponds to the *thermal confinement duration* during which heat flow does not significantly change the absorbed energy content of the thermal relaxation volume of the irradiated tissue (e.g., t_{min} is 18 microseconds in the spectral region from 0.4 to 1.05 μm and is 50 microseconds between 1.050 and 1.400 μm).

tonometry: Measurement of the pressure (tension) of the eyeball.

transmission: Passage of radiation through a medium.

transmittance: The ratio of transmitted power to incident power.

tunable dye laser: A laser with an active medium that is a liquid dye, pumped by another laser or flashlamps, to produce various colors of light. The color of light may be tuned by adjusting optical tuning elements or changing the dye used.

tunable laser: A laser system that can be *tuned* to emit laser light over a continuous range of wavelengths or frequencies.

ultraviolet radiation: Electromagnetic radiation with wavelengths shorter than those of visible radiation; for the purpose of this standard, 0.18–0.4 μm.

uncontrolled area: An area where the occupancy and activity of those within is not subject to control and supervision for the purpose of protection from radiation hazards.

viewing portal: An opening in a system that allows the user to observe the chamber. All viewing portals and display screens included as an integral part of a laser system must incorporate a suitable means to maintain the laser radiation at the viewing position at or below the applicable MPE (eye safe) for all conditions of operation and maintenance. It is essential that the materials used for viewing portals and display screens do not support combustion or release toxic vapors following exposure to laser radiation.

viewing window: Visually transparent parts of enclosures that contain laser processes. It may be possible to observe the laser processes through the viewing windows.

visible radiation (light): The term is used to describe electromagnetic radiation that can be detected by the human eye. This term is commonly used to describe wavelengths that lie in the range 0.4–0.7 μm.

watt (W): The unit of power or radiant flux. 1 watt = 1 joule per second.

watt/cm²: A unit of irradiance used in measuring the amount of power per area of absorbing surface or per area of CW laser beam.

wave: A sinusoidal undulation or vibration; a form of movement by which all radiant electromagnetic energy travels.

wavelength: The length of the light wave, usually measured from crest to crest, that determines its color. Common units of measurement are the micrometer (micron), the nanometer, and (earlier) the angstrom unit.

window: A piece of glass (or other material) with plane parallel sides that admits light into or through an optical system and excludes dirt and moisture.

work practices: Procedures used to accomplish a task.

YAG (yttrium aluminum garnet): A widely used solid-state crystal that is composed of yttrium and aluminum oxides, which is doped with a small amount of the rare-earth neodymium.

Z-cavity: A term referring to the shape of the optical layout of the tubes and resonator inside a laser.

Index